Java EE
核心框架实战
第 2 版

高洪岩 著

人民邮电出版社
北京

图书在版编目（CIP）数据

Java EE核心框架实战 / 高洪岩著. -- 2版. -- 北京：人民邮电出版社，2017.9
ISBN 978-7-115-46361-6

Ⅰ．①J… Ⅱ．①高… Ⅲ．①JAVA语言—程序设计 Ⅳ．①TP312.8

中国版本图书馆CIP数据核字(2017)第182196号

内 容 提 要

本书的宗旨是提高读者学习 Java EE 的效率，增强其项目实战能力。为此，本书摒弃了软件公司中不常用或不实用的技术，而是采用近 200 个开发案例，为读者讲解了开发商业软件的必备知识，帮助读者进行"精要"式的学习，汲取 Java EE 的思想，正确地进行项目实战。

本书涵盖了 MyBatis 3、Struts 2、Ajax、JSON、jQuery、Spring 4 MVC、Hibernate 5、Spring 4 等主流 Java EE 框架的核心开发技术，介绍了 MVC 框架的原理实现、上传、下载、数据验证、国际化、多模块分组开发、转发/重定向、JSON 的解析；将 Ajax 及 JSON 和 MVC 框架进行整合开发；ORM 框架的 CURD 操作以及 MyBatis 和 Hibernate 中的映射文件使用。本书还使用大量篇幅介绍了 Spring 4 中的核心技术 DI 与 AOP，以及企业中常用框架的整合开发，框架包含 Struts 2、Spring 4 MVC、MyBatis 3、Hibernate 5、Spring 4 整合开发等内容。

本书语言简洁，示例丰富，可帮助读者迅速掌握使用主流开源 Java EE 框架进行开发所需的各种技能。本书适合具有一定 Java 编程基础的读者，以及使用 Java 进行软件开发、测试的从业人员阅读。

◆ 著　　　高洪岩
责任编辑　傅道坤
责任印制　焦志炜

◆ 人民邮电出版社出版发行　北京市丰台区成寿寺路 11 号
邮编 100164　电子邮件 315@ptpress.com.cn
网址 http://www.ptpress.com.cn
固安县铭成印刷有限公司印刷

◆ 开本：787×1092　1/16
印张：33.75　　　　　　　　　2017年9月第 2 版
字数：813 千字　　　　　　　　2025年2月河北第 6 次印刷

定价：99.00 元

读者服务热线：(010)81055410　印装质量热线：(010)81055316
反盗版热线：(010)81055315

前言

感谢读者的厚爱。本书第 1 版在市场上反响强烈,很多读者向作者提出了若干建设性的意见,同时由于 IT 技术发展飞快,需要将这些技术进行更新;本书的第 2 版本加入了常用框架最新版本的新技术,完善了每个框架的学习案例,目的是让读者学习到更多的知识,这也是作者写作的初衷。

作者多年从事软件开发相关的工作,但从近几年用人单位对人才招聘的要求来看,越来越趋向于"实战性",也就是要求员工进入软件公司后能立即融入开发的任务中,快速地为软件公司创造巨大的经济利益。

我的很多学员建议能否出一本内容精悍而不失实用价值的主流 Java EE 开源框架图书,使其只包含主流框架最重要、最核心、最常用的内容。这样读者就可以尽快上手,并沿着这个核心指导出一些方向,自行在工作和学习中不断拓展和深掘,这就是我写这本书的主要目的。

的确,他们的建议非常有道理的,因为 Java EE 世界非常庞大,以至于世界上没有任何一本书能把它讲得巨细靡遗。要想学好 Java 语言或 Java EE 框架并吸取其中丰富的编码技巧、设计模式、代码优化,将他们熟练地综合应用在软件项目中,并没有捷径。所以,就有了本书。本书不但涵盖了学习主流 Java EE 框架所须掌握的核心技术,也涵盖了使用它们进行项目实战的必备知识。本书的主旨就是希望让读者尽快掌握开源 Java EE 框架的核心内容,正确进行项目实战,汲取 Java EE 的思想,并最终将这种思想活用到实际工作中。

现在主流的 Java EE 框架还是 MyBatis、Hibernate、Struts 2、Spring MVC、Spring,让初入 Java EE 框架的学习者从零开始到最终掌握这几个框架,一直是我的写作目标。有些 Java EE 开源的框架的确能极大改善开发效率,但由于使用的人不多,所以覆盖面比较窄,而软件公司在招聘时的技术需求大多数情况下却是"大众化"的,这就要求应聘者在面试前就有主流 Java EE 框架的学习或使用经验。如果读者找不到合适的教材,导致在学习某一项技术时根本摸不清哪些知识点是常用的、哪些是不常用的,就会大大降低学习者的学习效率,分散大量宝贵的注意力。

本书面向的读者

首先,本书适合所有 Java 的程序员阅读。Java EE 作为 Java 开源世界的主流框架,

Java 程序员没有理由不学习它们。其次，本书适合那些希望学习 Java EE 框架编程的在校学生。由于学校的功课很多，而一本大部头的框架书籍需要花费大量的时间去研读。在学习效率上，本书可以快速带领他们进入 Java EE 框架开发的殿堂，同时又不会遗漏掉应该掌握的核心技能。

本书的结构

第 1 章将会介绍基于 SQL 映射的 MyBatis 框架，可以使用此框架操作主流的数据库，并介绍 MyBatis 核心 API 的使用，采用自定义封装法来简化 MyBatis 的操作代码，进而加快开发效率。

第 2 章主要讲解了 MyBatis 映射有关的知识，包括<sql>、<resultMap>、<choose>、<set>、<foreach>等常用标签；DB 连接信息存储到 Properties 文件的读取；使用 JDBC 数据源；别名 typeAliases 的配置；CLOB 字段的读取以及分页等必备知识。

第 3 章在开始就为读者介绍了一个微型 MVC 框架的设计，体会 MVC 框架的工作原理；并介绍了基于 Struts 2 的有刷新验证及文本信息的国际化，转发/重定向的操作，多模块分组开发的实现，使用松/紧耦版的 API 进行实用开发。

第 4 章讲解了如何使用 Struts 2 框架实现上传、下载（支持中文）的功能，并且支持多文件上传。

在第 5 章中，读者将学习在 Web 开发中的主流技术 Ajax，还要学习 JSON 技术，并将 JSON 技术结合 Ajax 和 Struts 2 实现前台和后台数据通信的功能，掌握不同格式 JSON 的解析技术。Ajax 是学习 Java Web 必须要走的路。

第 6、7、8 章详细介绍主流 ORM 框架 Hibernate 4 的使用，包括核心 API 的使用和 HQL 语言的使用。

第 9 章介绍 Java EE 中的 JPA 规范，现主流的 ORM 框架都支持 JPA，比如 Hibernate 和 OpenJPA 等，所以掌握 JPA 也是考验一个程序员对 Java EE 的使用程度的一个标准。

第 10 章开始介绍 Spring 4 中的 IOC 和 AOP 技术，深入了解动态代理的实现与原理。

第 11 章又以 Struts 2+Hibernate 5+Spring 4 进行整合，这也是软件公司中非常常用的整合搭配。

在第 12 章中，读者将学习到最流行的 Spring 4 MVC 框架，体会使用此框架开发一个经典登录功能时使用的技术点，限制 form 提交方式，还要掌握分组分模块开发的技术，重定向/转发的使用，JSON+Ajax+Spring 4 MVC 联合开发，上传/下载的实现，以及数据验证功能的使用。

第 13 章对 Spring 4 MVC 框架进行详细介绍，包括 XML 配置文件的处理，业务层 Service 的注入，ModelAndView 对象的使用，以及 HttpSession 在 Spring 4 MVC 中的使用。

第 14 章讲解特别常用的 Spring 4 MVC+MyBatis 3+Spring 4 整合。本章以当前最具实战的组合框架来讲解整合的过程，而不囿于某一个框架本身，而且整合后的项目代码写法更加统一，便于维护与扩展。

如何使用本书

首先需要声明，本书不是 Java Web 的入门教程，学习本书之前，读者先要对 Java Web 中的 JSP、Servlet 等 Web 技术有所了解，尽量能完整地使用 JSP 或 Servlet 开发一个小型项目，再阅读本书，那样会发现代码的分层更加明确，结构更加清楚。

软件开发实践才是硬道理，当读者认为阅读本书不成问题时，通常是因为读者的英语能力还不错，可以识别书中大多数英文的意思，但软件开发并不是完全考验英语能力，而是设计、排错，以及拥有更多的经验，所以请拿起手中的键盘，练习一下吧！

请尽量在自己的计算机中执行本书的所有代码，仅仅看书和自己动手存在天壤之别，动手运行代码和看书结合起来才能更深刻地理解框架的各项功能。

代码下载

读者可从 box.ptpress.com.cn/y/46361 下载本书源代码。

与作者联系

由于 Java EE 内容涵盖面宽广，涉及的知识点非常多，并加之作者水平有限，错误之处在所难免，请广大读者在不足之处给予赐教和斧正。读者可以通过 QQ（279377921）与我联系，期待与您进行交流。

目录

第 1 章 MyBatis 3 操作数据库 1

- 1.1 MyBatis 介绍 1
- 1.2 MyBatis 基本使用 4
  - 1.2.1 使用 XML 配置文件创建 SqlSessionFactory 对象 5
  - 1.2.2 SqlSessionFactoryBuilder 和 SqlSessionFactory 类信息 7
  - 1.2.3 使用 MyBatis Generator 工具逆向操作 Oracle 数据库 7
  - 1.2.4 使用 MyBatis Generator 工具逆向操作 MySQL 数据库 14
- 1.3 使用 MyBatis 实现 CURD-2 种数据库（Oracle-MySQL） 16
  - 1.3.1 针对 Oracle 的 CURD 17
  - 1.3.2 针对 MySQL 的 CURD 23
  - 1.3.3 SQL 映射文件中 namespace 命名空间的作用 24
- 1.4 MyBatis 核心对象的生命周期与封装 26
  - 1.4.1 创建 GetSqlSessionFactory.java 类 27
  - 1.4.2 创建 GetSqlSession.java 类 28
  - 1.4.3 创建 DBOperate.java 类 29
  - 1.4.4 创建 userinfoMapping.xml 映射文件 29
  - 1.4.5 创建连接数据库 mybatis-config.xml 配置文件 30
  - 1.4.6 创建名称为 test 的 Servlet 对象 30
  - 1.4.7 添加记录及异常回滚的测试 31
  - 1.4.8 删除记录 33
  - 1.4.9 更改记录 34
  - 1.4.10 查询单条记录 35
  - 1.4.11 查询多条记录 36

第 2 章 MyBatis 3 必备开发技能 37

- 2.1 使用 getMapper()面向接口编程 37

2.1.1　接口-SQL 映射的对应关系　37
　　2.1.2　创建 Userinfo.java 实体类　38
　　2.1.3　创建 UserinfoMapper.java 接口　38
　　2.1.4　创建 SQL 映射文件 UserinfoMapper.xml　39
　　2.1.5　增加记录 insert 的操作代码　39
　　2.1.6　查询全部 selectAll 的操作代码　41
　　2.1.7　查询单条记录 selectById 的操作代码　41
　　2.1.8　修改记录 updateById 的操作代码　42
　　2.1.9　删除记录 deleteById 的操作代码　43
　2.2　使用 typeAliases 配置别名　43
　　2.2.1　使用 typeAlias 单独配置别名　44
　　2.2.2　使用 package 批量配置别名　44
　2.3　使用 properties 文件保存数据库信息　46
　　2.3.1　在 mybatis-config.xml 配置文件中读取 properties 文件中的参数　46
　　2.3.2　将数据库信息封装进 properties 属性文件中　47
　2.4　与数据源 DataSource 有关的操作　49
　　2.4.1　配置多个 environment 环境　49
　　2.4.2　什么是 JNDI 以及如何从 JNDI 获得数据源 DataSource　50
　　2.4.3　如何在 MyBatis 中使用数据源 DataSource　54
　　2.4.4　在 MyBatis 中使用 HikariCP 连接池　57
　2.5　多数据库执行不同 SQL 语句的支持　60
　2.6　多种获取 Mapper 的方式　61
　2.7　MyBatis3 的 SQL 映射文件与动态 SQL　62
　　2.7.1　<resultMap>标签　62
　　2.7.2　<sql>标签　64
　　2.7.3　使用${}拼接 SQL 语句　65
　　2.7.4　插入 null 值时的处理第 1 种方法 jdbcType　67
　　2.7.5　插入 null 值时的处理第 2 种方法<if>　68
　　2.7.6　<where>标签　69
　　2.7.7　<choose>标签的使用　71
　　2.7.8　<set>标签的使用　72
　　2.7.9　<foreach>标签的使用　73
　　2.7.10　使用<bind>标签对 like 语句进行适配　77
　2.8　插入超大的字符串 String 文本内容　81
　2.9　对查询的数据进行分页　84
　2.10　批处理 SQL 语句　86

第 3 章　Struts 2 必备开发技能　88

　3.1　使用 Struts 2 进行登录功能的开发　88

3.1.1 为什么要使用 MVC 89
3.1.2 准备 jar 文件 94
3.1.3 创建 Web 项目、添加 jar 文件及配置 web.xml 文件 96
3.1.4 创建控制层 Controller 文件-Login.java 96
3.1.5 创建业务逻辑层 Model 文件-UserinfoService.java 97
3.1.6 创建视图层 View 文件-login.jsp 98
3.1.7 添加核心配置文件 struts.xml 及解释 98
3.1.8 添加 ok.jsp 和 no.jsp 登录结果文件 99
3.1.9 运行项目 99
3.1.10 Struts 2 的拦截器 101
3.1.11 Struts 2 的数据类型自动转换 106

3.2 MVC 框架的开发模型 112

3.2.1 基础知识准备 1——XML 文件的 CURD 113
3.2.2 基础知识准备 2——Java 的反射 120
3.2.3 实现 MVC 模型——自定义配置文件 122
3.2.4 实现 MVC 模型——ActionMapping.java 封装<action>信息 123
3.2.5 实现 MVC 模型——ResultMapping.java 封装<result>信息 124
3.2.6 实现 MVC 模型——管理映射信息的 ActionMappingManager.java 对象 124
3.2.7 实现 MVC 模型——创建反射 Action 的 ActionManager.java 对象 126
3.2.8 实现 MVC 模型——创建核心控制器 ActionServlet.java 127
3.2.9 实现 MVC 模型——创建 Action 接口及控制层 Controller 实现类 129
3.2.10 实现 MVC 模型——创建视图层 V 对应的 JSP 文件 130
3.2.11 实现 MVC 模型——在 web.xml 中配置核心控制器 131
3.2.12 实现 MVC 模型——运行效果 131

3.3 Struts 2 的刷新验证功能 131

3.3.1 Action 接口 132
3.3.2 Validateable 和 ValidationAware 接口 133
3.3.3 TextProvider 和 LocaleProvider 接口 134
3.3.4 使用 ActionSupport 实现有刷新的验证 134

3.4 对 Struts 2 有刷新验证的示例进行升级 137

3.4.1 加入 xml 配置来屏蔽自动生成的 table/tr/td 代码 137
3.4.2 解决"出错信息不能自动显示"的问题 139

3.5 用<s:actionerror>标签显示全部出错信息 142

3.6 出错信息进行传参及国际化 144

3.6.1 创建 info_en_US.properties 和 info_zh_CN.properties 属性文件 144
3.6.2 在 JSP 文件中显示国际化的静态文本 147
3.6.3 在 JSP 文件中显示国际化的静态文本时传递参数 149
3.6.4 在 Action 中使用国际化功能 149

3.7 用实体类封装 URL 中的参数——登录功能的 URL 封装 151

- 3.8 Struts 2 中的转发操作 153
  - 3.8.1 Servlet 中的转发操作 153
  - 3.8.2 Struts 2 中的转发操作 154
- 3.9 由 Action 重定向到 Action——无参数 157
  - 3.9.1 什么样的情况下使用重定向 157
  - 3.9.2 新建起始控制层 Login.java 157
  - 3.9.3 新建目的控制层 List.java 157
  - 3.9.4 在 struts.xml 文件中配置重定向的重点 158
  - 3.9.5 新建显示列表的 JSP 文件 159
- 3.10 由 Action 重定向到 Action——有参数 159
  - 3.10.1 什么样的情况下需要重定向传递参数 159
  - 3.10.2 新建起始控制层 Login.java 文件 159
  - 3.10.3 更改 struts.xml 配置文件 160
  - 3.10.4 新建目的控制层 List.java 文件 161
  - 3.10.5 用 JSTL 和 EL 在 JSP 文件中打印数据 162
- 3.11 让 Struts 2 支持多模块多配置文件开发 164
  - 3.11.1 新建 4 个模块的控制层 164
  - 3.11.2 新建 3 个模块的配置文件 165
  - 3.11.3 使用 include 标记导入多个配置文件 167
  - 3.11.4 创建各模块使用的 JSP 文件 167
  - 3.11.5 运行各模块的效果 168
- 3.12 在 Action 中有多个业务方法时的处理 169
- 3.13 自定义全局 result 171
  - 3.13.1 新建全局 result 实例和控制层代码 171
  - 3.13.2 声明全局的 result 对象 172
  - 3.13.3 部属项目并运行 172
- 3.14 在 Action 中使用 servlet 的 API（紧耦版） 173
  - 3.14.1 将数据放到不同的作用域中 173
  - 3.14.2 从不同作用域中取值 174
- 3.15 在 Action 中使用 Servlet 的 API（松耦版） 175
  - 3.15.1 新建控制层 175
  - 3.15.2 新建 JSP 视图 176

第 4 章 Struts 2 文件的上传与下载 178

- 4.1 使用 Struts 2 进行单文件上传 178
  - 4.1.1 Struts 2 上传功能的底层依赖 178
  - 4.1.2 新建上传文件的 JSP 文件 178
  - 4.1.3 新建上传文件的控制层 Register.java 文件 179

4.1.4 Action 中 File 实例的命名规则　180
4.1.5 设置上传文件的大小　180
4.1.6 设计 struts.xml 配置文件　180
4.1.7 成功上传单个文件　181

4.2 使用 Struts 2 进行多文件上传　181
4.2.1 新建上传多个文件的 JSP　182
4.2.2 设计上传的控制层代码　182
4.2.3 成功上传多个文件　184

4.3 使用属性驱动形式的文件上传　185
4.3.1 创建上传多个文件的 JSP　185
4.3.2 设计上传文件的控制层　186
4.3.3 新建上传文件的封装类　187
4.3.4 将 JSP 文件中 s:file 标签的 name 属性进行更改　188
4.3.5 以属性驱动方式成功上传多个文件　189

4.4 用 Struts 2 实现下载文件的功能（支持中文文件名与 IE 和 FireFix 兼容）　190
4.4.1 新建下载文件的 JSP 文件　190
4.4.2 新建下载文件的控制层文件　191
4.4.3 更改 struts.xml 配置文件　192
4.4.4 成功下载中文文件名的文件　192

第 5 章　JSON、Ajax 和 jQuery 与 Struts 2 联合使用　193

5.1 JSON 介绍　193
5.2 用 JSON 创建对象　194
5.2.1 JSON 创建对象的语法格式　194
5.2.2 在 JSP 中用 JSON 创建一个对象　194
5.2.3 运行效果　194

5.3 用 JSON 创建字符串的限制　195
5.3.1 需要转义的特殊字符　195
5.3.2 在 JSP 中对 JSON 特殊字符进行转义　195
5.3.3 运行效果　196

5.4 用 JSON 创建数字类型语法格式　196
5.4.1 在 JSP 中用 JSON 创建数字类型　196
5.4.2 运行效果　197

5.5 用 JSON 创建数组对象的语法格式　197
5.5.1 JSON 创建一个数组对象　198
5.5.2 运行效果　198

5.6 用 JSON 创建嵌套的对象类型　198
5.7 将对象转成 JSON 字符串　200

5.7.1 什么情况下需要将对象转成 JSON 字符串　200
5.7.2 使用 stringify 方法将对象转成 JSON 字符串　200

5.8 将对象转成 JSON 字符串提交到 Action 并解析（以 post 方式提交）　201
5.8.1 在 JSP 中创建 JSON 和 Ajax 对象　201
5.8.2 用 Action 控制层接收通过 Ajax 传递过来的 JSON 字符串　202
5.8.3 运行效果　203
5.8.4 在控制台输出的数据　204

5.9 将对象转成 JSON 字符串提交到 Action 并解析(get 方式提交)　204
5.9.1 新建创建 JSON 字符串的 JSP 文件　204
5.9.2 新建接收 JSON 字符串的 Action 控制层　205
5.9.3 运行结果　206
5.9.4 在控制台输出的数据　206

5.10 将数组转成 JSON 字符串提交到 Action 并解析(get 和 post 方式提交)　207
5.10.1 在服务器端用 get 方法解析 JSON 字符串　208
5.10.2 在服务器端用 post 方法解析 JSON 字符串　209
5.10.3 运行结果　210
5.10.4 在控制台输出的数据　210

5.11 使用 Ajax 调用 Action 并生成 JSON 再传递到客户端(get 和 post 方式提交)　210
5.11.1 新建具有 Ajax 提交功能的 JSP　211
5.11.2 在 Action 控制层创建 List 中存 String　213
5.11.3 在 Action 控制层创建 List 中存 Bean　214
5.11.4 在 Action 控制层创建 Map 中存放的 String　215
5.11.5 在 Action 控制层创建 Map 中存放的 Bean　216
5.11.6 单击不同的 button 按钮调用不同的 Action　217

5.12 jQuery、JSON 和 Struts 2　218
5.12.1 jQuery 框架的 Ajax 功能介绍　218
5.12.2 用 jQuery 的 Ajax 功能调用远程 action（无返回结果）　219
5.12.3 jQuery 的 Ajax 方法结构　220
5.12.4 用 jQuery 的 Ajax 功能调用远程 action（有返回结果）　221
5.12.5 用 jQuery 的 Ajax 功能调用远程 action 并且传递 JSON 格式参数(有返回值)　223
5.12.6 用 jQuery 解析从 action 返回 List 中存 String 的 JSON 字符串　226

5.13 在服务器端解析复杂结构的 JSON 对象　228

第 6 章 用 Hibernate 5 操作数据库　230
6.1 Hibernate 概述与优势　230
6.2 持久层、持久化与 ORM　231
6.3 用 MyEclipse 开发第一个 Hibernate 示例　233

6.3.1　在 MyEclipse 中创建 MyEclipse Database Explorer 数据库连接　233
6.3.2　创建 Web 项目并添加 Hibernate 框架　235
6.3.3　开始 Hibernate 逆向　239
6.3.4　数据访问层 DAO 与实体类 entity 的代码分析　241
6.3.5　使用 Hibernate 进行持久化　242
6.3.6　映射文件 Userinfo.hbm.xml 的代码分析　243
6.3.7　查询—修改—删除的操作代码　245
6.3.8　其他类解释　247

第 7 章　Hibernate 5 核心技能　248

7.1　工厂类 HibernateSessionFactory.java 中的静态代码块　248
7.2　SessionFactory 介绍　249
7.3　Session 介绍　249
7.4　使用 Session 实现 CURD 功能　250
　　7.4.1　数据表 userinfo 结构与映射文件　250
　　7.4.2　创建 SessionFactory 工厂类　251
　　7.4.3　添加记录　251
　　7.4.4　查询单条记录　252
　　7.4.5　更改记录　253
　　7.4.6　删除记录　253
7.5　Hibernate 使用 JNDI 技术　254
　　7.5.1　备份 Tomcat/conf 路径下的配置文件　254
　　7.5.2　更改配置文件 context.xml　254
　　7.5.3　更改配置文件 web.xml　254
　　7.5.4　添加 Hibernate 框架配置的关键步骤　255
　　7.5.5　逆向工程　255
　　7.5.6　支持 JNDI 的 hibernate.cfg.xml 配置文件内容　255
　　7.5.7　创建查询数据的 Servlet　256
　　7.5.8　部属项目验证结果　256
7.6　缓存与实体状态　256
　　7.6.1　Hibernate 的 OID 与缓存　256
　　7.6.2　Hibernate 中的对象状态：瞬时状态、持久化状态和游离状态　258
7.7　双向一对多在 MyEclipse 中的实现　258
　　7.7.1　添加主表记录　261
　　7.7.2　添加子表记录　262
　　7.7.3　更改主表数据　262
　　7.7.4　更改子表数据　262
　　7.7.5　删除子表数据　263

7.7.6 删除主表 main 数据　263
7.7.7 通过主表获取子表数据　264
7.8 Hibernate 备忘知识点　265
7.9 对主从表结构中的 HashSet 进行排序　267
7.10 延迟加载与 load()和 get()的区别　267
7.10.1 主从表表结构的设计　267
7.10.2 对省表和市表内容的添充　268
7.10.3 更改映射文件　268
7.10.4 新建测试用的 Servlet 对象　268
7.10.5 更改映射文件 Sheng.hbm.xml　269
7.11 Hibernate 对 Oracle 中 CLOB 字段类型的读处理　270
7.12 Hibernate 中的 inverse 与 cascade 的测试　270

第 8 章 Hibernate 5 使用 HQL 语言进行检索　275

8.1 Hibernate 的检索方式　275
8.1.1 HQL 表别名　276
8.1.2 HQL 对结果进行排序与 list()和 iterator()方法的区别　278
8.1.3 HQL 索引参数绑定　281
8.1.4 HQL 命名参数绑定与安全性　282
8.1.5 HQL 方法链的使用　284
8.1.6 HQL 中的 uniqueResult 方法的使用　284
8.1.7 HQL 中的 where 子句与查询条件　285
8.1.8 查询日期——字符串格式　287
8.1.9 查询日期——数字格式　288
8.1.10 分页的处理　289
8.1.11 HQL 中的聚集函数：distinct-count-min-max-sum-avg　290
8.1.12 HQL 中的分组查询　292

第 9 章 JPA 核心技能　294

9.1 什么是 JPA 以及为什么要使用 JPA　294
9.2 搭建 JPA 开发环境与逆向　295
9.3 分析逆向出来的 Java 类　300
9.4 使用 IUserinfoDAO.java 接口中的方法　301
9.4.1 方法 public void save(Userinfo entity)的使用　302
9.4.2 方法 public Userinfo findById(Long id)的使用　303
9.4.3 方法 public List<Userinfo> findByProperty(String propertyName, final Object value, final int... rowStartIdxAndCount)的使用　304
9.4.4 方法 public List<Userinfo> findByUsername(Object username, int... rowStartIdxAndCount)

9.4.5 方法 public List<Userinfo> findByPassword(Object password, int... rowStartIdxAndCount) 的使用 304

9.4.6 方法 public List<Userinfo> findByAge(Object age, int... rowStartIdxAndCount)的使用 305

9.4.7 方法 public List<Userinfo> findAll(final int... rowStartIdxAndCount)的使用 305

9.4.8 方法 public Userinfo update(Userinfo entity)的使用 305

9.4.9 方法 public void delete(Userinfo entity)的使用 306

9.5 JPA 核心接口介绍 306

 9.5.1 类 Persistence 306

 9.5.2 JPA 中的事务类型 307

 9.5.3 接口 EntityManagerFactory 308

 9.5.4 接口 EntityManager 308

9.6 实体类的状态 308

9.7 使用原生 JPA 的 API 实现 1 个添加记录的操作 309

9.8 从零开始搭建 JPA 开发环境 309

9.9 EntityManager 核心方法的使用 311

 9.9.1 方法 void persist(Object entity)保存一条记录 311

 9.9.2 <T> T merge(T entity)方法和<T> T find(Class<T> entityClass, Object primaryKey)方法 311

 9.9.3 方法 void remove(Object entity) 312

 9.9.4 getReference(Class<T>, Object)方法 312

 9.9.5 createNativeQuery(string)方法 315

 9.9.6 clear()和 contains(Object)方法 317

 9.9.7 createQuery(String)方法 319

9.10 双向一对多的 CURD 实验 319

 9.10.1 逆向 Main.java 和 Sub.java 实体类 319

 9.10.2 创建 Main 322

 9.10.3 创建 Sub 322

 9.10.4 更新 Main 323

 9.10.5 更新 Sub 323

 9.10.6 删除 Main 时默认将 Sub 也一同删除 324

 9.10.7 从 Main 加载 Sub 时默认为延迟加载 324

9.11 JPQL 语言的使用 325

 9.11.1 参数索引式查询 325

 9.11.2 命名式参数查询 326

 9.11.3 JPQL 支持的运算符与聚合函数与排序 326

 9.11.4 is null 为空运算符的使用 327

 9.11.5 查询指定字段的示例 327

 9.11.6 JPQL 语言对日期的判断 329

 9.11.7 JPQL 语言中的分页功能 331

第 10 章　Spring 4 的 DI 与 AOP　332

10.1　Spring 介绍　332
10.2　依赖注入　333
10.3　DI 容器　333
10.4　AOP 的介绍　334
10.5　Spring 的架构　334
10.6　一个使用传统方式保存数据功能的测试　335
10.7　在 Spring 中创建 JavaBean　336
  10.7.1　使用 xml 声明法创建对象　337
  10.7.2　使用 Annotation 注解法创建对象　340
10.8　DI 的使用　350
  10.8.1　使用 xml 声明法注入对象　350
  10.8.2　使用注解声明法注入对象　352
  10.8.3　多实现类的歧义性　353
  10.8.4　使用@Autowired 注解向构造方法参数注入　356
  10.8.5　在 set 方法中使用@Autowired 注解　357
  10.8.6　使用@Bean 向工厂方法的参数传参　358
  10.8.7　使用@Autowired(required = false)的写法　358
  10.8.8　使用@Bean 注入多个相同类型的对象时出现异常　360
  10.8.9　使用@Bean 对 JavaBean 的 id 重命名　361
  10.8.10　对构造方法进行注入　362
  10.8.11　使用 p 命名空间对属性进行注入　368
  10.8.12　Spring 上下文环境的相关知识　370
  10.8.13　使用 Spring 的 DI 方式保存数据功能的测试　375
  10.8.14　BeanFactory 与 ApplicationContext　377
  10.8.15　注入 null 类型　377
  10.8.16　注入 Properties 类型　378
  10.8.17　在 DI 容器中创建 Singleton 单例和 Prototype 多例的 JavaBean 对象　379
  10.8.18　Spring 中注入外部属性文件的属性值　381
10.9　面向切面编程 AOP 的使用　383
  10.9.1　AOP 的原理之代理设计模式　384
  10.9.2　与 AOP 相关的必备概念　391
  10.9.3　面向切面编程 AOP 核心案例　395
  10.9.4　Strust 2、Spring 4 整合及应用 AOP 切面　432

第 11 章　Struts 2+Hibernate 5+Spring 4 整合　436

11.1　目的　436

11.2 创建数据库环境 436
11.2.1 新建数据表 userinfo 436
11.2.2 创建序列对象 437
11.3 新建整合用的 Web 项目 437
11.4 添加 Struts 2 框架支持 437
11.4.1 添加 Struts 2 框架 437
11.4.2 在 web.xml 文件中注册 Struts 2 的过滤器 438
11.4.3 在项目的 src 目录下创建 struts.xml 配置文件 438
11.5 添加 Hibernate 5 框架支持 439
11.6 添加 Spring 4 框架支持 440
11.7 创建 spring-dao.xml 文件 440
11.8 创建 spring-service.xml 文件 440
11.9 创建 spring-controller.xml 文件 441
11.10 创建 applicationContext.xml 文件 441
11.11 在 web.xml 文件中注册 Spring 监听器 442
11.12 加 Spring 4 框架后的 Web 项目结构 443
11.13 创建 Hibernate 中的实体类与映射文件 444
11.14 创建 Hibernate 5 的 DAO 类 445
11.15 创建 UserinfoService.java 服务对象 445
11.16 新建一个操作 userinfo 表数据的 Controller 控制层 446
11.17 测试成功的结果 447
11.18 测试回滚的结果 448

第 12 章 Spring 4 MVC 核心技能 450

12.1 Spring 4 MVC 介绍 450
12.1.1 Spring 4 MVC 核心控制器 451
12.1.2 基于注解的 Spring 4 MVC 开发 452
12.2 Spring 4 MVC 第一个登录测试 452
12.2.1 添加 Spring 4 MVC 的依赖 jar 文件 452
12.2.2 在 web.xml 中配置核心控制器 453
12.2.3 新建 springMVC-servlet.xml 配置文件 453
12.2.4 新建相关的 JSP 文件 453
12.2.5 新建控制层 Java 类文件 454
12.2.6 部署项目并运行 455
12.2.7 第一个示例的总结 456
12.2.8 Spring MVC 取参还能更加方便 456
12.3 执行控制层与限制提交的方式 457
12.3.1 新建控制层 ListUsername.java 文件 457

12.3.2　新建登录及显示数据的 JSP 文件　458
12.3.3　部署项目并测试　458
12.4　解决多人开发路径可能重复问题　460
12.4.1　错误的情况　460
12.4.2　解决办法　461
12.5　在控制层中处理指定的提交 get 或 post 方式　463
12.5.1　控制层代码　463
12.5.2　新建 JSP 文件并运行　464
12.6　控制层重定向到控制层——无参数传递　465
12.6.1　新建控制层 Java 文件　465
12.6.2　创建 JSP 文件并运行项目　466
12.7　控制层重定向到控制层——有参数传递　467
12.7.1　创建两个控制层 Java 文件　467
12.7.2　部署项目并运行　468
12.8　匹配 URL 路径执行指定控制层　468
12.8.1　新建控制层文件　468
12.8.2　部署项目并运行　469
12.9　在服务器端取得 JSON 字符串并解析——方式 1　470
12.9.1　在 web.xml 中配置字符编码过滤器　470
12.9.2　新建 JSP 文件　471
12.9.3　新建控制层 Java 文件　472
12.9.4　添加依赖的 jar 包文件　472
12.9.5　运行项目　473
12.10　在服务器端取得 JSON 字符串并解析——方式 2　473
12.10.1　新建封装 JSON 对象属性的实体类　473
12.10.2　新建控制层　474
12.10.3　在配置文件中添加<mvc:annotation-driven />注解　474
12.10.4　新建 JSP 文件　475
12.10.5　添加 jacksonJSON 解析处理类库并运行　475
12.10.6　解析不同格式的 JSON 字符串示例　476
12.11　将 URL 中的参数转成实体的示例　478
12.11.1　新建控制层文件　478
12.11.2　新建登录用途的 JSP 文件　479
12.11.3　在 web.xml 中注册编码过滤器　479
12.11.4　运行结果　479
12.12　在控制层返回 JSON 对象示例　479
12.12.1　新建控制层文件　480
12.12.2　新建 JSP 文件　480
12.12.3　部署项目并运行　481

12.13 在控制层传回 JSON 字符串示例　482
　12.13.1 新建控制层文件　482
　12.13.2 新建 JSP 文件及在配置文件中注册 utf-8 编码处理　482
　12.13.3 运行项目　483
12.14 在控制层取得 HttpServletRequest 和 HttpServletResponse 对象　483
　12.14.1 新建控制层　484
　12.14.2 JSP 文件中的 EL 代码及运行结果　484
　12.14.3 直接使用 HttpServletResopnse 对象输出响应字符　484
12.15 通过 URL 参数访问指定的业务方法　486
　12.15.1 新建控制层文件 List.java　486
　12.15.2 运行结果　487
12.16 Spring 4 MVC 单文件上传——写法 1　487
　12.16.1 新建控制层　487
　12.16.2 在配置文件 springMVC-servlet.xml 中声明上传请求　488
　12.16.3 创建前台 JSP 文件　489
　12.16.4 程序运行结果　489
12.17 Spring 4 MVC 单文件上传——写法 2　489
12.18 Spring 4 MVC 多文件上传　490
　12.18.1 新建控制层及 JSP 文件　490
　12.18.2 运行结果　491
12.19 Spring 4 MVC 支持中文文件名的文件下载　491
12.20 控制层返回 List 对象及实体的效果　493
　12.20.1 新建控制层文件　493
　12.20.2 新建 JSP 文件　493
　12.20.3 更改 springMVC-servlet.xml 配置文件　494
　12.20.4 程序运行结果　494
12.21 控制层 ModelMap 对象　495
　12.21.1 新建控制层　495
　12.21.2 JSP 文件代码　496
　12.21.3 运行效果　496
12.22 Spring 4 MVC 提交的表单进行手动数据验证　497
　12.22.1 创建控制层文件　497
　12.22.2 创建 JSP 文件　497
　12.22.3 运行结果　498

第 13 章　Spring 4 MVC 必备知识　499

13.1　web.xml 中的不同配置方法　499

13.1.1 存放于 src 资源路径中　499
13.1.2 指定存放路径　500
13.1.3 指定多个配置文件　500

13.2 路径中添加通配符的功能　501

13.3 业务逻辑层在控制层中进行注入　502
13.3.1 新建业务逻辑层　502
13.3.2 创建控制层文件　502
13.3.3 设计 springMVC-servlet.xml 配置文件　502
13.3.4 程序运行结果　503
13.3.5 多个实现类的情况　503

13.4 对象 ModelAndView 的使用　504
13.4.1 创建控制层及 JSP 文件　504
13.4.2 程序运行结果　505

13.5 控制层返回 void 数据的情况　505
13.5.1 创建控制层及 index.jsp 文件　505
13.5.2 更改配置文件　506
13.5.3 部署项目运行程序　506

13.6 使用 Spring 4 MVC 中的注解来操作 HttpSession 中的对象　507
13.6.1 创建控制层文件 PutGetSession.java　507
13.6.2 创建显示不同作用域中的值的 JSP 文件　508
13.6.3 部署项目并运行程序　508

第 14 章　Spring 4 MVC+MyBatis 3+Spring 4 整合　509

14.1 准备 Spring 4 的 JAR 包文件　509
14.2 准备 MyBatis 的 JAR 包文件　510
14.3 准备 MyBatis 3 与 Spring 4 整合的 JAR 文件　510
14.4 创建 Web 项目　510
14.5 配置 web.xml 文件　511
14.6 配置 springMVC-servlet.xml 文件　512
14.7 配置 MyBatis 配置文件　513
14.8 创建 MyBatis 与映射有关文件　513
14.9 配置 applicationContext.xml 文件　514
14.10 创建 Service 对象　516
14.11 创建 Controller 对象　516
14.12 测试正常的效果　517
14.13 测试回滚的效果　517

第 1 章　MyBatis 3 操作数据库

本章讲解的是 MyBatis 3 框架，此框架的主要作用就是更加便携地操作数据库，比如可以将 ResultSet 对象返回的数据自动封装进 Entity 实体类或 List 中，可以把 SQL 语句配置到 XML 文件中，也就是将 SQL 语句与*.java 文件进行分离，有利于代码的后期维护，也使代码的分层更加明确。另外由于 MyBatis 框架是使用 SQL 语句对数据库进行操作的，所以可以单独对 SQL 语句进行优化，以提高查询效率，这点与使用 Hibernate 框架相比有很大的优势，这也是为什么现阶段大部分的软件企业逐步用 MyBatis 替换掉 Hibernate 框架的主要原因。

MyBatis 是一个操作数据库的框架，那什么是框架？框架就是软件功能的半成品。框架提供了一个软件项目中通用的功能，将大多数常见的功能进行封装，无需自己重复开发，框架增加了开发及运行效率。

需要说明的是，MyBatis 并不是一个独立的技术，它内部操作数据库的原理还是使用 JDBC，只是对 JDBC 进行了轻量级的封装，便于程序员更方便地设计代码去操作数据库。

在本章中，读者应该着重掌握如下内容：
- 使用基于 Eclipse 的 MyBatis Generator 插件执行 CURD（Create、Update、Read、Delete）增删改查操作，以增加对 MyBatis API 的认识；
- 使用 MyBatis 框架操作常用的数据库，比如 Oracle、MySQL 等；
- 理解 MyBatis 框架中核心对象的生命周期，以便对 CURD 的操作进行封装；
- MyBatis 结合 ThreadLocal 类进行 CURD 的封装。

1.1　MyBatis 介绍

为什么要使用 MyBatis 框架呢？这个答案真的有很多，作者认为其中最具有代表性的就是在使用传统的 JDBC 代码时，需要写上必要的 DAO 层代码，在 DAO 层的查询代码中将数据表中的数据封装到自定义的实体类中。这样的代码写法在软件开发的过程中非常不便，因为大部分的代码都是重复冗余的，几乎每个 DAO 类都在做同样的事情，但 MyBatis 解决了这类问题。使用 MyBatis 做查询时可以自动将数据表中的数据记录封装到实体类或 Map 中，再将它们放入 List 进行返回，这么常见而且有利于提高开发效率的功能 MyBatis 都可以自由方便地处理。从此观点来看，使用 MyBatis 框架去开发应用软件是非常方便快捷，MyBatis 框架很有使用的必要性。

MyBatis 是一个"持久化 ORM 框架"，持久化是指内存中的数据保存到硬盘上。ORM 是对象关系映射（Object Relation Mapping），ORM 的概念可以从 2 个方面来介绍：

- 从大的方面来说，1 个类对应表中的 1 行；
- 从小的方面来说，1 个类中的属性对应 1 个表中的列。

也就是在使用 MyBatis 框架时，可以将 Java 类转化成数据表中的记录，或者将数据表中的记录转化成 Java 类，内部的技术原理其实就是 JDBC+反射。这些功能都是由 ORM 框架来进行处理的，MyBatis 可以实现这样的功能。MyBatis 有不同的语言版本，比如.Net 和 Java 都有对应的类库，它有大多数 ORM 框架都具有的功能，比如可以自定义 SQL 语句、调用存储过程、进行高级映射等。

下面来看一下 ORM 框架的技术原理，使用如下的代码进行实现。

创建类 Userinfo.java 代码如下：

```java
package entity;

public class Userinfo {

    private long id;
    private String username;
    private String password;

    public Userinfo() {
    }

    public Userinfo(long id, String username, String password) {
        super();
        this.id = id;
        this.username = username;
        this.password = password;
    }

    //get 和 set 方法忽略
}
```

运行类代码如下：

```java
package test;

import java.lang.reflect.Field;
import java.util.ArrayList;
import java.util.List;

import entity.Userinfo;

public class Test {

    public static void main(String[] args) throws IllegalArgumentException, IllegalAccessException {
        Userinfo userinfo = new Userinfo();
        userinfo.setId(100);
        userinfo.setUsername("中国");
        userinfo.setPassword("大中国");

        List valueList = new ArrayList();

        String sql = "insert into ";
        sql = sql + userinfo.getClass().getSimpleName().toLowerCase();

        String colSql = "";
```

```java
        String valueSql = "";
        Field[] fieldArray = userinfo.getClass().getDeclaredFields();
        for (int i = 0; i < fieldArray.length; i++) {
            Field eachField = fieldArray[i];
            eachField.setAccessible(true);
            String fieldName = eachField.getName();
            colSql = colSql + "," + fieldName;
            valueSql = valueSql + ",?";
            Object value = fieldArray[i].get(userinfo);
            valueList.add(value);
        }
        colSql = colSql.substring(1);
        valueSql = valueSql.substring(1);
        colSql = "(" + colSql + ")";
        valueSql = "(" + valueSql + ")";
        sql = sql + colSql + " " + valueSql;
        System.out.println(sql);
        System.out.println();
        for (int i = 0; i < valueList.size(); i++) {
            Object value = valueList.get(i);
            System.out.println(value);
        }

    }

}
```

程序运行后在控制台打印信息如下：

insert into userinfo(id,username,password) (?,?,?)

100
中国
大中国

这就是将一个 Userinfo.java 对象转成添加到 userinfo 数据表的 SQL 语句，也是 ORM 框架的原理。

但更严格来讲，MyBatis 是一种"半自动化"的 ORM 映射框架，它应该算作 SQL 映射框架（SQL mapper framework），官方也是这么介绍的，将 MyBatis 称为"半自动化的 ORM 框架"。这是因为 MyBatis 操作数据库时还是使用原始的 SQL 语句，这些 SQL 语句还是需要程序员自己来进行设计，这就是半自动化，只不过把重复的 JDBC 代码进行了封装与简化。在使用的方式上和全自动的 ORM 框架 Hibernate 有着非常大的区别，MyBatis 是以 SQL 语句为映射基础，而 Hibernate 是彻底的基于实体类与表进行映射，基本是属于全自动化的 ORM 映射框架。但正是 MyBatis 属于半自动化的 ORM 映射框架这个特性，所以可以将 SQL 语句灵活多变的特性溶入到项目开发中。

另外，如果使用 MyBatis 框架，还可以省略大多数的 JDBC 代码，因为它把常用的 JDBC 操作都进行了封装，进而可以加快开发效率。MyBatis 可以使用 XML 或 Annotations 注解的方式将数据表中的记录映射成 1 个 Map 或 Java POJOs 实体对象。但还是推荐使用 XML 的方式，该方式也是 MyBatis 官方推荐的。

由于 MyBatis 是第三方的框架，javaee.jar 中并不包含它的 API，所以得单独进行下载。打开网页后看到的界面如图 1-1 所示。

图 1-1　MyBatis 官方网站

继续操作，单击 Download Latest 链接后打开的界面如图 1-2 所示。

图 1-2　准备下载 MyBatis 框架

到这一步已经把 MyBatis 框架从官网下载到本地，然后就可以使用它的 jar 文件进行开发了。下载的 zip 压缩文件中包含开发时要用到的 PDF 文档、source 源代码和 jar 文件等资源。

1.2　MyBatis 基本使用

开门见山永远是快速学习一门技术最好的方式。

MyBatis 框架的核心是 SqlSessionFactory 对象，从 SqlSessionFactory 类的名称来看，它是创建 SqlSession 对象的工厂。但 SqlSessionFactory 对象的创建来自于 SqlSessionFactoryBuilder 类，也就是使用 SqlSessionFactoryBuilder 类创建出 SqlSessionFactory 对象。

这 3 者之间的创建关系为：SqlSessionFactoryBuilder 创建出 SqlSessionFactory，SqlSessionFactory 创建出 SqlSession。

使用 SqlSessionFactoryBuilder 类创建 SqlSessionFactory 对象的方式可以来自于 1 个 XML 配置文件，还可以来自于 1 个实例化的 Configuration 对象，由于使用 XML 方式创建 SqlSessionFactory 对象在使用上比较广泛，所以在下面的小节就介绍这些内容。

1.2.1 使用 XML 配置文件创建 SqlSessionFactory 对象

创建名称为 mybatis1 的 Web 项目。

想要使用 MyBatis 实现 CURD 的操作，必须先要创建出 SqlSessionFactory 对象。

根据 XML 文件中的配置创建 SqlSessionFactory 对象的核心 Test.java 类的代码如下。

```java
package test;

import java.io.IOException;
import java.io.InputStream;

import org.apache.ibatis.io.Resources;
import org.apache.ibatis.session.SqlSessionFactory;
import org.apache.ibatis.session.SqlSessionFactoryBuilder;

public class Test {
    public static void main(String[] args) {
        try {
            String resource = "mybatis-config.xml";
            InputStream inputStream = Resources.getResourceAsStream(resource);
            SqlSessionFactory sqlSessionFactory = new SqlSessionFactoryBuilder()
                    .build(inputStream);
            System.out.println(sqlSessionFactory);
        } catch (IOException e) {
            // TODO Auto-generated catch block
            e.printStackTrace();
        }
    }
}
```

上面代码的主要作用就是取得 SqlSessionFactory 工厂对象。其中 mybatis-config.xml 配置文件连接数据库的内容如下。

```xml
<?xml version="1.0" encoding="UTF-8" ?>
<!DOCTYPE configuration
PUBLIC "-//mybatis.org//DTD Config 3.0//EN"
"http://mybatis.org/dtd/mybatis-3-config.dtd">
<configuration>
    <environments default="development">
        <environment id="development">
            <transactionManager type="JDBC" />
            <dataSource type="POOLED">
                <property name="driver" value="${driver}" />
                <property name="url" value="${url}" />
                <property name="username" value="${username}" />
```

```xml
            <property name="password" value="${password}" />
        </dataSource>
    </environment>
  </environments>
</configuration>
```

配置文件 mybatis-config.xml 主要作用就是如何连接数据库，包含连接数据库所用到的 username 和 password 及 url 等参数，但并没有实质的属性值，而是使用${xxxx}做为替代。因为在获取 SqlSessionFactory 工厂对象时，不需要提供这些具体的参数值。

添加最新版的 jar 包的项目结构如图 1-3 所示。

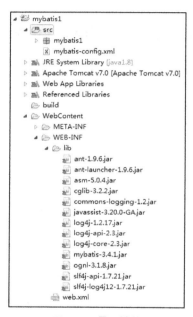

图 1-3　项目结构

来测试一下代码是否能正常创建出 SqlSessionFactory 类的实例，运行程序后并没有出现异常，输出信息如图 1-4 所示。

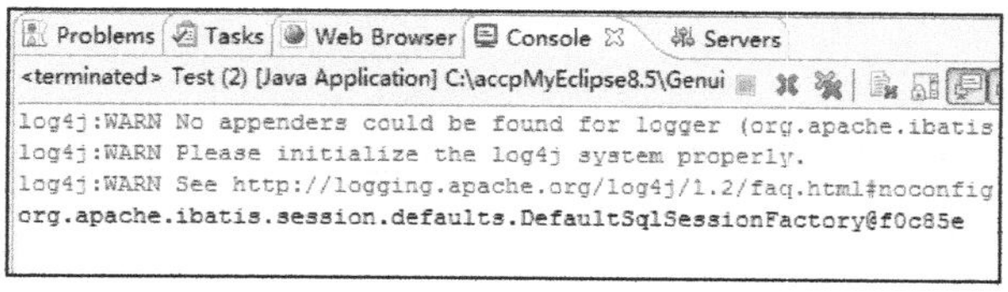

图 1-4　输出 SqlSessionFactory 对象

DefaultSqlSessionFactory.java 是 SqlSessionFactory.java 接口的实现类，具有实现关系，效果如图 1-5 所示。

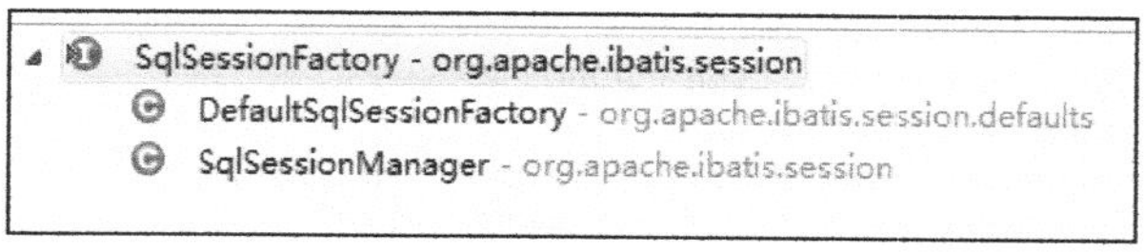

图 1-5　实现关系

到此，SqlSessionFactory 对象来自于 XML 配置文件创建成功。

1.2.2　SqlSessionFactoryBuilder 和 SqlSessionFactory 类信息

虽然现在正在学习新的知识，包括要熟悉新的类名，新的方法名称，新的包名等，但 MyBatis 在 API 的设计结构上是相当简洁的，大部分都是重载的方法。我们来看一看 SqlSessionFactoryBuilder 类的结构，效果如图 1-6 所示。

SqlSessionFactory 类的结构如图 1-7 所示。

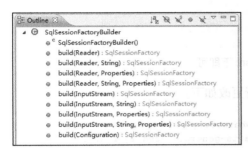

图 1-6　SqlSessionFactoryBuilder 类结构　　图 1-7　SqlSessionFactory 类结构

从图 1-6 和图 1-7 可以看到，两者的类结构中基本全是重载的方法，主要就是通过 build() 方法取得 SqlSessionFactory 对象，使用 openSession()方法取得 SqlSession 对象，对象 SqlSession 主要的作用是对数据库进行 CURD 操作。

1.2.3　使用 MyBatis Generator 工具逆向操作 Oracle 数据库

使用 MyBatis 实现操作数据库要使用到 SQL 映射文件，但该文件的配置代码比较复杂，这种情况也存在于 Hibernate 框架中，所以为了加快开发效率，MyBatis 官方提供 1 个 Eclipse 插件，名称为 MyBatis Generator，该插件主要的功能就是生成 SQL 映射文件。此插件需要在 Eclipse 中在线安装，安装的过程请参考搜索引擎提供的资料，下面来看此插件的使用。

新建名为 GeneratorOracle 的 Web 项目，然后在 Java 项目的 src 结点下单击鼠标右键新建 1 个 MyBatis 的 Generator 配置文件，如图 1-8 所示。

单击 "Next" 按钮出现如图 1-9 所示界面。

在图 1-9 界面中不需要更改配置，保持默认设置即可，单击 "Finish" 按钮完成 Generator 配置文件的创建。

图 1-8　创建生成 ORM 的配置 XML 文件

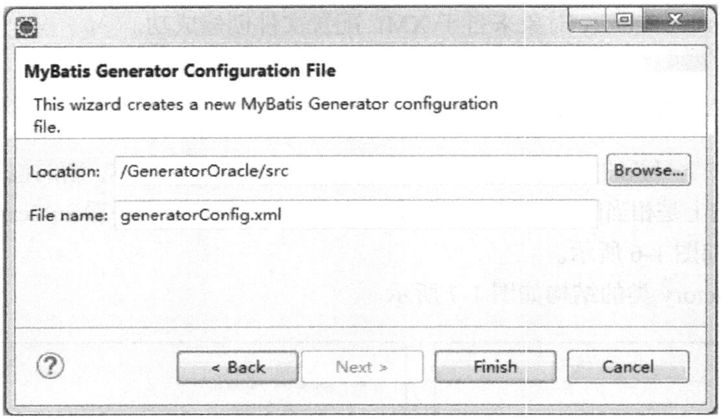

图 1-9　将文件放入 src 下即可

对生成的 generatorConfig.xml 配置文件代码进行更改如下。

```
<?xml version="1.0" encoding="UTF-8"?>
<!DOCTYPE generatorConfiguration PUBLIC "-//mybatis.org//DTD MyBatis Generator
Configuration 1.0//EN" "http://mybatis.org/dtd/mybatis-generator-config_1_0.dtd">
<generatorConfiguration>
    <context id="context1">
        <jdbcConnection connectionURL="jdbc:oracle:thin:@localhost:1521:orcl"
            driverClass="oracle.jdbc.OracleDriver" password="123" userId="y2" />
        <javaModelGenerator targetPackage="sqlmapping"
            targetProject="GeneratorOracle" />
        <sqlMapGenerator targetPackage="sqlmapping"
            targetProject="GeneratorOracle" />
        <javaClientGenerator targetPackage="sqlmapping"
            targetProject="GeneratorOracle" type="XMLMAPPER" />
        <table schema="y2" tableName="userinfo">
        </table>
    </context>
</generatorConfiguration>
```

配置文件 generatorConfig.xml 是 MyBatis Generator 插件中必备的文件，通过此文件可以将数据表的结构逆向出对应的 Java 类以及存储 SQL 语句的 SQL 映射文件，然后用 MyBatis 的 API 就可以对这些 Java 类进行操作，从而演变成对数据表的增删改查操作。

数据表 userinfo 的表结构如图 1-10 所示。

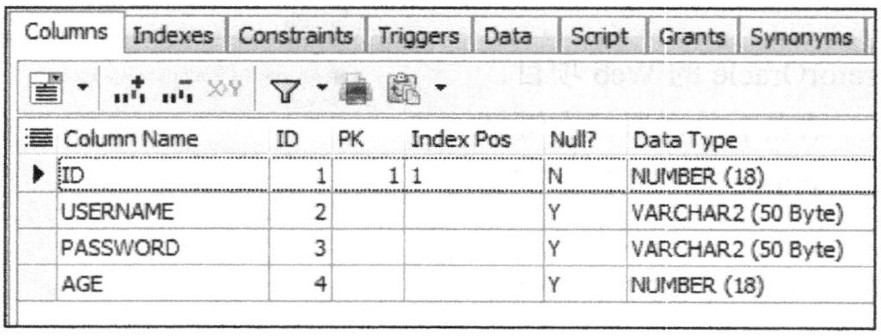

Column Name	ID	PK	Index Pos	Null?	Data Type
ID	1	1	1	N	NUMBER (18)
USERNAME	2			Y	VARCHAR2 (50 Byte)
PASSWORD	3			Y	VARCHAR2 (50 Byte)
AGE	4			Y	NUMBER (18)

图 1-10　userinfo 数据表结构

配置文件 generatorConfig.xml 准备就绪后单击图 1-11 中的菜单。

1.2 MyBatis 基本使用

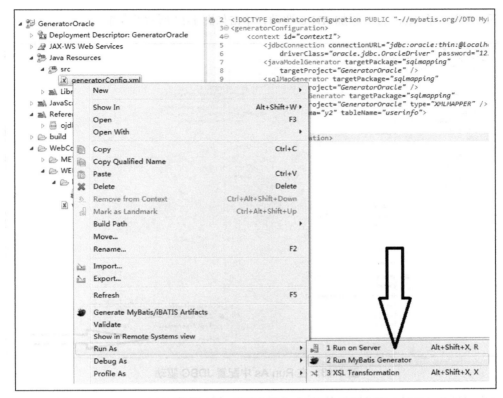

图 1-11 根据 XML 配置文件生成 ORM 映射文件

这里却出现了异常情况，效果如图 1-12 所示。

图 1-12 找不到 JDBC 驱动

下一步就要在 Run As 中添加 JDBC 驱动，单击如图 1-13 所示的菜单。

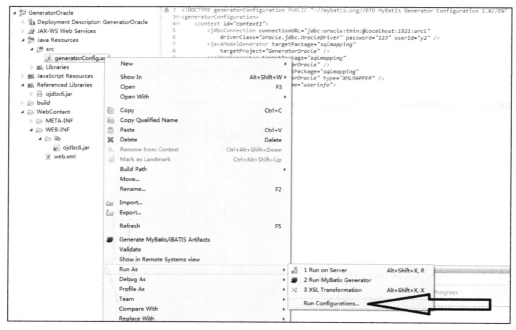

图 1-13　在 Run As 中配置 JDBC 驱动

弹出配置界面如图 1-14 所示，按（1）、（2）、（3）、（4）的步骤将 ojdbc6.jar 添加到 classpath 中。

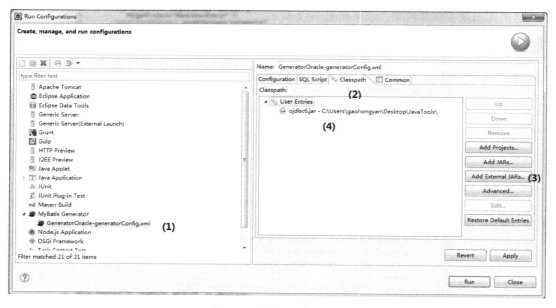

图 1-14　添加驱动 jar 包

驱动 jar 包添加成功后单击右下角的 Run 按钮开始逆向，逆向后的项目结构如图 1-15 所示。文件 UserinfoMapper.java 中出现了异常，原因是并没有添加 MyBatis 的 jar 包，添加成功后 Java 文件不再出现红叉，如图 1-16 所示。

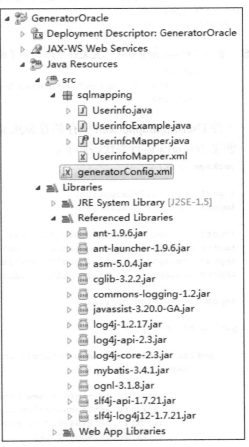

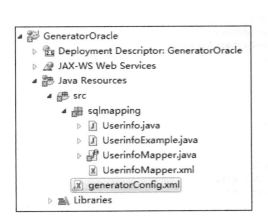

图 1-15　成功生成 ORM 映射文件　　　　图 1-16　MyEclipse 中 Web 项目结构

至此 MyBatis 操作数据库的基础文件已经准备完毕。

下面开始在 Oracle 数据库的 userinfo 数据表中添加 1 条记录。

在 src 中创建 mybatis-config.xml 文件，代码如下。

```xml
<?xml version="1.0" encoding="UTF-8" ?>
<!DOCTYPE configuration
PUBLIC "-//mybatis.org//DTD Config 3.0//EN"
"http://mybatis.org/dtd/mybatis-3-config.dtd">
<configuration>
    <environments default="development">
        <environment id="development">
            <transactionManager type="JDBC" />
            <dataSource type="POOLED">
                <property name="driver" value="oracle.jdbc.OracleDriver" />
                <property name="url" value="jdbc:oracle:thin:@localhost:1521:orcl" />
                <property name="username" value="y2" />
                <property name="password" value="123" />
            </dataSource>
        </environment>
    </environments>
```

```xml
        <mappers>
            <mapper resource="sqlmapping/UserinfoMapper.xml" />
        </mappers>
</configuration>
```

注意：在 mybatis-config.xml 配置文件中添加如下配置。

```xml
        <mappers>
            <mapper resource="sqlmapping/UserinfoMapper.xml" />
        </mappers>
```

文件 UserinfoMapper.xml 中存储着 SQL 语句，所以在配置文件中需要关联。

创建 Test.java 运行类代码如下。

```java
package test;

import java.io.IOException;
import java.io.InputStream;

import org.apache.ibatis.io.Resources;
import org.apache.ibatis.session.SqlSession;
import org.apache.ibatis.session.SqlSessionFactory;
import org.apache.ibatis.session.SqlSessionFactoryBuilder;

import sqlmapping.Userinfo;

public class Test {

    public static void main(String[] args) {
        try {

            Userinfo userinfo = new Userinfo();
            userinfo.setUsername("usernameValue");
            userinfo.setPassword("passwordValue");
            userinfo.setAge(100L);

            String resource = "mybatis-config.xml";
            InputStream inputStream = Resources.getResourceAsStream(resource);
            SqlSessionFactory sqlSessionFactory = new SqlSessionFactoryBuilder().build(inputStream);
            SqlSession sqlSession = sqlSessionFactory.openSession();
            sqlSession.insert("insert", userinfo);
            sqlSession.commit();
            sqlSession.close();

        } catch (IOException e) {
            // TODO Auto-generated catch block
            e.printStackTrace();
        }

    }

}
```

方法 main() 中的代码是一个经典的 insert 数据表的功能，从代码中可以看到 MyBatis 用最精简的 API 就可以完全地控制数据表中的记录，可见 MyBatis 不管是在学习还是开发等方面成本都是比较低的。

程序运行后在控制台打印出的异常信息如下。

Caused by: java.sql.SQLException: Error setting driver on UnpooledDataSource. Cause: java.lang.ClassNotFoundException: Cannot find class: oracle.jdbc.OracleDriver。

异常信息提示该项目中并没有添加 JDBC 的驱动 jar 文件，添加 ojdbc6.jar 并将其添加进 classpath 中再运行 Test.java 时依然出现了异常提示，如图 1-17 所示。

```
log4j:WARN No appenders could be found for logger (org.apache.ibatis.logging.LogFactory).
log4j:WARN Please initialize the log4j system properly.
log4j:WARN See http://logging.apache.org/log4j/1.2/faq.html#noconfig for more info.
Exception in thread "main" org.apache.ibatis.exceptions.PersistenceException:
### Error updating database.  Cause: java.sql.SQLIntegrityConstraintViolationException: ORA-01400: 无法将NULL 插入 ("Y2"."USERINFO"."ID")
### The error may involve sqlmapping.UserinfoMapper.insert-Inline
### The error occurred while setting parameters
### SQL: insert into Y2.USERINFO (ID, USERNAME, PASSWORD,        AGE)     values (?, ?, ?,        ?)
### Cause: java.sql.SQLIntegrityConstraintViolationException: ORA-01400: 无法将NULL 插入 ("Y2"."USERINFO"."ID")
    at org.apache.ibatis.exceptions.ExceptionFactory.wrapException(ExceptionFactory.java:30)
    at org.apache.ibatis.session.defaults.DefaultSqlSession.update(DefaultSqlSession.java:200)
    at org.apache.ibatis.session.defaults.DefaultSqlSession.insert(DefaultSqlSession.java:185)
    at test.Test.main(Test.java:26)
Caused by: java.sql.SQLIntegrityConstraintViolationException: ORA-01400: 无法将NULL 插入 ("Y2"."USERINFO"."ID")
```

图 1-17　不能对 id 主键列赋 null 空值

Oracle 数据库的主键并不是自增的，在 insert 语句中需要结合序列来实现添加记录的功能，逆向生成错误的 insert 语句如图 1-18 所示。

```xml
141  <insert id="insert" parameterType="sqlmapping.Userinfo">
142    <!--
143      WARNING - @mbg.generated
144      This element is automatically generated by MyBatis Generator, do not modify.
145      This element was generated on Tue Oct 11 20:31:06 CST 2016.
146    -->
147    insert into Y2.USERINFO (ID, USERNAME, PASSWORD,
148      AGE)
149    values (#{id,jdbcType=DECIMAL}, #{username,jdbcType=VARCHAR}, #{password,jdbcType=VARCHAR},
150      #{age,jdbcType=DECIMAL})
151  </insert>
```

图 1-18　SQL 语句中没有使用序列

这说明 sqlmapping 包中的文件都是错误的，所以需要连同 sqlmapping 包与下面的所有文件一同进行删除，因为要重新进行逆向了。

在逆向之前需要更改 generatorConfig.xml 配置文件中的配置，添加<generatedKey>标签后的配置代码如下。

```xml
<table schema="y2" tableName="userinfo">
    <generatedKey column="id" sqlStatement="select idauto.nextval from dual"
        identity="false" />
</table>
```

标签<generatedKey>的主要作用就是在生成 insert 的 SQL 语句时使用名称为 idauto 序列的 nextval 值来做为主键 id 值，identity="false"属性说明主键并不是自增的，而是由序列生成的。

如果想同时逆向多个表，则在 generatorConfig.xml 文件中写入多个<table>标签即可，示例代码如下。

```xml
<?xml version="1.0" encoding="UTF-8"?>
<!DOCTYPE generatorConfiguration PUBLIC "-//mybatis.org//DTD MyBatis Generator Configuration 1.0//EN" "http://mybatis.org/dtd/mybatis-generator-config_1_0.dtd">
```

```xml
<generatorConfiguration>
    <context id="context1">
        <jdbcConnection connectionURL="jdbc:oracle:thin:@localhost:1521:orcl"
            driverClass="oracle.jdbc.OracleDriver" password="123" userId="y2" />
        <javaModelGenerator targetPackage="mapping"
            targetProject="GeneratorOracle" />
        <sqlMapGenerator targetPackage="mapping"
            targetProject="GeneratorOracle" />
        <javaClientGenerator targetPackage="mapping"
            targetProject="GeneratorOracle" type="XMLMAPPER" />
        <table schema="y2" tableName="userinfo">
            <generatedKey column="id" sqlStatement="select idauto.nextval from dual"
                identity="false" />
        </table>
        <table schema="y2" tableName="A">
            <generatedKey column="id" sqlStatement="select idauto.nextval from dual"
                identity="false" />
        </table>
        <table schema="y2" tableName="B">
            <generatedKey column="id" sqlStatement="select idauto.nextval from dual"
                identity="false" />
        </table>
    </context>
</generatorConfiguration>
```

配置文件 generatorConfig.xml 准备结束后重新进行逆向操作，在 sqlmapping 包中生成最新版正确的*.java 和*.xml 文件，最新版的 insert 语句如图 1-19 所示。

```
<insert id="insert" parameterType="sqlmapping.Userinfo">
    <!--
      WARNING - @mbg.generated
      This element is automatically generated by MyBatis Generator, do not modify.
      This element was generated on Tue Oct 11 20:57:56 CST 2016.
    -->
    <selectKey keyProperty="id" order="BEFORE" resultType="java.lang.Long">
        select idauto.nextval from dual
    </selectKey>
    insert into Y2.USERINFO (ID, USERNAME, PASSWORD,
        AGE)
    values (#{id,jdbcType=DECIMAL}, #{username,jdbcType=VARCHAR}, #{password,jdbcType=VARCHAR},
        #{age,jdbcType=DECIMAL})
</insert>
```

图 1-19 新版的 insert 语句

再次运行 Test.java 类，成功在数据表中添加了 1 条记录，如图 1-20 所示。

ID	USERNAME	PASSWORD	AGE
1445434	usernameValue	passwordValue	100

图 1-20 成功添加 1 条记录

1.2.4 使用 MyBatis Generator 工具逆向操作 MySQL 数据库

MySQL 数据库中的 userinfo 数据表结构如图 1-21 所示。

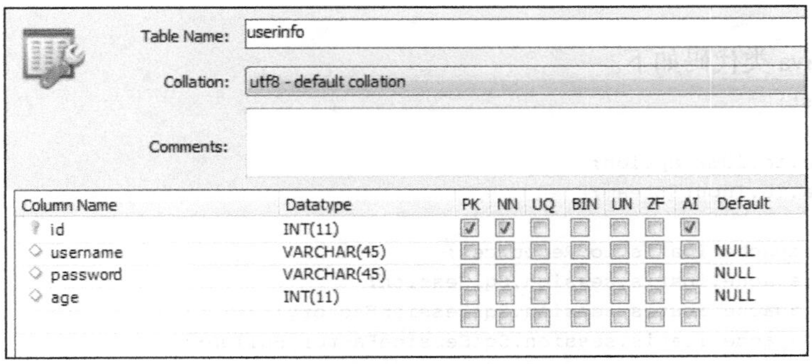

图 1-21 数据表 userinfo 结构

创建名称为 GeneratorMySQL 的项目，更改配置文件 generatorConfig.xml 的代码如下。

```xml
<?xml version="1.0" encoding="UTF-8"?>
<!DOCTYPE generatorConfiguration PUBLIC "-//mybatis.org//DTD MyBatis Generator Configuration 1.0//EN" "http://mybatis.org/dtd/mybatis-generator-config_1_0.dtd">
<generatorConfiguration>
    <context id="context1">
        <jdbcConnection connectionURL="jdbc:mysql://localhost:3306/y2"
            driverClass="com.mysql.jdbc.Driver" password="123" userId="root" />
        <javaModelGenerator targetPackage="sqlmapping"
            targetProject="GeneratorMySQL" />
        <sqlMapGenerator targetPackage="sqlmapping"
            targetProject="GeneratorMySQL" />
        <javaClientGenerator targetPackage="sqlmapping"
            targetProject="GeneratorMySQL" type="XMLMAPPER" />
        <table schema="y2" tableName="userinfo">
            <generatedKey column="id" sqlStatement="mysql" identity="true" />
        </table>
    </context>
</generatorConfiguration>
```

对 generatorConfig.xml 文件进行逆向。

在 src 中创建 mybatis-config.xml 文件，代码如下。

```xml
<?xml version="1.0" encoding="UTF-8" ?>
<!DOCTYPE configuration
PUBLIC "-//mybatis.org//DTD Config 3.0//EN"
"http://mybatis.org/dtd/mybatis-3-config.dtd">
<configuration>
    <environments default="development">
        <environment id="development">
            <transactionManager type="JDBC" />
            <dataSource type="POOLED">
                <property name="driver" value="com.mysql.jdbc.Driver" />
                <property name="url" value="jdbc:mysql://localhost:3306/y2" />
                <property name="username" value="root" />
                <property name="password" value="123" />
            </dataSource>
        </environment>
    </environments>
    <mappers>
        <mapper resource="sqlmapping/UserinfoMapper.xml" />
    </mappers>
</configuration>
```

运行 Test.java 类代码如下。

```java
package test;

import java.io.IOException;
import java.io.InputStream;

import org.apache.ibatis.io.Resources;
import org.apache.ibatis.session.SqlSession;
import org.apache.ibatis.session.SqlSessionFactory;
import org.apache.ibatis.session.SqlSessionFactoryBuilder;

import sqlmapping.Userinfo;

public class Test {

    public static void main(String[] args) {
        try {

            Userinfo userinfo = new Userinfo();
            userinfo.setUsername("usernameValue");
            userinfo.setPassword("passwordValue");
            userinfo.setAge(888);

            String resource = "mybatis-config.xml";
            InputStream inputStream = Resources.getResourceAsStream(resource);
            SqlSessionFactory sqlSessionFactory = new SqlSessionFactoryBuilder().build(inputStream);
            SqlSession sqlSession = sqlSessionFactory.openSession();
            sqlSession.insert("insert", userinfo);
            sqlSession.commit();
            sqlSession.close();

        } catch (IOException e) {
            // TODO Auto-generated catch block
            e.printStackTrace();
        }

    }

}
```

运行 Test.java 类后在 userinfo 数据表添加了新的记录，效果如图 1-22 所示。

id	username	password	age
16	usernameValue	passwordValue	888
NULL	NULL	NULL	NULL

图 1-22　数据表 userinfo 中的新记录

1.3　使用 MyBatis 实现 CURD-2 种数据库（Oracle-MySQL）

前面都是使用 MyBatis Generator 工具生成实体和 SQL 映射文件，并不能从基础上掌握 MyBatis 框架的使用，本小节将从 0 起步开始研究如何使用 MyBatis 框架实现经典功能 CURD，

1.3　使用 MyBatis 实现 CURD-2 种数据库（Oracle-MySQL）

主要针对 2 种主流数据库。

1.3.1　针对 Oracle 的 CURD

MyBatis 框架针对每一种数据库的操作都大同小异，本节将用示例的方式演示 Oracle 数据库的 CURD 操作。

1．准备开发环境

（1）创建数据表。

创建 userinfo 数据表，表结构如图 1-23 所示。

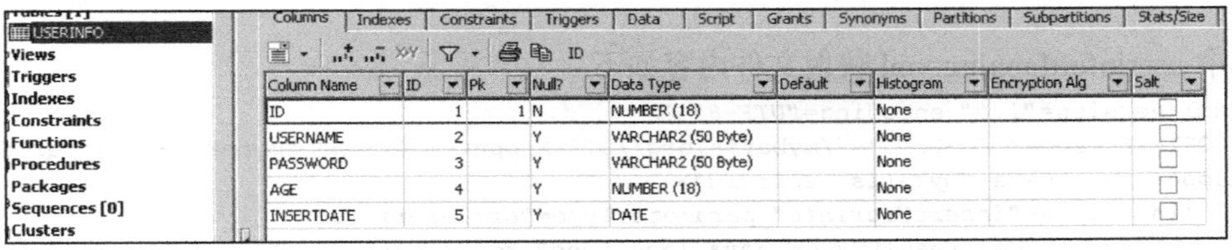

图 1-23　userinfo 数据表结构

（2）准备 generatorConfig.xml 逆向配置文件。
（3）创建名称为 mybatis_curd_oracle 的 Web 项目。
（4）创建 Userinfo.java 实体类，实体类 Userinfo.java 类结构如图 1-24 所示。
（5）项目结构如图 1-25 所示。

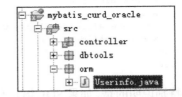

图 1-24　Userinfo.java 类结构　　　图 1-25　Eclipse 中的 orm 包中有实体

（6）在 Web 项目 mybatis_curd_oracle 中的 src 路径下创建连接数据库的配置文件 mybatis-config.xml，代码如下。

```
<?xml version="1.0" encoding="UTF-8" ?>
<!DOCTYPE configuration
PUBLIC "-//mybatis.org//DTD Config 3.0//EN"
"mybatis-3-config.dtd">
<configuration>
```

```xml
<environments default="development">
    <environment id="development">
        <transactionManager type="JDBC" />
        <dataSource type="POOLED">
            <property name="driver" value="oracle.jdbc.driver.OracleDriver" />
            <property name="url" value="jdbc:oracle:thin:@localhost:1522:accp11g" />
            <property name="username" value="ghy" />
            <property name="password" value="123" />
        </dataSource>
    </environment>
</environments>
<mappers>
    <mapper resource="userinfoMapping.xml" />
</mappers>
</configuration>
```

其中 userinfoMapping.xml 映射文件内容如下。

```xml
<?xml version="1.0" encoding="UTF-8" ?>
<!DOCTYPE mapper PUBLIC "-//mybatis.org//DTD Mapper 3.0//EN" "mybatis-3-mapper.dtd">
<mapper namespace="mybatis.testcurd">
    <insert id="insertUserinfo" parameterType="orm.Userinfo">
        <selectKey resultType="java.lang.Long" keyProperty="id"
            order="BEFORE">
            select idauto.nextval from dual
        </selectKey>
        insert into
        userinfo(id,username,password,age,insertDate)
        values(#{id},#{username},#{password},#{age},#{insertdate})
    </insert>
    <select id="getUserinfoById" parameterType="int" resultType="orm.Userinfo">
        select * from
        userinfo where id=#{id}
    </select>
    <delete id="deleteUserinfoById" parameterType="int">
        delete from
        userinfo where id=#{id}
    </delete>
    <select id="getAllUserinfo" resultType="orm.Userinfo">
        select * from userinfo
    </select>
    <update id="updateUserinfoById" parameterType="orm.Userinfo">
        update userinfo
        set
        username=#{username},password=#{password},age=#{age},insertDate=#{insertdate}
        where id=#{id}
    </update>
</mapper>
```

需要特别说明的是，如果 SQL 语句有一些特殊符号，则必须使用如下的格式进行设计 SQL 语句。

```
<![CDATA[ sql 语句 ]]>
```

其中配置代码如下所示。

```xml
<selectKey resultType="java.lang.Long" keyProperty="id"
    order="BEFORE">
    select idauto.nextval from dual
</selectKey>
```

标记<selectKey>的 order="BEFORE"属性含义是 select 语句比 insert 语句先执行，resultType 属性值为"java.lang.Long"是将序列返回的数字转成 Long 类型，keyProperty="id"的作用是将这个 Long 值放入 parameterType 的 Userinfo 的 id 属性里面。此段配置的主要功能是根据序列对象生成一个主键 id 值，并且此值还可以在代码中获取，也就是插入一条记录后使用程序代码就可以获取刚才插入记录的 id 值。

属性 parameterType 定义参数类型，属性 resultType 定义返回值的类型。

继续创建测试用的 Servlet 对象，完整的项目结构如图 1-26 所示。

在图中可以发现 src 路径下有两个 dtd 文件，这是为了在开发 XML 配置或 XML 映射文件时实现代码自动提示功能。

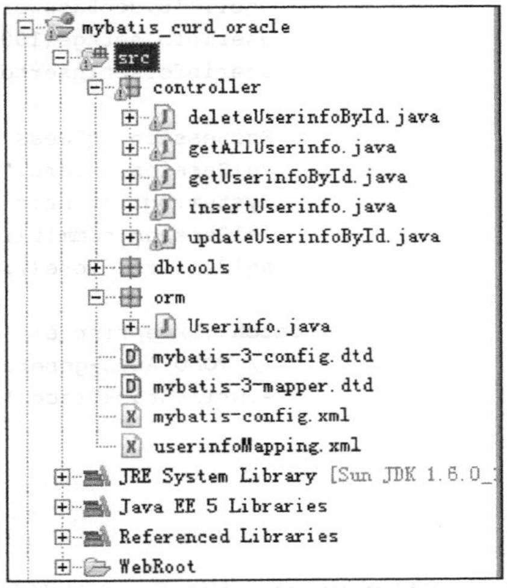

图 1-26 完整项目结构

2. 创建获取 SqlSession 对象工具类

核心代码如下。

```java
package dbtools;

import java.io.IOException;
import java.io.InputStream;

import org.apache.ibatis.io.Resources;
import org.apache.ibatis.session.SqlSession;
import org.apache.ibatis.session.SqlSessionFactory;
import org.apache.ibatis.session.SqlSessionFactoryBuilder;

public abstract class GetSqlSession {

    public static SqlSession getSqlSession() throws IOException {
        String resource = "mybatis-config.xml";
        InputStream inputStream = Resources.getResourceAsStream(resource);
        SqlSessionFactory sqlSessionFactory = new SqlSessionFactoryBuilder()
                .build(inputStream);
        SqlSession sqlSsession = sqlSessionFactory.openSession();
        return sqlSsession;
    }

}
```

3. 插入多条记录

创建 Servlet，核心代码如下。

```java
public class insertUserinfo extends HttpServlet {

    public void doGet(HttpServletRequest request, HttpServletResponse response)
            throws ServletException, IOException {

        try {
```

```
                Userinfo userinfo = new Userinfo();
                userinfo.setUsername("高洪岩");
                userinfo.setPassword("岩洪高");
                userinfo.setAge(100L);
                userinfo.setInsertdate(new Date());

                SqlSession sqlSession = GetSqlSession.getSqlSession();
                sqlSession.insert("mybatis.testcurd.insertUserinfo", userinfo);
                System.out.println(userinfo.getId());
                sqlSession.commit();
                sqlSession.close();

            } catch (Exception e) {
                // TODO Auto-generated catch block
                e.printStackTrace();
            }

        }
    }
```

变量 SqlSession 的 insert 方法的第 1 个参数是 userinfoMapping.xml 映射文件<insert>标签的 id 值，还要加上 namespace 命名空间的前缀，映射文件部分代码如下。

```
<?xml version="1.0" encoding="UTF-8" ?>
<!DOCTYPE mapper PUBLIC "-//mybatis.org//DTD Mapper 3.0//EN" "mybatis-3-mapper.dtd">
<mapper namespace="mybatis.testcurd">
    <insert id="insertUserinfo" ……>
```

在代码中可以看到获取已经插入数据表中记录的主键值。

执行 Servlet 后在控制台输出如图 1-27 所示。

Oracle 数据库中的 userinfo 数据表内容如图 1-28 所示。

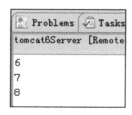

图 1-27　插入多条记录

图 1-28　出现 3 条数据记录

4．根据 id 值查询记录

创建 Servlet，核心代码如下。

```
public class getUserinfoById extends HttpServlet {

    public void doGet(HttpServletRequest request, HttpServletResponse response)
            throws ServletException, IOException {
        try {
            SqlSession sqlSession = GetSqlSession.getSqlSession();
            Userinfo userinfo = sqlSession.selectOne(
                    "mybatis.testcurd.getUserinfoById", 7);
            System.out.println(userinfo.getId());
            System.out.println(userinfo.getUsername());
```

```
            System.out.println(userinfo.getPassword());
            System.out.println(userinfo.getAge());
            System.out.println(userinfo.getInsertdate().toLocaleString());
            sqlSession.commit();
            sqlSession.close();
        } catch (Exception e) {
            // TODO Auto-generated catch block
            e.printStackTrace();
        }
    }
}
```

程序运行后的效果如图 1-29 所示。

图 1-29 打印 id 是 7 的信息

5. 查询所有记录

创建 Servlet，核心代码如下。

```
public class getAllUserinfo extends HttpServlet {

    public void doGet(HttpServletRequest request, HttpServletResponse response)
            throws ServletException, IOException {

        try {

            SqlSession sqlSession = GetSqlSession.getSqlSession();
            List<Userinfo> listUserinfo = sqlSession
                    .selectList("mybatis.testcurd.getAllUserinfo");
            for (int i = 0; i < listUserinfo.size(); i++) {
                Userinfo userinfo = listUserinfo.get(i);
                System.out.println(userinfo.getId() + " "
                        + userinfo.getUsername() + " " + userinfo.getPassword()
                        + " " + userinfo.getAge() + " "
                        + userinfo.getInsertdate().toLocaleString());

            }
            sqlSession.commit();
            sqlSession.close();
        } catch (Exception e) {
            // TODO Auto-generated catch block
            e.printStackTrace();
        }
    }

}
```

程序运行后打印出 3 条记录，如图 1-30 所示。

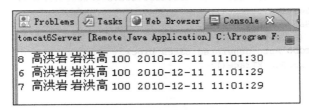

图 1-30 打印 3 条记录信息

6. 更新记录

创建 Servlet，核心代码如下。

```
public class updateUserinfoById extends HttpServlet {

    public void doGet(HttpServletRequest request, HttpServletResponse response)
            throws ServletException, IOException {

        try {
            SqlSession sqlSession = GetSqlSession.getSqlSession();
            Userinfo userinfo = sqlSession.selectOne(
                    "mybatis.testcurd.getUserinfoById", 7);
            userinfo.setUsername("最新版高洪岩");
            sqlSession.update("mybatis.testcurd.updateUserinfoById", userinfo);

            sqlSession.commit();
            sqlSession.close();
        } catch (Exception e) {
            // TODO Auto-generated catch block
            e.printStackTrace();
        }
    }

}
```

程序运行后数据表 userinfo 中的记录被更新，如图 1-31 所示。

	ID	USERNAME	PASSWORD	AGE	INSERTDATE
	8	高洪岩	岩洪高	100	2010-12-11 11:01:30
▶	6	高洪岩	岩洪高	100	2010-12-11 11:01:29
	7	最新版高洪岩	岩洪高	100	2010-12-11 11:01:29

图 1-31 userinfo 数据表内容被更新

7. 删除记录

创建 Servlet，核心代码如下。

```
public class deleteUserinfoById extends HttpServlet {

    public void doGet(HttpServletRequest request, HttpServletResponse response)
            throws ServletException, IOException {

        try {
            SqlSession sqlSession = GetSqlSession.getSqlSession();
            sqlSession.delete("mybatis.testcurd.deleteUserinfoById", 6);
            sqlSession.commit();
            sqlSession.close();
        } catch (Exception e) {
            // TODO Auto-generated catch block
            e.printStackTrace();
        }
    }

}
```

程序运行后，将 id 为 6 的记录删除了，如图 1-32 所示。

ID	USERNAME	PASSWORD	AGE	INSERTDATE
8	高洪岩	岩洪高	100	2010-12-11 11:01:30
7	最新版高洪岩	岩洪高	100	2010-12-11 11:01:29

图 1-32　不存在 id 为 6 的记录

至此，针对 Oracle 数据库的 CURD 操作到此结束。

1.3.2　针对 MySQL 的 CURD

本章节也将从零开始，使用 MyBatis 操作 MySQL 数据库。

1．准备开发环境

（1）在 Eclipse 中创建名称为 mybatis_curd_mysql 的项目。

（2）创建 Userinfo.java 实体类。

（3）创建 mybatis-config.xml 配置文件，代码如下。

```
<?xml version="1.0" encoding="UTF-8" ?>
<!DOCTYPE configuration
PUBLIC "-//mybatis.org//DTD Config 3.0//EN"
"mybatis-3-config.dtd">
<configuration>
    <environments default="development">
        <environment id="development">
            <transactionManager type="JDBC" />
            <dataSource type="POOLED">
                <property name="driver" value="com.mysql.jdbc.Driver" />
                <property name="url" value="jdbc:mysql://localhost:3307/ghydb" />
                <property name="username" value="root" />
                <property name="password" value="123" />
            </dataSource>
        </environment>
    </environments>
    <mappers>
        <mapper resource="userinfoMapping.xml" />
    </mappers>
</configuration>
```

（4）创建 SQL 映射文件 userinfoMapping.xml，代码如下。

```
<?xml version="1.0" encoding="UTF-8" ?>
<!DOCTYPE mapper PUBLIC "-//mybatis.org//DTD Mapper 3.0//EN" "mybatis-3-mapper.dtd">
<mapper namespace="mybatis.testcurd">
    <insert id="insertUserinfo" parameterType="orm.Userinfo"
        useGeneratedKeys="true" keyProperty="id">
        insert into
        userinfo(username,password,age,insertDate)
        values(#{username},#{password},#{age},#{insertdate})
    </insert>
    <select id="getUserinfoById" parameterType="int" resultType="orm.Userinfo">
        select * from
        userinfo where id=#{id}
```

```xml
        </select>
        <delete id="deleteUserinfoById" parameterType="int">
            delete from
            userinfo where id=#{id}
        </delete>
        <select id="getAllUserinfo" resultType="orm.Userinfo">
            select * from userinfo
        </select>
        <update id="updateUserinfoById" parameterType="orm.Userinfo">
            update userinfo
            set
            username=#{username},password=#{password},age=#{age},insertDate=#{insertdate}
            where id=#{id}
        </update>
</mapper>
```

MySQL 的主键是自增的，所以不需要序列这种机制。

2．增加一条记录并且返回主键值

创建 Servlet 对象，插入记录的代码和 Oracle 数据库对应的 Servlet 代码一致，运行后在控制台打印 4 条记录的主键 ID 值，效果如图 1-33 所示。

数据表 userinfo 中的内容如图 1-34 所示。

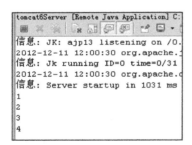

图 1-33　向 MySQL 数据库中插入 4 条记录　　图 1-34　数据表 userinfo 中的 4 条记录

3．其他业务方法的测试

其他的业务方法代码和操作 Oracle 数据库的大体一致，并且已经成功运行，详细代码请参看随书下载的源代码。

1.3.3　SQL 映射文件中 namespace 命名空间的作用

当多个 SQL 映射文件中的 id 值一样时，在使用 SqlSession 操作数据库时会出现异常。

创建测试用的 namespaceError 项目，创建 userinfoMapping.xml 映射文件代码如下。

```xml
<mapper namespace="AAAAA">
    <insert id="insertUserinfo" parameterType="orm.Userinfo">
        <selectKey resultType="java.lang.Long" keyProperty="id"
            order="BEFORE">
            select idauto.nextval from dual
        </selectKey>
        insert into
```

```xml
            userinfo(id,username,password,age)
            values(#{id},#{username},#{password},#{age})
    </insert>
</mapper>
```

再创建名称为 userinfoMappingZZZZZZZZ.xml 映射文件，代码如下。

```xml
<mapper namespace="BBBBB">
    <insert id="insertUserinfo" parameterType="orm.Userinfo">
        <selectKey resultType="java.lang.Long" keyProperty="id"
            order="BEFORE">
            select idauto.nextval from dual
        </selectKey>
        insert into
        userinfo(id,username,password,age)
        values(#{id},#{username},#{password},#{age})
    </insert>
</mapper>
```

将这 2 个 SQL 映射文件使用 `<mapper>` 标签注册到 mybatis-config.xml 配置文件中，代码如下。

```xml
<configuration>
    <environments default="development">
        <environment id="development">
            <transactionManager type="JDBC" />
            <dataSource type="POOLED">
                <property name="driver" value="oracle.jdbc.driver.OracleDriver" />
                <property name="url" value="jdbc:oracle:thin:@localhost:1521:orcl" />
                <property name="username" value="y2" />
                <property name="password" value="123" />
            </dataSource>
        </environment>
    </environments>
    <mappers>
        <mapper resource="userinfoMapping.xml" />
        <mapper resource="userinfoMappingZZZZZZZZ.xml" />
    </mappers>
</configuration>
```

在 2 个 SQL 映射文件中都有 insert 的 id 属性值为 insertUserinfo 的配置代码，执行如下 Java 代码。

```java
public class insertUserinfo {

    public static void main(String[] args) {
        try {
            Userinfo userinfo = new Userinfo();
            userinfo.setUsername("高洪岩");
            userinfo.setPassword("岩洪高");
            userinfo.setAge(100L);
            userinfo.setInsertdate(new Date());

            SqlSession sqlSession = GetSqlSession.getSqlSession();
            sqlSession.insert("insertUserinfo", userinfo);
            System.out.println(userinfo.getId());
            sqlSession.commit();
            sqlSession.close();
```

```
            } catch (Exception e) {
                e.printStackTrace();
            }
        }
    }
```

执行后出现了异常，异常信息如下。

java.lang.IllegalArgumentException: insertUserinfo is ambiguous in Mapped Statements collection (try using the full name including the namespace, or rename one of the entries)

异常信息提示 insertUserinfo 并不是确定的名字，是模糊的，可以尝试使用全名称的方式来调用此 SQL 映射，所谓的"全名称"方式就是指 sqlId 之前写上 namspace 命名空间，对 namespace 命名空间的命名可以写上表的名称，或者是业务的名称，这样有助于区分重复的 sqlId。

更改后的 Java 代码如下。

```
public class insertUserinfoOK {

    public static void main(String[] args) {
        try {
            Userinfo userinfo = new Userinfo();
            userinfo.setUsername("高洪岩");
            userinfo.setPassword("岩洪高");
            userinfo.setAge(100L);

            SqlSession sqlSession = GetSqlSession.getSqlSession();
            sqlSession.insert("BBBBB.insertUserinfo", userinfo);
            System.out.println(userinfo.getId());
            sqlSession.commit();
            sqlSession.close();
        } catch (Exception e) {
            e.printStackTrace();
        }
    }

}
```

执行 insert() 方法时在第 1 个参数加上了命名空间 BBBBB，使用小数点（.）做为间隔，再执行程序就不会出现异常了。

1.4 MyBatis 核心对象的生命周期与封装

在前面对 2 种主流数据库实现基本的 CURD 后，对 MyBatis 核心对象在使用上应该不再陌生，在本中将会继续介绍一下这些核心对象的生命周期。

对象的生命周期也就是对象从创建到销毁的过程，但在此过程中，如果实现的代码质量不太优质，那么很容易造成程序上的错误或效率的降低。

- SqlSessionFactoryBuilder 对象可以被 JVM 虚拟机所实例化、使用或者销毁。一旦你使用 SqlSessionFactoryBuilder 对象创建了 SqlSessionFactory 后，SqlSessionFactoryBuilder 类就不需要存在了，也就是不需要保持此对象的状态，可以随意的任由 JVM 销毁，因

- 此 SqlSessionFactoryBuilder 对象的最佳使用范围是方法之内，也就是说可以在方法内部声明 SqlSessionFactoryBuilder 对象来创建 SqlSessionFactory 对象。
- SqlSessionFactory 对象是由 SqlSessionFactoryBuilder 对象创建而来。一旦 SqlSessionFactory 类的实例被创建，该实例应该在应用程序执行期间都存在，根本不需要在每一次操作数据库时都重新创建它，所以应用它的最佳方式就是写一个单例模式，或使用 Spring 框架来实现单例模式对 SqlSessionFactory 对象进行有效的管理。
- SqlSession 对象是由 SqlSessionFactory 类创建，需要注意的是，每个线程都应该有它自己的 SqlSession 实例。SqlSession 的实例不能共享使用，因为它是线程不安全的，所以千万不要在 Servlet 中声明该对象的 1 个实例变量。因为 Servlet 是单例的，声明成实例变量会造成线程安全问题，也绝不能将 SqlSession 对象放在一个类的静态字段甚至是实例字段中，还不可以将 SqlSession 对象放在 HttpSession 会话或 ServletContext 上下文中。在接收到 HTTP 请求时，可以打开了 1 个 SqlSession 对象操作数据库，然后返回响应，就可以关闭它了。关闭 SqlSession 很重要，你应该确保使用 finally 块来关闭它，下面的示例就是一个确保 SqlSession 对象正常关闭的基本模式代码。

```java
public class insertUserinfo extends HttpServlet {
    public void doGet(HttpServletRequest request, HttpServletResponse response)
            throws ServletException, IOException {
        SqlSession sqlSession = GetSqlSession.getSqlSession();
        try {
            // sqlSession curd code
            sqlSession.commit();
        } catch (Exception e) {
            sqlSession.rollback();
            e.printStackTrace();
        } finally {
            sqlSession.close();
        }
    }
}
```

1.4.1 创建 GetSqlSessionFactory.java 类

根据前面学习过的生命周期的知识，在后面的章节将对 MyBatis 核心代码进行封装，这样更有助于对数据 CURD 的方便性，创建 Web 项目，名称为 mybatis_threadlocal。

创建 GetSqlSessionFactory.java 类，完整代码如下。

```java
package dbtools;

import java.io.IOException;
import java.io.InputStream;

import org.apache.ibatis.io.Resources;
import org.apache.ibatis.session.SqlSessionFactory;
import org.apache.ibatis.session.SqlSessionFactoryBuilder;

public class GetSqlSessionFactory {

    private static SqlSessionFactory sqlSessionFactory;
```

```java
    private GetSqlSessionFactory() {
    }

    synchronized public static SqlSessionFactory getSqlSessionFactory() {
        try {
            if (sqlSessionFactory == null) {
                String resource = "mybatis-config.xml";
                InputStream inputStream = Resources
                        .getResourceAsStream(resource);
                sqlSessionFactory = new SqlSessionFactoryBuilder()
                        .build(inputStream);
            } else {
            }
        } catch (IOException e) {
            // TODO Auto-generated catch block
            e.printStackTrace();
        }

        return sqlSessionFactory;
    }
}
```

在 GetSqlSessionFactory.java 类中使用单例设计模式来取得 SqlSessionFactory 对象。

1.4.2 创建 GetSqlSession.java 类

创建 GetSqlSession.java 类的核心代码如下。

```java
package dbtools;

import org.apache.ibatis.session.SqlSession;

public class GetSqlSession {

    private static ThreadLocal<SqlSession> tl = new ThreadLocal<SqlSession>();

    public static SqlSession getSqlSession() {
        SqlSession sqlSession = tl.get();
        if (sqlSession == null) {
            sqlSession = GetSqlSessionFactory.getSqlSessionFactory()
                    .openSession();
            tl.set(sqlSession);
        } else {

        }
        System.out.println("获得的 sqlSession 对象的 hashCode: " + sqlSession.hashCode());
        return sqlSession;
    }

    public static void commit() {
        if (tl.get() != null) {
            tl.get().commit();
            tl.get().close();
            tl.set(null);
            System.out.println("提交了");
```

```java
        }
    }
    public static void rollback() {
        if (tl.get() != null) {
            tl.get().rollback();
            tl.get().close();
            tl.set(null);
            System.out.println("回滚了");
        }
    }
}
```

1.4.3 创建 DBOperate.java 类

创建 DBOperate.java 类核心代码如下。

```java
package dbtools;

import java.util.List;
import java.util.Map;

import org.apache.ibatis.session.SqlSession;

public class DBOperate {

    public int insert(String sqlId, Map valueMap) {
        SqlSession sqlSession = GetSqlSession.getSqlSession();
        return sqlSession.insert(sqlId, valueMap);
    }

    public int delete(String sqlId, Map valueMap) {
        SqlSession sqlSession = GetSqlSession.getSqlSession();
        return sqlSession.delete(sqlId, valueMap);
    }

    public int update(String sqlId, Map valueMap) {
        SqlSession sqlSession = GetSqlSession.getSqlSession();
        return sqlSession.update(sqlId, valueMap);
    }

    public List<Map> select(String sqlId, Map valueMap) {
        SqlSession sqlSession = GetSqlSession.getSqlSession();
        return sqlSession.selectList(sqlId, valueMap);
    }

}
```

所有 CURD 的参数值都用 Map 对象进行封装，所以看一下 SQL 映射文件中的代码吧。

1.4.4 创建 userinfoMapping.xml 映射文件

创建 userinfoMapping.xml 映射文件的代码如下。

```xml
<?xml version="1.0" encoding="UTF-8" ?>
<!DOCTYPE mapper PUBLIC "-//mybatis.org//DTD Mapper 3.0//EN" "mybatis-3-mapper.dtd">
```

```xml
<mapper namespace="mybatis.testcurd">
    <insert id="insertUserinfo" parameterType="map"
        useGeneratedKeys="true" keyProperty="id">
        insert into
        userinfo(username,password,age,insertDate)
        values(#{username},#{password},#{age},#{insertdate})
    </insert>
    <select id="getUserinfoById" parameterType="map" resultType="map">
        select * from
        userinfo where id=#{id}
    </select>
    <delete id="deleteUserinfoById" parameterType="map">
        delete from
        userinfo where id=#{id}
    </delete>
    <select id="getAllUserinfo" resultType="map">
        select * from userinfo
    </select>
    <update id="updateUserinfoById" parameterType="map">
        update userinfo
        set
        username=#{username},password=#{password},age=#{age},insertDate=#{insertdate}
        where id=#{id}
    </update>
</mapper>
```

1.4.5 创建连接数据库 mybatis-config.xml 配置文件

创建连接数据库的 mybatis-config.xml 配置文件，代码如下。

```xml
<?xml version="1.0" encoding="UTF-8" ?>
<!DOCTYPE configuration
PUBLIC "-//mybatis.org//DTD Config 3.0//EN"
"mybatis-3-config.dtd">
<configuration>
    <environments default="development">
        <environment id="development">
            <transactionManager type="JDBC" />
            <dataSource type="POOLED">
                <property name="driver"
                    value="自定义的值" />
                <property name="url"
                    value="自定义的值" />
                <property name="username" value="自定义的值" />
                <property name="password" value="自定义的值" />
            </dataSource>
        </environment>
    </environments>
    <mappers>
        <mapper resource="userinfoMapping.xml" />
    </mappers>
</configuration>
```

1.4.6 创建名称为 test 的 Servlet 对象

该对象的主要作用就是测试在 1 个请求中多次获取 SqlSession 对象是不是 1 个，核心代码如下。
```java
public class test extends HttpServlet {
    public void doGet(HttpServletRequest request, HttpServletResponse response)
            throws ServletException, IOException {
        GetSqlSession.getSqlSession();
        GetSqlSession.getSqlSession();
        GetSqlSession.getSqlSession();
        GetSqlSession.getSqlSession();
        GetSqlSession.getSqlSession();
    }
}
```

程序运行后，在控制台输出信息如图 1-35 所示。

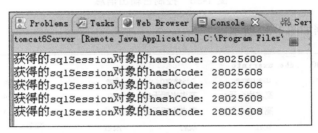

图 1-35　获得的 SqlSession 对象是 1 个

1.4.7 添加记录及异常回滚的测试

添加记录及异常回滚的测试，核心代码如下。

```java
public class insert extends HttpServlet {

    public void doGet(HttpServletRequest request, HttpServletResponse response)
            throws ServletException, IOException {

        try {
            HashMap valueMap1 = new HashMap();
            valueMap1.put("username", "高洪岩今天 1");
            valueMap1.put("password", "高洪岩明天 1");
            valueMap1.put("age", 100);
            valueMap1.put("insertdate", new Date());

            HashMap valueMap2 = new HashMap();
            valueMap2.put("username", "高洪岩今天 2");
            valueMap2.put("password", "高洪岩明天 2");
            valueMap2.put("age", 100);
            valueMap2.put("insertdate", new Date());

            DBOperate dbo = new DBOperate();
            dbo.insert("insertUserinfo", valueMap1);
            dbo.insert("insertUserinfo", valueMap2);
        } catch (Exception e) {
            e.printStackTrace();
            GetSqlSession.rollback();
        } finally {
            GetSqlSession.commit();
        }
```

        }
    }

程序运行后，在控制台输出信息如图 1-36 所示。

图 1-36　控制台输出信息

数据表中的数据如图 1-37 所示。

id	username	password	age	insertDate
39	高洪岩今天1	高洪岩明天1	100	2010-12-14 10:09:41.297
40	高洪岩今天2	高洪岩明天2	100	2010-12-14 10:09:41.297
NULL	NULL	NULL	NULL	NULL

图 1-37　成功添加 2 条记录

上面的步骤证明添加多条记录成功，userinfo 数据表中有 2 条记录。

再来测试异常回滚的情况，更改部分代码如下。

```
HashMap valueMap2 = new HashMap();
valueMap2.put("username",
    "高洪岩今天2_123456789_123456789_123456789_123456789_123456789");
valueMap2.put("password", "高洪岩明天2");
valueMap2.put("age", 100);
valueMap2.put("insertdate", new Date());
```

程序运行后在控制台输出异常信息如下。

```
获得的 sqlSession 对象的 hashCode: 24442607
获得的 sqlSession 对象的 hashCode: 24442607
org.apache.ibatis.exceptions.PersistenceException:
### Error updating database.  Cause: com.microsoft.sqlserver.jdbc.SQLServerException: 将截断字符串或二进制数据。
### The error may involve mybatis.testcurd.insertUserinfo-Inline
### The error occurred while setting parameters
### SQL: insert into   userinfo(username,password,age,insertDate)   values(?,?,?,?)
### Cause: com.microsoft.sqlserver.jdbc.SQLServerException: 将截断字符串或二进制数据。
    at org.apache.ibatis.exceptions.ExceptionFactory.wrapException(ExceptionFactory.java:23)
    at org.apache.ibatis.session.defaults.DefaultSqlSession.update(DefaultSqlSession.java:147)
    at org.apache.ibatis.session.defaults.DefaultSqlSession.insert(DefaultSqlSession.java:134)
    at dbtools.DBOperate.insert(DBOperate.java:12)
    at controller.insert.doGet(insert.java:36)
    at javax.servlet.http.HttpServlet.service(HttpServlet.java:690)
rotocol.java:624)
```

                at org.apache.tomcat.util.net.JIoEndpoint$Worker.run(JIoEndpoint.java:445)
                at java.lang.Thread.run(Thread.java:619)
        Caused by: com.microsoft.sqlserver.jdbc.SQLServerException: 将截断字符串或二进制数据。
                at
        org.apache.ibatis.session.defaults.DefaultSqlSession.update(DefaultSqlSession.java:145)
                ... 17 more
        回滚了

通过上面的信息可以得知，程序出现异常，并且已经回滚，那 userinfo 数据表中是否还是 2 条记录呢？查看一下，其内容如图 1-38 所示。

图 1-38　成功回滚后还是 2 条记录

在 userinfo 数据表中多增加几条记录，便于后面的测试，新增的记录如图 1-39 所示。

图 1-39　userinfo 表中的多条记录

1.4.8　删除记录

删除记录的核心代码如下。

```java
public class delete extends HttpServlet {

    public void doGet(HttpServletRequest request, HttpServletResponse response)
            throws ServletException, IOException {

        try {
            HashMap valueMap1 = new HashMap();
            valueMap1.put("id", 44);

            DBOperate dbo = new DBOperate();
            dbo.delete("deleteUserinfoById", valueMap1);
        } catch (Exception e) {
            e.printStackTrace();
            GetSqlSession.rollback();
        } finally {
```

```
            GetSqlSession.commit();
        }
    }
}
```

程序运行后,userinfo 数据表中的记录如图 1-40 所示。

id	username	password	age	insertDate
39	高洪岩今天1	高洪岩明天1	100	2010-12-14 10:09:41.297
40	高洪岩今天2	高洪岩明天2	100	2010-12-14 10:09:41.297
43	高洪岩今天1	高洪岩明天1	100	2010-12-14 10:14:24.517
45	高洪岩今天1	高洪岩明天1	100	2010-12-14 10:14:26.250
46	高洪岩今天2	高洪岩明天2	100	2010-12-14 10:14:26.250
NULL	NULL	NULL	NULL	NULL

图 1-40 成功删除 id 为 44 的记录

1.4.9 更改记录

更改记录的核心代码如下。

```java
public class update extends HttpServlet {

    public void doGet(HttpServletRequest request, HttpServletResponse response)
            throws ServletException, IOException {
        try {
            HashMap valueMap1 = new HashMap();
            valueMap1.put("id", 45);
            valueMap1.put("username", "高洪岩今天 3new");
            valueMap1.put("password", "高洪岩明天 3new");
            valueMap1.put("age", 100);
            valueMap1.put("insertdate", new Date());

            DBOperate dbo = new DBOperate();
            dbo.update("updateUserinfoById", valueMap1);
        } catch (Exception e) {
            e.printStackTrace();
            GetSqlSession.rollback();
        } finally {
            GetSqlSession.commit();
        }
    }
}
```

程序运行后,userinfo 表中的数据如图 1-41 所示。

1.4 MyBatis 核心对象的生命周期与封装

图 1-41 userinfo 表中的记录

1.4.10 查询单条记录

查询单条记录的核心代码如下。

```java
public class getUserinfoById extends HttpServlet {

    public void doGet(HttpServletRequest request, HttpServletResponse response)
            throws ServletException, IOException {

        try {
            HashMap valueMap1 = new HashMap();
            valueMap1.put("id", 39);

            DBOperate dbo = new DBOperate();
            List<Map> list = dbo.select("getUserinfoById", valueMap1);
            for (int i = 0; i < list.size(); i++) {
                Map rowMap = list.get(i);
                System.out.println(rowMap.get("id") + "_"
                        + rowMap.get("username") + "_" + rowMap.get("password")
                        + "_" + rowMap.get("age") + "_"
                        + rowMap.get("insertDate"));
            }
        } catch (Exception e) {
            e.printStackTrace();
            GetSqlSession.rollback();
        } finally {
            GetSqlSession.commit();
        }

    }

}
```

程序运行后,在控制台输出如图 1-42 所示的结果。

图 1-42 控制台输出信息

1.4.11 查询多条记录

查询多条记录的核心代码如下。

```java
public class getAllUserinfo extends HttpServlet {

    public void doGet(HttpServletRequest request, HttpServletResponse response)
            throws ServletException, IOException {

        try {
            DBOperate dbo = new DBOperate();
            List<Map> list = dbo.select("getAllUserinfo", null);
            for (int i = 0; i < list.size(); i++) {
                Map rowMap = list.get(i);
                System.out.println(rowMap.get("id") + "_"
                    + rowMap.get("username") + "_" + rowMap.get("password")
                    + "_" + rowMap.get("age") + "_"
                    + rowMap.get("insertDate"));
            }
        } catch (Exception e) {
            e.printStackTrace();
            GetSqlSession.rollback();
        } finally {
            GetSqlSession.commit();
        }

    }

}
```

程序运行后，控制台的输出如图 1-43 所示。

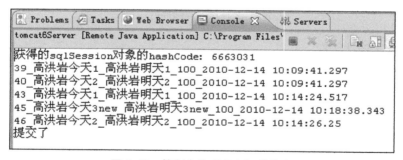

图 1-43　控制台输出多条记录信息

ORM 框架 MyBatis 介绍到这里，读者应该能熟练地使用它进行数据库的 CURD 操作，并且对核心 API 在使用上有一个了解。

第 2 章 MyBatis 3 必备开发技能

本章内容是学习 MyBatis 框架必须要掌握的技能，这些技能在使用 MyBatis 框架时使用率是非常得多。可以这样说，MyBatis 大部分核心功能都在本章中进行介绍。

2.1 使用 getMapper() 面向接口编程

前面知识点中的代码都是使用如下程序对数据库进行操作。

```
sqlSession.insert("sqlId");
sqlSession.delete("sqlId");
sqlSession.update("sqlId");
sqlSession.selectList("sqlId");
```

对方法 insert、delete、update 和 selectList 传入 sqlId 来达到调用 SQL 语句而对数据库的操作目的，完完全全是在面向 String 字符串类型的 sqlId 编程，虽然能达到实现操作数据库的目的，但这种代码写法却是不规范的。理想中规范的写法不是面向 sqlId 编程，而是面向接口编程，新版的 MyBatis 提供了"接口-SQL 映射"的功能，使程序员完全面向接口进行编程，相比 sqlId 的方式，使用"接口-SQL 映射"在代码规范上更上一个台阶。

2.1.1 接口-SQL 映射的对应关系

"接口-SQL 映射"的对应关系如图 2-1 所示。

"接口-SQL 映射"的原理如下：

SQL 映射文件 UserinfoMapper.xml 中的 namespace 属性值 sqlmapping.UserinfoMapper 代表该映射对应的就是 sqlmapping 包中的 UserinfoMapper.java 接口，而<insert>标签的 id 属性值 insertUserinfo 就是 UserinfoMapper.java 接口中的 public void insertUserinfo(Userinfo userinfo)方法，<insert> 标签的 parameterType 属性值 sqlmapping.Userinfo 就是 public void insertUserinfo(Userinfo userinfo)方法参数类型，只要它们一一对应，就能实现"接口-SQL 映射"，程序员完全以面向接口的方式设计软件。

下面来看一下详细代码吧！

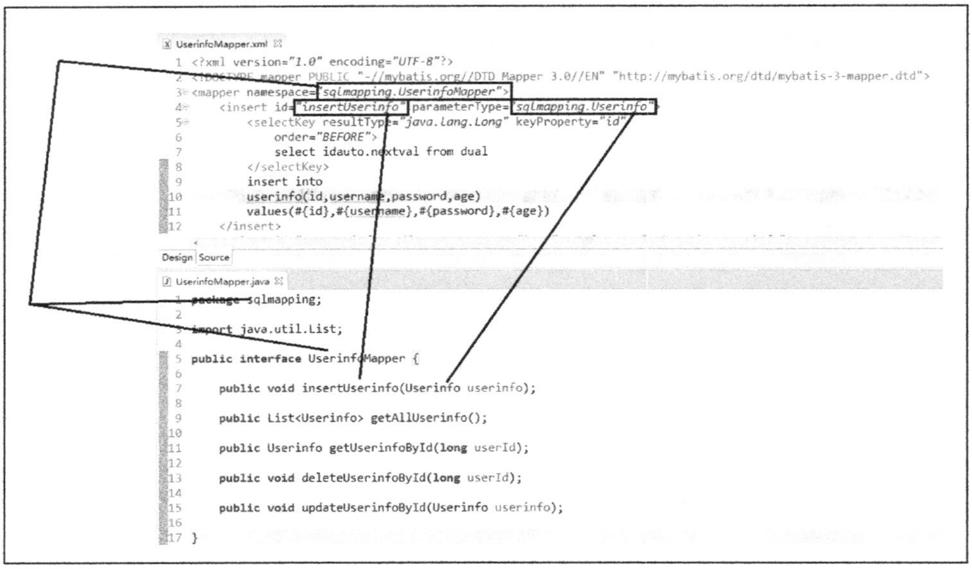

图 2-1 接口-SQL 映射原理

2.1.2 创建 Userinfo.java 实体类

创建名称为 getMapperTest 的项目。

创建实体类 Userinfo.java，代码如下。

```java
package sqlmapping;

public class Userinfo {

    private Long id;
    private String username;
    private String password;
    private Long age;

    //省略 set 和 get 方法

}
```

2.1.3 创建 UserinfoMapper.java 接口

程序员需要使用该接口进行操作数据库，示例代码如下。

```java
package sqlmapping;

import java.util.List;

public interface UserinfoMapper {

    public void insertUserinfo(Userinfo userinfo);

    public List<Userinfo> getAllUserinfo();
```

```
    public Userinfo getUserinfoById(long userId);

    public void deleteUserinfoById(long userId);

    public void updateUserinfoById(Userinfo userinfo);

}
```

2.1.4 创建 SQL 映射文件 UserinfoMapper.xml

最为关键的 SQL 映射文件 UserinfoMapper.xml 中的代码和前面章节中的代码大体一样,具体如下。

```xml
<?xml version="1.0" encoding="UTF-8"?>
<!DOCTYPE mapper PUBLIC "-//mybatis.org//DTD Mapper 3.0//EN"
"http://mybatis.org/dtd/mybatis-3-mapper.dtd">
<mapper namespace="sqlmapping.UserinfoMapper">
    <insert id="insertUserinfo" parameterType="sqlmapping.Userinfo">
        <selectKey resultType="java.lang.Long" keyProperty="id"
            order="BEFORE">
            select idauto.nextval from dual
        </selectKey>
        insert into
        userinfo(id,username,password,age)
        values(#{id},#{username},#{password},#{age})
    </insert>
    <select id="getUserinfoById" parameterType="long" resultType="sqlmapping.Userinfo">
        select * from
        userinfo where id=#{id}
    </select>
    <delete id="deleteUserinfoById" parameterType="long">
        delete from
        userinfo where id=#{id}
    </delete>
    <select id="getAllUserinfo" resultType="sqlmapping.Userinfo">
        select * from userinfo
    </select>
    <update id="updateUserinfoById" parameterType="sqlmapping.Userinfo">
        update userinfo
        set
        username=#{username},password=#{password},age=#{age}
        where id=#{id}
    </update>
</mapper>
```

SQL 与接口的映射对应关系一定要配置成功,不然程序就会出现异常。

2.1.5 增加记录 insert 的操作代码

增加 insert 的操作代码如下。

```
package test;

import java.io.IOException;
import java.io.InputStream;
```

```java
import org.apache.ibatis.io.Resources;
import org.apache.ibatis.session.SqlSession;
import org.apache.ibatis.session.SqlSessionFactory;
import org.apache.ibatis.session.SqlSessionFactoryBuilder;

import sqlmapping.Userinfo;
import sqlmapping.UserinfoMapper;

public class InsertUserinfo {

    public static void main(String[] args) {
        try {
            Userinfo userinfo = new Userinfo();
            userinfo.setUsername("高洪岩");
            userinfo.setPassword("大中国");
            userinfo.setAge(100L);

            String resource = "mybatis-config.xml";
            InputStream inputStream = Resources.getResourceAsStream(resource);
            SqlSessionFactory sqlSessionFactory = new SqlSessionFactoryBuilder().build(inputStream);
            SqlSession sqlSession = sqlSessionFactory.openSession();

            UserinfoMapper userinfoMapper = sqlSession.getMapper(UserinfoMapper.class);

            userinfoMapper.insertUserinfo(userinfo);

            sqlSession.commit();
            sqlSession.close();

            System.out.println("createId=" + userinfo.getId());
        } catch (IOException e) {
            // TODO Auto-generated catch block
            e.printStackTrace();
        }
    }
}
```

接口 UserinfoMapper.java 并不能直接使用,必须得有接口 UserinfoMapper.java 的实现类才可以执行任务。MyBatis 动态地创建出了接口 UserinfoMapper.java 的实现类,示例代码如下。

```
UserinfoMapper userinfoMapper = sqlSession.getMapper(UserinfoMapper.class);
System.out.println(userinfoMapper);
```

程序运行结果如图 2-2 所示。

```
<terminated> InsertUserinfo (3) [Java Application] C:\java1.8\bin\javaw.
log4j:WARN No appenders could be found for log
log4j:WARN Please initialize the log4j system
log4j:WARN See http://logging.apache.org/log4j
org.apache.ibatis.binding.MapperProxy@d2cc05a
```

图 2-2 对象 userinfoMapper 是代理类

控制台输出来的"org.apache.ibatis.binding.MapperProxy"类就是被 MyBatis 动态创建出来的，而且实现了接口 UserinfoMapper.java，所以才可以对 UserinfoMapper 对象进行赋值，属于多态关系。

2.1.6 查询全部 selectAll 的操作代码

查询全部 selectAll 的操作代码如下。

```java
public class GetAllUserinfo {

    public static void main(String[] args) {
        try {

            String resource = "mybatis-config.xml";
            InputStream inputStream = Resources.getResourceAsStream(resource);
            SqlSessionFactory sqlSessionFactory = new SqlSessionFactoryBuilder().build(inputStream);
            SqlSession sqlSession = sqlSessionFactory.openSession();
            UserinfoMapper userinfoMapper = sqlSession.getMapper(UserinfoMapper.class);

            List<Userinfo> listUserinfo = userinfoMapper.getAllUserinfo();
            for (int i = 0; i < listUserinfo.size(); i++) {
                Userinfo userinfo = listUserinfo.get(i);
                System.out.println(userinfo.getId() + " " + userinfo.getUsername() + " " + userinfo.getPassword() + " "
                        + userinfo.getAge());

            }

            sqlSession.commit();
            sqlSession.close();

        } catch (IOException e) {
            // TODO Auto-generated catch block
            e.printStackTrace();
        }

    }

}
```

2.1.7 查询单条记录 selectById 的操作代码

查询单条记录 selectById 的操作代码如下。

```java
public class GetUserinfoById {

    public static void main(String[] args) {
        try {

            String resource = "mybatis-config.xml";
            InputStream inputStream = Resources.getResourceAsStream(resource);
            SqlSessionFactory sqlSessionFactory = new SqlSessionFactoryBuilder().build(inputStream);
            SqlSession sqlSession = sqlSessionFactory.openSession();
```

```java
            UserinfoMapper userinfoMapper = sqlSession.getMapper(UserinfoMapper.class);

            Userinfo userinfo = userinfoMapper.getUserinfoById(1445436L);

            System.out.println(userinfo.getId() + " " + userinfo.getUsername() + " " + userinfo.getPassword() + " "
                    + userinfo.getAge());

            sqlSession.commit();
            sqlSession.close();

        } catch (IOException e) {
            // TODO Auto-generated catch block
            e.printStackTrace();
        }

    }

}
```

2.1.8 修改记录 updateById 的操作代码

修改记录 updateById 的操作代码如下。

```java
public class UpdateUserinfoById {

    public static void main(String[] args) {
        try {
            String resource = "mybatis-config.xml";
            InputStream inputStream = Resources.getResourceAsStream(resource);
            SqlSessionFactory sqlSessionFactory = new SqlSessionFactoryBuilder().build(inputStream);
            SqlSession sqlSession = sqlSessionFactory.openSession();
            UserinfoMapper userinfoMapper = sqlSession.getMapper(UserinfoMapper.class);

            Userinfo userinfo = userinfoMapper.getUserinfoById(1445436L);

            userinfo.setUsername("新的账号");
            userinfo.setPassword("新的密码");
            userinfo.setAge(888L);

            userinfoMapper.updateUserinfoById(userinfo);

            sqlSession.commit();
            sqlSession.close();

        } catch (IOException e) {
            // TODO Auto-generated catch block
            e.printStackTrace();
        }

    }

}
```

2.1.9　删除记录 deleteById 的操作代码

删除记录 deleteById 的操作代码如下。

```java
public class DeleteUserinfoById {

    public static void main(String[] args) {
        try {
            String resource = "mybatis-config.xml";
            InputStream inputStream = Resources.getResourceAsStream(resource);
            SqlSessionFactory sqlSessionFactory = new SqlSessionFactoryBuilder().build(inputStream);
            SqlSession sqlSession = sqlSessionFactory.openSession();
            UserinfoMapper userinfoMapper = sqlSession.getMapper(UserinfoMapper.class);

            userinfoMapper.deleteUserinfoById(1445434L);

            sqlSession.commit();
            sqlSession.close();

        } catch (IOException e) {
            // TODO Auto-generated catch block
            e.printStackTrace();
        }

    }

}
```

程序员一直在使用接口进行编程，不再使用 sqlId 进行软件设计。

2.2　使用 typeAliases 配置别名

在执行 select 查询或 insert 添加的 SQL 语句时，都要在 parameterType 或 resultType 属性中写上完整的实体类路径，路径中需要包含完整的包名，示例代码如下。

```xml
<insert id="insertUserinfo" parameterType="sqlmapping.Userinfo">
    <selectKey resultType="java.lang.Long" keyProperty="id"
        order="BEFORE">
        select idauto.nextval from dual
    </selectKey>
    insert into
    userinfo(id,username,password,age)
    values(#{id},#{username},#{password},#{age})
</insert>
<select id="getUserinfoById" parameterType="long" resultType="sqlmapping.Userinfo">
    select * from
    userinfo where id=#{id}
</select>
```

如果包名嵌套层级较多，则会出现大量冗余的配置代码，这时可以在 mybatis-config.xml 配置文件中使用<typeAliases>标签来简化。

2.2.1 使用 typeAlias 单独配置别名

本节的示例代码在名称为 typeAliasTest 的项目中，在 mybatis-config.xml 配置文件中添加 <typeAlias>标签，代码如下。

```xml
<configuration>
    <typeAliases>
        <typeAlias alias="userinfo" type="sqlmapping.Userinfo" />
    </typeAliases>
    <environments default="development">
        <environment id="development">
            <transactionManager type="JDBC" />
            <dataSource type="POOLED">
                <property name="driver" value="oracle.jdbc.OracleDriver" />
                <property name="url" value="jdbc:oracle:thin:@localhost:1521:orcl" />
                <property name="username" value="y2" />
                <property name="password" value="123" />
            </dataSource>
        </environment>
    </environments>
    <mappers>
        <mapper resource="sqlmapping/UserinfoMapper.xml" />
    </mappers>
</configuration>
```

其中配置代码<typeAlias alias="userinfo" type="sqlmapping.Userinfo" />中的 type 属性值就是完整的实体类包路径，而属性 alias 就是实体类的别名，这个别名可以在 SQL 映射文件中进行使用，从而简化了配置代码，示例代码如下。

```xml
<mapper namespace="sqlmapping.UserinfoMapper">
    <insert id="insertUserinfo" parameterType="userinfo">
        <selectKey resultType="java.lang.Long" keyProperty="id"
            order="BEFORE">
            select idauto.nextval from dual
        </selectKey>
        insert into
        userinfo(id,username,password,age)
        values(#{id},#{username},#{password},#{age})
    </insert>
    <select id="getAllUserinfo" resultType="userinfo">
        select * from
        userinfo
    </select>
</mapper>
```

使用类型别名 typeAlias 后，即可以正常地操作数据库，配置代码也不再出现冗余了。
在引用别名时是不区分大小写的，比如如下代码也能正确得到运行。

```
parameterType="USERinfo"
resultType="USERINFo"
```

2.2.2 使用 package 批量配置别名

使用<typeAlias>虽然可以实现配置别名，但如果实体类的数量较多，则极易出现

<typeAlias>配置爆炸,这种情况可以通过使用<package>标签来解决,它的原理就是扫描指定包下的类,这些类都被自动赋于了与类同名的别名,不区分大小写,别名中不包含包名。

创建名称为 typeAliasPackageTest 的项目,更改 mybatis-config.xml 配置文件,代码如下。

```xml
<configuration>
    <typeAliases>
        <package name="sqlmapping" />
    </typeAliases>
    <environments default="development">
        <environment id="development">
            <transactionManager type="JDBC" />
            <dataSource type="POOLED">
                <property name="driver" value="oracle.jdbc.OracleDriver" />
                <property name="url" value="jdbc:oracle:thin:@localhost:1521:orcl" />
                <property name="username" value="y2" />
                <property name="password" value="123" />
            </dataSource>
        </environment>
    </environments>
    <mappers>
        <mapper resource="sqlmapping/UserinfoMapper.xml" />
    </mappers>
</configuration>
```

在<typeAliases>标签中使用<package>子标签来扫描包中的类而自动创建出类的别名。
SQL 映射文件也需要进行更改,代码如下。

```xml
<mapper namespace="sqlmapping.UserinfoMapper">
    <insert id="insertUserinfo" parameterType="usERInFO">
        <selectKey resultType="java.lang.Long" keyProperty="id"
            order="BEFORE">
            select idauto.nextval from dual
        </selectKey>
        insert into
        userinfo(id,username,password,age)
        values(#{id},#{username},#{password},#{age})
    </insert>
    <select id="getAllUserinfo" resultType="Userinfo">
        select * from
        userinfo
    </select>
</mapper>
```

在 SQL 映射文件中使用配置 parameterType="usERInFO"也能正确地操作数据库,说明使用<package>方式定义的别名是不区分大写的。

另外在使用<package>包扫描时,代码如下。

```xml
<typeAliases>
    <package name="entity1" />
    <package name="entity2" />
</typeAliases>
```

如果在不同的包中出现相同实体类名的情况下,在 MyBatis 解析 XML 配置文件时就会出现异常信息。

```
Caused by: org.apache.ibatis.type.TypeException: The alias 'Userinfo' is already mapped
to the value 'entity1.Userinfo'.
```

此结论在项目 typeAliasPackageTest2 中得到验证。

2.3 使用 properties 文件保存数据库信息

配置文件 mybatis-config.xml 承载的配置信息过多，所以可以将连接数据库的信息转移到 properties 属性文件中，这样也便于代码的后期维护与管理。

2.3.1 在 mybatis-config.xml 配置文件中读取 properties 文件中的参数

在前面章节示例中，连接数据库时的具体参数是直接在 mybatis-config.xml 文件中进行定义的，比如 url、username 和 password 这些信息，MyBatis 还支持将这些参数值写入*.properties 属性文件中，然后使用 Java 代码对这些参数值进行传递。

更改 mybatis-config.xml 配置文件的部分代码如下。

```xml
<dataSource type="POOLED">
    <property name="driver" value="${driver}" />
    <property name="url" value="${url}" />
    <property name="username" value="${username}" />
    <property name="password" value="${password}" />
</dataSource>
```

在 src 中创建 db.properties 属性文件，内容如下。

```
url=jdbc:sqlserver://localhost:1079;databaseName=ghydb
driver=com.microsoft.sqlserver.jdbc.SQLServerDriver
username=sa
password=
```

使用如下 Java 代码即可对类似${driver}的表达式传值了。

```java
InputStream isRef = GetSqlSession.class
        .getResourceAsStream("/db.properties");
Properties prop = new Properties();
prop.load(isRef);

String resource = "mybatis-config.xml";
InputStream inputStream = Resources.getResourceAsStream(resource);
SqlSessionFactory sqlSessionFactory = new SqlSessionFactoryBuilder()
        .build(inputStream, prop);
SqlSession sqlSession = sqlSessionFactory.openSession();
```

在 SqlSessionFactoryBuilder 类的 build()方法中将配置 inputStream 流与属性对象 prop 进行关联，就可以根据${username}表达式读取 properties 属性文件中的同名 key 对应的值了，从而成功读取连接数据库的 4 个必要参数。

本章节使用 Java 代码的方式来传入 4 个必要变量，此种用法可以保护 properties 属性文件中的内容具有十足的安全性，比如可以在 properties 属性文件中保存经过加密后的 password 值，然后在 Java 代码中对 properties 属性文件中的 password 值再进行进行解密即可，示例伪代码如下。

```java
Properties prop = new Properties();
prop.setProperty("driver", 解密的方法(prop.getProperty("driver")));
prop.setProperty("url", 解密的方法(prop.getProperty("url")));
```

2.3 使用 properties 文件保存数据库信息

```
            prop.setProperty("username", 解密的方法(prop.getProperty("username")));
            prop.setProperty("password", 解密的方法(prop.getProperty("password")));
```

本实验的源代码在项目 properties_encode_decode 中。
完整示例代码如下。

```java
public class SelectAll {

    // 加密算法
    public static String encode(String password) {
        return password = "_" + password;
    }

    // 解密算法
    public static String decode(String password) {
        return password.substring(1);
    }

    public static void main(String[] args) throws IOException {
        String configFile = "mybatis-config.xml";
        String dbInfoFile = "dbinfo.properties";

        InputStream configStream = Resources.getResourceAsStream(configFile);
        InputStream dbinfoStream = Resources.getResourceAsStream(dbInfoFile);

        Properties prop = new Properties();
        prop.load(dbinfoStream);
        String password = decode(prop.getProperty("password"));
        prop.setProperty("password", password);

        SqlSessionFactoryBuilder builder = new SqlSessionFactoryBuilder();
        SqlSessionFactory factory = builder.build(configStream, prop);
        SqlSession session = factory.openSession();
        UserinfoMapping mapping = session.getMapper(UserinfoMapping.class);
        System.out.println("数据库信息来自于 properties 文件！");

        List<Userinfo> listUserinfo = mapping.getAllUserinfo();
        for (int i = 0; i < listUserinfo.size(); i++) {
            Userinfo eachUserinfo = listUserinfo.get(i);
            System.out.println(eachUserinfo.getId() + " " + eachUserinfo.getUsername()
 + " "
                    + eachUserinfo.getPassword() + " " + eachUserinfo.getAge() + " " +
eachUserinfo.getInsertdate());
        }
        session.commit();
        session.close();
    }

}
```

2.3.2 将数据库信息封装进 properties 属性文件中

如果不要求 properties 属性文件信息的安全性，还可以在 mybatis-config.xml 配置文件中直接引用指定的 properties 属性文件，再根据固定的 key 找到对应的 value，也能达到连接数据库的目的，此种方法不需要使用 Java 代码来操作 properties 属性文件。

创建名称为 propertiesSaveDBInfo 的项目，添加 dbinfo.properties 文件，代码如下。

```
driver=oracle.jdbc.OracleDriver
url=jdbc:oracle:thin:@localhost:1521:orcl
username=y2
password=123
```

配置文件 mybatis-config.xml 的代码如下。

```xml
<configuration>
    <properties resource="dbinfo.properties"></properties>
    <typeAliases>
        <typeAlias alias="userinfo" type="sqlmapping.Userinfo" />
    </typeAliases>
    <environments default="development">
        <environment id="development">
            <transactionManager type="JDBC" />
            <dataSource type="POOLED">
                <property name="driver" value="${driver}" />
                <property name="url" value="${url}" />
                <property name="username" value="${username}" />
                <property name="password" value="${password}" />
            </dataSource>
        </environment>
    </environments>
    <mappers>
        <mapper resource="sqlmapping/UserinfoMapper.xml" />
    </mappers>
</configuration>
```

添加了配置代码<properties resource="*dbinfo.properties*"></properties>来使用 dbinfo.properties 文件中的连接数据库信息。

运行类代码如下。

```java
public class GetAllUserinfo {

    public static void main(String[] args) {
        try {
            String resource = "mybatis-config.xml";
            InputStream inputStream = Resources.getResourceAsStream(resource);
            SqlSessionFactory sqlSessionFactory = new SqlSessionFactoryBuilder().build(inputStream);
            SqlSession sqlSession = sqlSessionFactory.openSession();
            UserinfoMapper userinfoMapper = sqlSession.getMapper(UserinfoMapper.class);

            List<Userinfo> listUserinfo = userinfoMapper.getAllUserinfo();
            for (int i = 0; i < listUserinfo.size(); i++) {
                Userinfo userinfo = listUserinfo.get(i);
                System.out.println(userinfo.getId() + " " + userinfo.getUsername() + " "
                        + userinfo.getPassword() + " "
                        + userinfo.getAge());
            }

            sqlSession.commit();
            sqlSession.close();

        } catch (IOException e) {
            e.printStackTrace();
```

```
            }
        }
}
```

成功将数据表中的数据查询出来。

2.4 与数据源 DataSource 有关的操作

数据源 DataSource 可以提升操作数据库的速度,加快程序运行的效率。本节将介绍 MyBatis 与 DataSource 常见的联合使用方法。

2.4.1 配置多个 environment 环境

可以在 mybatis-config.xml 配置文件中创建多个数据源，方便切换欲操作的数据库。
创建名称为 moreEnvironment 的项目，并创建 dbinfo.properties 属性文件，内容如下。

```
oracle1.driver=oracle.jdbc.OracleDriver
oracle1.url=jdbc:oracle:thin:@localhost:1521:orcl
oracle1.username=y2
oracle1.password=123


oracle2.driver=oracle.jdbc.OracleDriver
oracle2.url=jdbc:oracle:thin:@localhost:1521:orcl
oracle2.username=y2
oracle2.password=errorPassword_errorPassword_###########
```

从属性文件中提供的内容可以发现是 2 个 Oracle 数据库的连接信息，其中第 2 个 Oracle 的连接密码是错误的。

配置文件 mybatis-config.xml 代码如下。

```xml
<?xml version="1.0" encoding="UTF-8" ?>
<!DOCTYPE configuration
PUBLIC "-//mybatis.org//DTD Config 3.0//EN"
"http://mybatis.org/dtd/mybatis-3-config.dtd">
<configuration>
    <properties resource="dbinfo.properties"></properties>
    <typeAliases>
        <typeAlias alias="userinfo" type="sqlmapping.Userinfo" />
    </typeAliases>
    <environments default="oracle2">
        <environment id="oracle1">
            <transactionManager type="JDBC" />
            <dataSource type="POOLED">
                <property name="driver" value="${oracle1.driver}" />
                <property name="url" value="${oracle1.url}" />
                <property name="username" value="${oracle1.username}" />
                <property name="password" value="${oracle1.password}" />
            </dataSource>
        </environment>
        <environment id="oracle2">
            <transactionManager type="JDBC" />
```

```xml
        <dataSource type="POOLED">
            <property name="driver" value="${oracle2.driver}" />
            <property name="url" value="${oracle2.url}" />
            <property name="username" value="${oracle2.username}" />
            <property name="password" value="${oracle2.password}" />
        </dataSource>
    </environment>
</environments>
<mappers>
    <mapper resource="sqlmapping/UserinfoMapper.xml" />
</mappers>
</configuration>
```

配置<environments default="oracle2">代码中的 default 属性值是 oracle2，代表要连接的是<environment id="oracle2">中的配置信息，也就是在配置代码<environments default="oracle2">中来切换欲操作数据库的实例，从而达到配置多数据源的目的。

2.4.2 什么是 JNDI 以及如何从 JNDI 获得数据源 DataSource

什么是 JNDI？这里暂时先不介绍 JNDI 的理论概念。在通常情况下，常用对象的组织结构是树形结构，如图 2-3 所示。

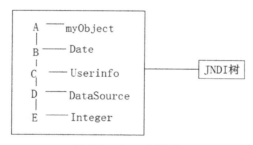

图 2-3 JNDI 树结构

为了更好地组织多个对象，可以把多个对象绑定到一个树中，使用结点名称对应结点值，来找到目标对象。这非常类似于 Map 结构，这个树称为"JNDI 树"。

如果每个厂商都不使用树形结构来组织数据，也就是说每个厂商对对象的组织方式都不一样，有的放入 List，有的放入 Map，有的创建自己的数据结构来保存对象，那么程序员需要掌握以上 3 种 API 来获取每个厂商提供的对象。

```
MyList.get(0)
MyMap.get(key)
自定义的数据结构.getObject()
```

这样的写法大大增加了程序员学习的成本，并且使得程序代码的移植性非常得不便。为了达到获取对象的 API 命名的统一性，Sun 发布了 JNDI 规范，让这些厂商遵循 JNDI 规范。使用了 JNDI 规范后，不管使用什么样的方式保存对象，程序员只需要使用 1 种方式获得对象，就可以在不同厂商的不同数据结构中获得指定的对象值，这个特点和使用 1 种 JDBC 接口就可以操作不同数据库的原理是一样的。

接口 DataSource 的主要作用就是获得 Connection 连接，是为了制定在不同容器产品中获得

Connection 的 API 进行规范化，而接口 JNDI 的主要作用就是制定在不同容器产品中操作 JNDI 树中的数据的 API 的规范化。简单总结：

- DataSource 规范是为了获得 Connection 规范化；
- JNDI 规范是为了获得 JNDI 树中数据的规范化。

下面来看一下在 Tomcat 容器中使用 JNDI 接口的 API 来操作 JNDI 树结点与对应值的代码吧。

创建 Web 项目 jdbcJNDITest。

在 JNDI 树中绑定值的代码如下。

```java
package jnditest;

import java.io.IOException;

import javax.naming.Context;
import javax.naming.InitialContext;
import javax.naming.NamingException;
import javax.servlet.ServletException;
import javax.servlet.http.HttpServlet;
import javax.servlet.http.HttpServletRequest;
import javax.servlet.http.HttpServletResponse;

public class test1 extends HttpServlet {
    protected void doGet(HttpServletRequest request, HttpServletResponse response)
            throws ServletException, IOException {
        try {
            Context context = new InitialContext();
            context.bind("username", "中国人");
            context.close();
        } catch (NamingException e) {
            e.printStackTrace();
        }
    }
}
```

从 JNDI 树中取得值的代码如下。

```java
public class test2 extends HttpServlet {
    protected void doGet(HttpServletRequest request, HttpServletResponse response)
            throws ServletException, IOException {
        try {
            Context context = new InitialContext();
            String value = (String) context.lookup("username");
            context.close();
            System.out.println(value);
        } catch (NamingException e) {
            e.printStackTrace();
        }
    }
}
```

结点值还可以是父子关系，代码如下。

```java
public class test3 extends HttpServlet {
    protected void doGet(HttpServletRequest request, HttpServletResponse response)
            throws ServletException, IOException {
        try {
            Context context = new InitialContext();
            context.bind("username.password", "中国人密码");
            context.close();
```

```
        } catch (NamingException e) {
            e.printStackTrace();
        }
    }
}
```

上面的代码可以在 Tomcat 中进行运行,并且是没有错误的,但在 weblogic 中以这样的代码方式创建父子结点就会出现错误,正确地在 weblogic 中创建父子结点的代码如下:

```
try {
    Context context = new InitialContext();
    context = context.createSubcontext("a");
    context = context.createSubcontext("b");
    context.bind("c", "cValue");
    context.close();
} catch (NamingException e) {
    e.printStackTrace();
}
```

取得节点是父子关系对应值的代码如下。

```
public class test4 extends HttpServlet {
    protected void doGet(HttpServletRequest request, HttpServletResponse response)
            throws ServletException, IOException {
        try {
            Context context = new InitialContext();
            String value = (String) context.lookup("username.password");
            context.close();
            System.out.println(value);
        } catch (NamingException e) {
            e.printStackTrace();
        }
    }
}
```

删除结点的代码如下。

```
public class test5 extends HttpServlet {
    protected void doGet(HttpServletRequest request, HttpServletResponse response)
            throws ServletException, IOException {
        try {
            Context context = new InitialContext();
            context.unbind("username");
            context.close();
        } catch (NamingException e) {
            e.printStackTrace();
        }
    }
}
```

更新 JNDI 结点值的代码如下。

```
public class test6 extends HttpServlet {
    protected void doGet(HttpServletRequest request, HttpServletResponse response)
            throws ServletException, IOException {
        try {
            Context context = new InitialContext();
            context.rebind("username", "新中国人");
            context.close();
        } catch (NamingException e) {
            e.printStackTrace();
```

        }
    }
}

上面的代码就是在 Tomcat 中操作 JNDI 树结点与对应的值,这些代码可以在不更改任何一行的情况下成功在其他容器中移植运行。这就是 JNDI 规范的优点:API 统一,代码移植性好。

使用 JNDI 规范中的 API 就可以从 JNDI 树中找到对应的对象。大多数 Web 容器都支持 JNDI 规范,因为 JNDI 规范属于 Java EE 技术体系之一。有了 JNDI 规范后,只要 Web 容器支持 JNDI 规范,那么在获得对象时使用的代码是一样的。

在 Web 开发中,如果使用 Tomcat 做为 Web 容器,则 Tomcat 内部默认提供了获得 Connection 连接对象的数据源 DataSource,可以使用相应的 API 来从 DataSource 中取得连接 Connection,并操作数据库。

数据源 DataSource 接口的信息如图 2-4 所示。

数据源 DataSource 接口在 Java Doc 中官方解释如下。

该工厂用于提供到此 DataSource 对象所表示的物理数据源的连接。作为 DriverManager 工具的替代项,DataSource 对象是获取连接 Connection 的首选方法。实现 DataSource 接口的对象通常在基于 JavaTM Naming and Directory Interface (JNDI) API 的命名服务中注册。

以上文字可以总结出 3 条:

图 2-4 数据源 DataSource 接口的信息

- 数据源 DataSource 是 DriverManager 工具的替代使用方式;
- 获得数据源 DataSource 要使用 JNDI;
- 通过 DataSource 可以获得 Connection 连接。

大多数 JavaWeb 容器内部的 DataSource 数据源提供的 Connection 都已经被池化,所谓的"池化"也就是预先在 Tomcat 中创建一个空间,在这个空间中存储一些已经连接到数据库的 Connection 连接,客户端在使用 Connection 时只需要 get 获取即可,省略了每次连接数据库时的 IP 寻址、username 和 password 验证等重复的环节,大大提高软件的运行效率,这个空间就是 Connection Pool 连接池。

那么 JNDI 与 DataSource 数据源以及 ConnectionPool 连接池之间到底是什么关系呢?效果如图 2-5 所示。

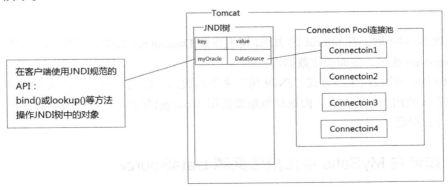

图 2-5 三者之间的关系

在 Tomcat 中获得数据源 DataSource 之前要在 Tomcat 的 conf 文件夹中的 context.xml 文件里配置 DataSource 数据源信息，Tomcat 启动时会读取这些信息，然后将 DataSource 对象放到 JNDI 树结点上，核心代码如下。

```
<Resource name="jdbc/myOracle" type="javax.sql.DataSource"
    driverClassName="oracle.jdbc.OracleDriver"
url="jdbc:oracle:thin:@localhost:1521:orcl"
    username="y2" password="123" maxActive="200" maxIdle="10" maxWait="-1" />
```

<Resource>中的 name 值 "jdbc/myOracle" 就是 JNDI 树节点的名称，对应的值就是数据源 DataSource。

然后在 tomcat 的 conf 文件夹中的 web.xml 文件里引用数据源，核心代码如下。

```
<resource-ref>
    <res-ref-name>jdbc/myOracle</res-ref-name>
    <res-type>javax.sql.DataSource</res-type>
</resource-ref>
```

先来看看若不使用 MyBatis 如何从数据源 DataSource 中获得 Connection。继续在 jdbcJNDITest 项目中进行测试，创建 Servlet 的代码如下。

```java
public class test7 extends HttpServlet {
    protected void doGet(HttpServletRequest request, HttpServletResponse response)
            throws ServletException, IOException {

        try {
            Context context = new InitialContext();
            DataSource ds = (DataSource) context.lookup("java:/comp/env/jdbc/myOracle");
            Connection connection = ds.getConnection();
            ResultSet rs = connection.prepareStatement("select * from userinfo order by id asc").executeQuery();
            while (rs.next()) {
                System.out.println(rs.getString("id") + " " + rs.getString("username"));
            }
            rs.close();
            connection.close();
        } catch (NamingException e) {
            e.printStackTrace();
        } catch (SQLException e) {
            e.printStackTrace();
        }
    }
}
```

执行此 Servlet 后，在 tomcat 的 JDNI 树上找到 DataSource 数据源对象，然后通过此对象获得 Connection 连接，进而操作数据库。

JNDI 以统一的数据组织方式 "JNDI 树" 来组织数据，以统一的方式对数据进行存取，比如存储数据要使用 bind()方法，而取得数据要使用 lookup()方法。它是访问数据，共享数据的一种 Java EE 规范。

2.4.3 如何在 MyBatis 中使用数据源 DataSource

MyBatis 中使用 DataSource 数据源的实验在项目 JNDIDataSourceTest 中进行测试，创建配

2.4 与数据源 DataSource 有关的操作

置文件 mybatis-config.xml 的代码如下。

```xml
<?xml version="1.0" encoding="UTF-8" ?>
<!DOCTYPE configuration
PUBLIC "-//mybatis.org//DTD Config 3.0//EN"
"http://mybatis.org/dtd/mybatis-3-config.dtd">
<configuration>
    <properties resource="dbinfo.properties">
    </properties>
    <environments default="jndi1">
        <environment id="jndi1">
            <transactionManager type="JDBC" />
            <dataSource type="JNDI">
                <property name="data_source" value="${jndiName}" />
            </dataSource>
        </environment>
        <environment id="jndi2">
            <transactionManager type="JDBC" />
            <dataSource type="JNDI">
                <property name="data_source" value="java:/comp/env/jdbc/myOracle" />
            </dataSource>
        </environment>
    </environments>
    <mappers>
        <mapper resource="mapping/UserinfoMapper.xml" />
    </mappers>
</configuration>
```

其中配置代码<transactionManager type="JDBC" />的主要作用就是程序员需要使用显式代码进行事务的提交，也就是程序员必须要调用 SqlSession 对象的 commit()方法才会更改数据库。

还有另外一个写法。

```xml
<transactionManager type="MANAGED" />
```

此写法的作用是每执行 1 条 SQL 语句后，事务进行自动提交。

创建 Servlet 类 Test1.java 代码如下。

```java
package controller;

import java.io.IOException;
import java.util.Date;

import javax.servlet.ServletException;
import javax.servlet.http.HttpServlet;
import javax.servlet.http.HttpServletRequest;
import javax.servlet.http.HttpServletResponse;

import org.apache.ibatis.session.SqlSession;

import dbtools.GetSqlSession;
import entity.Userinfo;
import mapping.UserinfoMapper;

public class test1 extends HttpServlet {

    protected void doGet(HttpServletRequest request, HttpServletResponse response)
            throws ServletException, IOException {
```

```
        Userinfo userinfo1 = new Userinfo();
        userinfo1.setUsername("中国1");
        userinfo1.setPassword("中国人1");
        userinfo1.setAge(100L);
        userinfo1.setInsertdate(new Date());

        Userinfo userinfo2 = new Userinfo();
        userinfo2.setUsername("中国2");
        userinfo2.setPassword("中国人2");
        userinfo2.setAge(100L);
        userinfo2.setInsertdate(new Date());

        SqlSession session = GetSqlSession.getSqlSession();
        UserinfoMapper mapper = session.getMapper(UserinfoMapper.class);
        mapper.insertUserinfo(userinfo1);
        mapper.insertUserinfo(userinfo2);
        // 此案例没有提交
        // session.commit();
        session.close();
    }
}
```

程序运行后在数据表 userinfo 中并没有添加任何一条记录，因为事务使用的是 <transactionManager type="JDBC" />显式处理。

更改配置文件 mybatis-config.xml 的代码如下。

```
<environments default="jndi1">
    <environment id="jndi1">
        <transactionManager type="MANAGED" />
        <dataSource type="JNDI">
            <property name="data_source" value="${jndiName}" />
        </dataSource>
    </environment>
```

再次执行 Test1.java 类，程序执行完毕后在数据表 userinfo 中添加了 2 条记录，每执行 1 条 SQL 语句后事务进行了自动提交。

以上测试的结果说明，在程序代码没有错误的情况下，配置代码<transactionManager type="MANAGED" />和<transactionManager type="JDBC" />的主要区别就是事务是否自动提交。

下面再来看一下在程序代码出现错误的情况下，这两者是如何处理事务的。

首先要将 userinfo 数据表中的全部记录删除。

更改配置文件 mybatis-config.xml 的代码如下。

```
<environments default="jndi1">
    <environment id="jndi1">
        <transactionManager type="JDBC" />
        <dataSource type="JNDI">
            <property name="data_source" value="${jndiName}" />
        </dataSource>
    </environment>
```

创建 Servlet 类 Test2.java 的代码如下。

```
public class test2 extends HttpServlet {
    protected void doGet(HttpServletRequest request, HttpServletResponse response)
```

2.4 与数据源 DataSource 有关的操作

```
            throws ServletException, IOException {
        Userinfo userinfo1 = new Userinfo();
        userinfo1.setUsername("中国1");
        userinfo1.setPassword("中国人1");
        userinfo1.setAge(100L);
        userinfo1.setInsertdate(new Date());

        Userinfo userinfo2 = new Userinfo();
        userinfo2.setUsername("中国2");
        userinfo2.setPassword(
               "中国人2中国人2中国人2中国人2中国人2中国人2中国人2中国人2中国人2中国人2中国人2中国人2中国人2中国人2中国人2中国人2中国人2中国人2中国人2中国人2中国人2中国人2中国人2中国人2中国人2中国人2中");
        userinfo2.setAge(100L);
        userinfo2.setInsertdate(new Date());

        SqlSession session = GetSqlSession.getSqlSession();
        UserinfoMapper mapper = session.getMapper(UserinfoMapper.class);
        mapper.insertUserinfo(userinfo1);
        mapper.insertUserinfo(userinfo2);
        // 此案例没有提交
        // session.commit();
        session.close();
    }

}
```

程序运行后，数据表 userinfo 中并没有添加任何一条记录，程序代码出现异常导致事务整体回滚，因为配置代码<transactionManager type="*JDBC*" />会使用 autocommit 值为 false 的配置，而配置代码<transactionManager type="*MANAGED*" />会使用 autocommit 值为 true 的配置。

继续更改配置文件 mybatis-config.xml，代码如下。

```
<environments default="jndi1">
    <environment id="jndi1">
        <transactionManager type="MANAGED" />
        <dataSource type="JNDI">
            <property name="data_source" value="${jndiName}" />
        </dataSource>
    </environment>
```

再次执行名称为 Test2.java 的 Servlet，程序执行完毕后，数据表 userinfo 中只出现 1 条记录，因为第 2 个 insert 操作失败。由于<transactionManager type="MANAGED" />的配置是每执行一个 SQL 就提交事务，所以第 2 条 insert 语句由于出现异常并没有成功添加到 userinfo 数据表中。

从 JNDI 树中根据 JNDI 结点名称 name 取得对应的 DataSource 数据源 value 值，再从 DataSource 中取得 Connection 连接对象。以上就是在 Web 环境中使用 JNDI 的过程，JNDI 的代码是需要程序员自己输入，而 MyBatis 封装了这个过程。

2.4.4 在 MyBatis 中使用 HikariCP 连接池

在第 1 章中使用 MyBatis 时，使用配置<dataSource type="POOLED">在 MyBatis 中创建 ConnectionPool 连接池，这个连接池是 MyBatis 自己创建的，在效率上得到了保障。MyBatis 还支持第三方的连接池，第三方的连接池在获得 Connection 时会更加高效，本章就将现阶段主流的 HikariCP 连接池与 MyBatis 进行整合。

创建名称为 use_HikariDataSource 的项目，类 MyDataSourceFactory.java 的代码如下。

```java
package mysourcefactory;

import java.util.Properties;
import javax.sql.DataSource;
import org.apache.ibatis.datasource.DataSourceFactory;
import com.zaxxer.hikari.HikariDataSource;

public class MyDataSourceFactory implements DataSourceFactory {

    Properties prop;

    @Override
    public DataSource getDataSource() {
        System.out.println("2 public DataSource getDataSource()");
        HikariDataSource ds = new HikariDataSource();
        ds.setDriverClassName(prop.getProperty("driver"));
        ds.setJdbcUrl(prop.getProperty("url"));
        ds.setUsername(prop.getProperty("username"));
        ds.setPassword(prop.getProperty("password"));
        ds.setMinimumIdle(10);
        ds.setMaximumPoolSize(10);
        return ds;
    }

    @Override
    public void setProperties(Properties arg0) {
        System.out.println("1 public void setProperties(Properties arg0) " + arg0.size());
        prop = arg0;
    }

}
```

配置文件 mybatis-config.xml 的代码如下。

```xml
<?xml version="1.0" encoding="UTF-8" ?>
<!DOCTYPE configuration
PUBLIC "-//mybatis.org//DTD Config 3.0//EN"
"http://mybatis.org/dtd/mybatis-3-config.dtd">
<configuration>
    <properties resource="dbinfo.properties"></properties>
    <typeAliases>
        <package name="entity" />
    </typeAliases>
    <environments default="jdbc2">
        <environment id="jdbc1">
            <transactionManager type="JDBC" />
            <dataSource type="mysourcefactory.MyDataSourceFactory">
                <property name="driver" value="${driver}" />
                <property name="url" value="${url}" />
                <property name="username" value="${username}" />
                <property name="password" value="${password}" />
            </dataSource>
        </environment>
        <environment id="jdbc2">
            <transactionManager type="JDBC" />
```

```xml
            <dataSource type="mysourcefactory.MyDataSourceFactory">
                <property name="driver" value="oracle.jdbc.OracleDriver" />
                <property name="url" value="jdbc:oracle:thin:@localhost:1521:orcl" />
                <property name="username" value="y2" />
                <property name="password" value="123" />
            </dataSource>
        </environment>
    </environments>
    <mappers>
        <mapper resource="mapping/userinfoMapping.xml" />
    </mappers>
</configuration>
```

运行类的代码如下。

```java
public class SelectAll {

    public static void main(String[] args) throws IOException, SQLException {
        String configFile = "mybatis-config.xml";

        InputStream configStream = Resources.getResourceAsStream(configFile);

        SqlSessionFactoryBuilder builder = new SqlSessionFactoryBuilder();
        SqlSessionFactory factory = builder.build(configStream);
        SqlSession session = factory.openSession();
        UserinfoMapping mapping = session.getMapper(UserinfoMapping.class);
        System.out.println("z13");

        List<Userinfo> listUserinfo = mapping.getAllUserinfo();
        for (int i = 0; i < listUserinfo.size(); i++) {
            Userinfo eachUserinfo = listUserinfo.get(i);
            System.out.println(eachUserinfo.getId() + " " + eachUserinfo.getUsername() + " "
                    + eachUserinfo.getPassword() + " " + eachUserinfo.getAge() + " " +
eachUserinfo.getInsertdate());
        }
        System.out.println();
        System.out.println();
        DataSource dataSource =
factory.getConfiguration().getEnvironment().getDataSource();
        for (int i = 0; i < 15; i++) {
            // 注意，不要打印 connection 对象的 hashCode() 值
            // 因为现在的 Connection 接口的实现类并不是：
            // oracle.jdbc.driver.T4CConnection
            // 而是 HikariProxyConnection
            // 多个 HikariProxyConnection 对象使用的是同一个
oracle.jdbc.driver.T4CConnection 对象
            Connection connection = dataSource.getConnection();
            System.out.println(connection);
            connection.close();
        }
        session.commit();
        session.close();
    }

}
```

程序运行后控制台输出信息如下。

```
1 public void setProperties(Properties arg0)  4
2 public DataSource getDataSource()
1542469 中国 大中国 100 null
```

```
1542470 中国 大中国 100 null
1542471 中国 中国人 100 Wed Nov 02 09:26:44 CST 2016
1542472 中国 中国人 100 Wed Nov 02 09:31:15 CST 2016
1542476 x xx 200 Mon Nov 07 09:39:37 CST 2016
1542477 a 我是中文密码! 100 Mon Nov 07 09:43:22 CST 2016
1542478 a 我是中文密码! 100 Mon Nov 07 09:50:06 CST 2016


HikariProxyConnection@1543148593 wrapping oracle.jdbc.driver.T4CConnection@3db346a8
HikariProxyConnection@2028555727 wrapping oracle.jdbc.driver.T4CConnection@3db346a8
HikariProxyConnection@591391158 wrapping oracle.jdbc.driver.T4CConnection@3db346a8
HikariProxyConnection@898557489 wrapping oracle.jdbc.driver.T4CConnection@3db346a8
HikariProxyConnection@247944893 wrapping oracle.jdbc.driver.T4CConnection@3db346a8
HikariProxyConnection@1014166943 wrapping oracle.jdbc.driver.T4CConnection@3db346a8
HikariProxyConnection@1625082366 wrapping oracle.jdbc.driver.T4CConnection@3db346a8
HikariProxyConnection@572593338 wrapping oracle.jdbc.driver.T4CConnection@3db346a8
HikariProxyConnection@384294141 wrapping oracle.jdbc.driver.T4CConnection@3db346a8
HikariProxyConnection@1024597427 wrapping oracle.jdbc.driver.T4CConnection@3db346a8
HikariProxyConnection@990355670 wrapping oracle.jdbc.driver.T4CConnection@3db346a8
HikariProxyConnection@296347592 wrapping oracle.jdbc.driver.T4CConnection@3db346a8
HikariProxyConnection@956420404 wrapping oracle.jdbc.driver.T4CConnection@3db346a8
HikariProxyConnection@349420578 wrapping oracle.jdbc.driver.T4CConnection@3db346a8
HikariProxyConnection@315932542 wrapping oracle.jdbc.driver.T4CConnection@3db346a8
```

从控制台打印的信息来看，使用的是同 1 个 Connection 连接。

2.5 多数据库执行不同 SQL 语句的支持

在设计软件系统时经常需要考虑多数据库的支持，也就是在 Java 代码与 SQL 映射代码都不变的情况下来执行不同的 SQL 语句，MyBatis 新版是支持这个功能的。

创建名称为 databaseIdTest 的项目，SQL 映射文件 UserinfoMapper.xml 的代码如下。

```xml
<?xml version="1.0" encoding="UTF-8"?>
<!DOCTYPE mapper PUBLIC "-//mybatis.org//DTD Mapper 3.0//EN"
"http://mybatis.org/dtd/mybatis-3-mapper.dtd">
<mapper namespace="sqlmapping.UserinfoMapper">
    <select id="getAllUserinfo" resultType="userinfo" databaseId="xxxOracle">
        select * from
        userinfo order by id asc
    </select>
    <select id="getAllUserinfo" resultType="userinfo" databaseId="yyyMYSQL">
        select * from
        userinfo order by id desc
    </select>
</mapper>
```

配置文件 mybatis-config.xml 的代码如下。

```xml
<?xml version="1.0" encoding="UTF-8" ?>
<!DOCTYPE configuration
PUBLIC "-//mybatis.org//DTD Config 3.0//EN"
"http://mybatis.org/dtd/mybatis-3-config.dtd">
<configuration>
    <typeAliases>
        <typeAlias alias="userinfo" type="sqlmapping.Userinfo" />
    </typeAliases>
```

```xml
<environments default="mysql">
    <environment id="oracle">
        <transactionManager type="JDBC" />
        <dataSource type="POOLED">
            <property name="driver" value="oracle.jdbc.OracleDriver" />
            <property name="url" value="jdbc:oracle:thin:@localhost:1521:orcl" />
            <property name="username" value="y2" />
            <property name="password" value="123" />
        </dataSource>
    </environment>
    <environment id="mysql">
        <transactionManager type="JDBC" />
        <dataSource type="POOLED">
            <property name="driver" value="com.mysql.jdbc.Driver" />
            <property name="url" value="jdbc:mysql://localhost:3306/y2" />
            <property name="username" value="root" />
            <property name="password" value="123" />
        </dataSource>
    </environment>
</environments>
<databaseIdProvider type="DB_VENDOR">
    <property name="Oracle" value="xxxOracle" />
    <property name="MySQL" value="yyyMYSQL" />
</databaseIdProvider>
<mappers>
    <mapper resource="sqlmapping/UserinfoMapper.xml" />
</mappers>

</configuration>
```

配置代码<property name="*Oracle*" value="*xxxOracle*" />中的 name 属性值是通过代码获得的：

```
System.out.println("getDatabaseProductName()=" +
factory.getConfiguration().getEnvironment().getDataSource()
            .getConnection().getMetaData().getDatabaseProductName());
```

属性 name 的值不能随意填写，并且还要注意大小写，因为是区分大小写的。属性 name 代表数据库的产品名称，而 value 属性代表这个数据库产品名称的别名。value 的属性可以随意命名，但尽量命名的有意义。然后在 SQL 映射文件中使用如下代码。

```xml
<select id="getAllUserinfo" resultType="userinfo" databaseId="xxxOracle">
    select * from
    userinfo order by id asc
</select>
<select id="getAllUserinfo" resultType="userinfo" databaseId="yyyMYSQL">
    select * from
    userinfo order by id desc
</select>
```

在不同的 SQL 映射上引用不同的数据库别名 xxxOracle 和 yyyMYSQL，这样就可以达到和 SQL 映射的 id 值一样，但在不同的数据库中可以执行不同的 SQL 语句的目的了。

使用不同的数据源可以执行不同的 SQL 语句，Oracle 是正序，而 MySQL 是倒序。

2.6 多种获取 Mapper 的方式

创建名称为 moreMapperTest 的项目。

在 mybatis-config.xml 配置文件中可以使用 4 种方式来获取 Mapper 映射，示例代码如下。

```xml
<!-- 第 1 种写法： 直接获取 SQL 映射 xml 文件-->
<mapper resource="sqlmapping/UserinfoMapper.xml" />

<!-- 第 2 种写法： 扫描包，注意接口的名称和 xml 的文件的主文件名必须一样-->
<package name="sqlmapping" />

<!-- 第 3 种写法： 指定接口名称，注意接口的名称和 xml 的文件的主文件名必须一样-->
<mapper class="sqlmapping.UserinfoMapper" />

<!-- 第 4 种写法： 使用绝对路径-->
<mapper
    url="file:\\\C:\Users\gaohongyan\workspace\moreMapperTest\src\sqlmapping\UserinfoMapper.xml" />
```

这 4 种方式建议不要混用。

2.7 MyBatis3 的 SQL 映射文件与动态 SQL

MyBatis 框架是基于 SQL 映射的，所以 SQL 映射文件在此框架中的位置非常重要。动态 SQL 是 MyBatis 提供的，它根据指定的条件来执行指定的 SQL 语句，它使 SQL 映射文件中的 SQL 语句在执行时具有动态性。SQL 映射文件与动态 SQL 被设计得非常简单。本节将和大家一起学习 SQL 映射文件中的常用实例的使用。

2.7.1 <resultMap>标签

如果数据表中字段的名称和 Java 实体类中属性的名称不一致，就要使用<resultMap>标签来做一个映射。

创建名称为 resultMapTest 的 Java 项目，映射配置文件 userinfoMapping.xml 的代码如下。

```xml
<?xml version="1.0" encoding="UTF-8" ?>
<!DOCTYPE mapper PUBLIC "-//mybatis.org//DTD Mapper 3.0//EN" "mybatis-3-mapper.dtd">
<mapper namespace="mybatis.testcurd">

    <resultMap type="entity.Userinfo" id="userinfo">
        <result column="id" property="idghy" />
        <result column="username" property="usernameghy" />
        <result column="password" property="passwordghy" />
        <result column="age" property="ageghy" />
        <result column="insertdate" property="insertdateghy" />
    </resultMap>

    <select id="getUserinfoAll" resultMap="userinfo">
        select * from
        userinfo
    </select>
</mapper>
```

在<select>标签中使用 resultMap 属性来引用<resultMap>的 id 属性值，形成映射关系。
实体类 Userinfo.java 的结构如图 2-6 所示。

图 2-6 类 Userinfo.java 结构

创建 Run.java 运行类，核心代码如下。

```java
public class Run {

    public static void main(String[] args) {

        try {
            String resource = "mybatis-config.xml";
            InputStream inputStream = Resources.getResourceAsStream(resource);
            SqlSessionFactory sqlSessionFactory = new SqlSessionFactoryBuilder()
                    .build(inputStream);
            SqlSession sqlSession = sqlSessionFactory.openSession();
            List<Userinfo> listUserinfo = sqlSession
                    .selectList("getUserinfoAll");
            for (int i = 0; i < listUserinfo.size(); i++) {
                Userinfo userinfo = listUserinfo.get(i);
                System.out.println(userinfo.getIdghy() + " "
                        + userinfo.getUsernameghy() + " "
                        + userinfo.getPasswordghy() + " "
                        + userinfo.getAgeghy() + " "
                        + userinfo.getInsertdateghy());
            }
        } catch (IOException e) {
            // TODO Auto-generated catch block
            e.printStackTrace();
        }

    }

}
```

程序运行后的结果如图 2-7 所示。

```
┌─────────────────────────────────────────────┐
│ 🔲 Problems  🔲 Tasks  🔲 Web Browser  🔲 Co│
│ <terminated> Run [Java Application] C:\MyEclipse│
│ 1  a1  aa  100  2010-12-19  14:17:30.0      │
│ 2  a2  aa  100  2010-12-19  14:17:30.0      │
│ 3  a3  aa  100  2010-12-19  14:17:30.0      │
│ 4  a4  aa  100  2010-12-19  14:17:30.0      │
│ 5  a5  aa  100  2010-12-19  14:17:30.0      │
│ 6  a6  aa  100  2010-12-19  14:17:30.0      │
│ 7  a7  aa  100  2010-12-19  14:17:30.0      │
│ 8  a8  aa  100  2010-12-19  14:17:30.0      │
└─────────────────────────────────────────────┘
```

图 2-7 成功输出数据表中的数据

2.7.2 \<sql\>标签

重复的 sql 语句永远不可避免，\<sql\>标签就是用来解决这个问题的。

创建名称为 sqlTest 的 Java 项目，映射配置文件 userinfoMapping.xml 的代码如下。

```xml
<?xml version="1.0" encoding="UTF-8" ?>
<!DOCTYPE mapper PUBLIC "-//mybatis.org//DTD Mapper 3.0//EN" "mybatis-3-mapper.dtd">
<mapper namespace="mybatis.testcurd">

    <sql id="userinfoField">id,username,password,age,insertdate</sql>

    <select id="getUserinfoAll" resultType="map">
        select
        <include refid="userinfoField" />
        from
        userinfo
    </select>

    <select id="getUserinfoById" resultType="map" parameterType="int">
        select
        <include refid="userinfoField" />
        from
        userinfo where id=#{0}
    </select>
</mapper>
```

上面代码中的 id、username、password、age 和 insertdate 这 5 个字段在映射文件中多处出现，所以可以将这 5 个字段封装进\<sql\>标签中，以减少配置的代码量。

创建 Run.java 运行测试类，代码如下。

```java
public class Run {

    public static void main(String[] args) {
        try {
            String resource = "mybatis-config.xml";
            InputStream inputStream = Resources.getResourceAsStream(resource);
            SqlSessionFactory sqlSessionFactory = new SqlSessionFactoryBuilder()
                    .build(inputStream);
            SqlSession sqlSession = sqlSessionFactory.openSession();
            List<Map> listUserinfo = sqlSession.selectList("getUserinfoAll");
            for (int i = 0; i < listUserinfo.size(); i++) {
```

```
                    Map map = listUserinfo.get(i);
                    System.out.println(map.get("ID") + " " + map.get("USERNAME")
                            + " " + map.get("PASSWORD") + " " + map.get("AGE")
                            + " " + map.get("INSERTDATE"));
                }
                System.out.println("");
                System.out.println("");
                listUserinfo = sqlSession.selectList("getUserinfoById", 5);
                for (int i = 0; i < listUserinfo.size(); i++) {
                    Map map = listUserinfo.get(i);
                    System.out.println(map.get("ID") + " " + map.get("USERNAME")
                            + " " + map.get("PASSWORD") + " " + map.get("AGE")
                            + " " + map.get("INSERTDATE"));
                }
        } catch (IOException e) {
            // TODO Auto-generated catch block
            e.printStackTrace();
        }
    }
}
```

程序运行后的结果如图 2-8 所示。

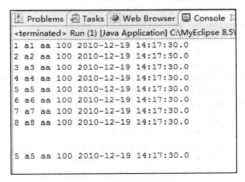

图 2-8　程序运行结果

本例的 sql 映射文件是从 map 里使用 get（字段名称）的形式取得字段对应的值，但字段名称在 Oracle 中是大写字母，所以这里的字段名称也必须要写成大写形式。为了支持方便的小写形式，可以在映射文件中定义的 SQL 语句中为字段另起一个别名即可：

```
select id "id",username "username",password "password",age "age"
```

还可以使用<sql>标签来进行声明，示例代码如下：

```
<sql id="userinfo4ColName">id "id",username "username",password "password",age "age"
</sql>
```

这样从 map 中就可以以小写的形式取得字段值了。

2.7.3　使用${}拼接 SQL 语句

在 MyBatis 中也支持将 SQL 语句当成变量传入。

新建名称为 sqlStringVar 的 Java 项目，映射文件 userinfoMapping.xml 的内容如下。

```xml
<?xml version="1.0" encoding="UTF-8" ?>
<!DOCTYPE mapper PUBLIC "-//mybatis.org//DTD Mapper 3.0//EN" "mybatis-3-mapper.dtd">
<mapper namespace="mybatis.testcurd">

    <select id="getUserinfo" parameterType="map" resultType="map">
        select
        id,username,password,age,insertdate
        from
        userinfo where id>#{id} order
        by ${orderSql}
    </select>

</mapper>
```

Java 类文件 Run.java 的代码如下。

```java
public class Run {

    public static void main(String[] args) {

        try {
            HashMap mapParam = new HashMap();
            mapParam.put("id", 5);
            mapParam.put("orderSql", "id desc");

            String resource = "mybatis-config.xml";
            InputStream inputStream = Resources.getResourceAsStream(resource);
            SqlSessionFactory sqlSessionFactory = new SqlSessionFactoryBuilder()
                    .build(inputStream);
            SqlSession sqlSession = sqlSessionFactory.openSession();
            List<Map> listUserinfo = sqlSession.selectList("getUserinfo",
                    mapParam);
            for (int i = 0; i < listUserinfo.size(); i++) {
                Map map = listUserinfo.get(i);
                System.out.println(map.get("ID") + " " + map.get("USERNAME")
                        + " " + map.get("PASSWORD") + " " + map.get("AGE")
                        + " " + map.get("INSERTDATE"));
            }
        } catch (IOException e) {
            // TODO Auto-generated catch block
            e.printStackTrace();
        }

    }

}
```

运行结果如图 2-9 所示。

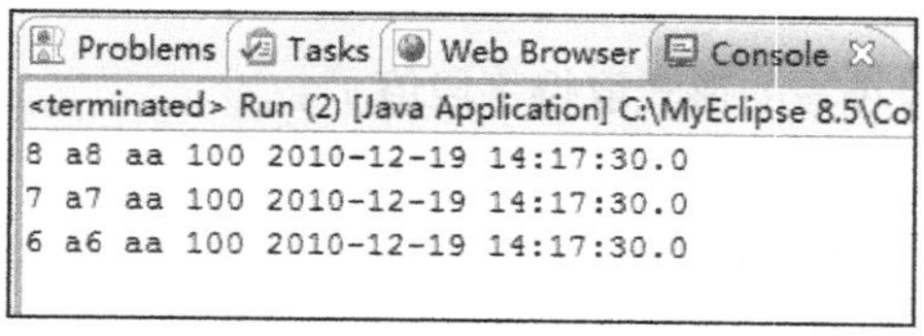

图 2-9　运行结果

2.7 MyBatis3 的 SQL 映射文件与动态 SQL

如果在 SQL 语句中出现 < 小于或者是 <= 小于等于的条件时，由于 < 符号是特殊的符号，所以要使用<![CDATA[]]>将其转化成为普通的字符串，而不是 XML 标记的左开始标签，示例代码如下。

```xml
<select id="getAllUserinfo" parameterType="map" resultType="map">
    select
    <include refid="userinfo4ColName"></include>
    from userinfo where id
    <![CDATA[<=]]>
    #{id}
    order by ${orderSQL}
</select>
```

2.7.4 插入 null 值时的处理第 1 种方法 jdbcType

创建名称为 dynSqlTest 的 Web 项目，创建名称为 test1 的 Servlet 对象，核心代码如下。

```java
public class test1 extends HttpServlet {

    public void doGet(HttpServletRequest request, HttpServletResponse response)
            throws ServletException, IOException {

        Userinfo userinfo = new Userinfo();
        userinfo.setUsername("中国");
        userinfo.setPassword(null);
        userinfo.setAge(200L);
        userinfo.setInsertdate(new Date());

        SqlSession sqlSessionRef = GetSqlSession.getSqlSession();
        sqlSessionRef.insert("insertUserinfo", userinfo);
        sqlSessionRef.commit();
        sqlSessionRef.close();

    }

}
```

在映射文件 userinfoMapping.xml 中添加如下配置。

```xml
<insert id="insertUserinfo" parameterType="orm.Userinfo">
    <selectKey keyProperty="id" order="BEFORE" resultType="java.lang.Long">
        select idauto.nextval from dual
    </selectKey>
    insert into userinfo(id,username,password,age,insertdate)
    values(#{id},#{username},#{password},#{age},#{insertdate})
</insert>
```

程序运行后出现如下异常信息。

Error updating database. Cause: org.apache.ibatis.type.TypeException: Error setting null for parameter #3 with JdbcType OTHER . Try setting a different JdbcType for this parameter or a different jdbcTypeForNull configuration property. Cause: java.sql.SQLException: 无效的列类型: 1111

从出错信息中可以看到，是用 null 值对 password 字段进行了设置，造成 MyBatis 无法识别出应该插入默认的数据类型值，这种情况可以通过设置映射的默认数据类型来解决。更改映射配置 userinfoMapping.xml 代码如下。

```xml
<?xml version="1.0" encoding="UTF-8" ?>
<!DOCTYPE mapper PUBLIC "-//mybatis.org//DTD Mapper 3.0//EN" "mybatis-3-mapper.dtd">
<mapper namespace="mybatis.testcurd">
    <insert id="insertUserinfo" parameterType="orm.Userinfo">
        <selectKey keyProperty="id" order="BEFORE" resultType="java.lang.Long">
            select idauto.nextval from dual
        </selectKey>
        insert into userinfo(id,username,password,age,insertdate)
        values(#{id,jdbcType=INTEGER},#{username,jdbcType=VARCHAR},#{password,jdbcType=VARCHAR},#{age,jdbcType=INTEGER},#{insertdate,jdbcType=TIMESTAMP})
    </insert>
</mapper>
```

这里在#{}格式中加入了数据类型的声明,这样可以明确地告诉 MyBatis 框架遇到 null 值该如何处理。再次运行程序,成功插入数据表,运行结果如图 2-10 所示。

| 192 | 中国 | 200 | 2010/12/26 8:58:45 |
| 193 | 中国 | 200 | 2010/12/26 8:58:47 |

图 2-10 成功插入数据表

2.7.5 插入 null 值时的处理第 2 种方法 <if>

通过在#{}格式中加入 jdbcType 即可避免插入 null 值时的异常,其实使用动态 SQL 标签也可以达到同样的效果。

继续在 dynSqlTest 项目中进行添加代码。

在映射文件 userinfoMapping.xml 添加如下代码。

```xml
<insert id="insertUserinfo2" parameterType="orm.Userinfo">
    <selectKey keyProperty="id" order="BEFORE" resultType="java.lang.Long">
        select idauto.nextval from dual
    </selectKey>
    <if test="password!=null">
        insert into userinfo(id,username,password,age,insertdate)
        values(#{id},#{username},#{password},#{age},#{insertdate})
    </if>
    <if test="password==null">
        insert into userinfo(id,username,age,insertdate)
        values(#{id},#{username},#{age},#{insertdate})
    </if>
</insert>
```

创建名称为 test2 的 Servlet,核心代码如下。

```java
public class test2 extends HttpServlet {

    public void doGet(HttpServletRequest request, HttpServletResponse response)
            throws ServletException, IOException {

        Userinfo userinfo1 = new Userinfo();
        userinfo1.setUsername("英国");
        userinfo1.setPassword(null);
        userinfo1.setAge(200L);
        userinfo1.setInsertdate(new Date());
```

```
            Userinfo userinfo2 = new Userinfo();
            userinfo2.setUsername("法国");
            userinfo2.setPassword("法国人");
            userinfo2.setAge(200L);
            userinfo2.setInsertdate(new Date());

            SqlSession sqlSessionRef = GetSqlSession.getSqlSession();
            sqlSessionRef.insert("insertUserinfo2", userinfo1);
            sqlSessionRef.insert("insertUserinfo2", userinfo2);
            sqlSessionRef.commit();
            sqlSessionRef.close();
    }
}
```

程序运行后，成功向数据表中插入 2 条记录，如图 2-11 所示。

| 198 | 英国 | | 200 | 2010/12/26 9:10:56 |
| 199 | 法国 | 法国人 | 200 | 2010/12/26 9:10:56 |

图 2-11　成功插入 2 条记录

标签 <if> 在判断字符串为空时还可以更加具体，示例代码如下。

```
<if test="password!=null and password!=''">
```

也就是 null 值与""空字符串一同进行判断。

2.7.6　<where> 标签

标签 where 的主要作用就是生成根据条件查询的 SQL 语句，可以用在 insert、delete、update 以及 select 语句中。

创建测试用的项目 whereTAG，创建映射文件 userinfoMapping.xml 的代码如下。

```xml
<?xml version="1.0" encoding="UTF-8" ?>
<!DOCTYPE mapper
PUBLIC "-//mybatis.org//DTD Mapper 3.0//EN"
"mybatis-3-mapper.dtd">
<mapper namespace="test">
    <select id="selectA" resultType="map">
        select * from userinfo
        <where>
            <if test="username!=null">username like #{username}</if>
            <if test="age!=null">or age = #{age}</if>
        </where>
    </select>
</mapper>
```

运行类 Test1.java 的代码如下。

```java
package controller;

import java.io.IOException;
import java.io.InputStream;
```

```java
import java.util.HashMap;
import java.util.List;
import java.util.Map;

import org.apache.ibatis.io.Resources;
import org.apache.ibatis.session.SqlSession;
import org.apache.ibatis.session.SqlSessionFactory;
import org.apache.ibatis.session.SqlSessionFactoryBuilder;

public class Test1 {

    public static void main(String[] args) {
        try {

            Map queryMap = new HashMap();
            queryMap.put("username", "%美%");
            queryMap.put("age", "999");

            InputStream configStream = Resources
                    .getResourceAsStream("mybatis-config.xml");
            SqlSessionFactory factory = new SqlSessionFactoryBuilder()
                    .build(configStream);
            SqlSession session = factory.openSession();

            List<Map> listMap = session.selectList("selectA", queryMap);
            for (int i = 0; i < listMap.size(); i++) {
                Map eachMap = listMap.get(i);
                System.out.println(eachMap.get("ID") + " "
                        + eachMap.get("USERNAME") + " "
                        + eachMap.get("PASSWORD") + " " + eachMap.get("AGE"));
            }
            session.commit();
            session.close();
        } catch (IOException e) {
            e.printStackTrace();
        }

    }
}
```

程序运行结果如图 2-12 所示。

图 2-12 根据 where 条件查询的结果

如果在代码中不向 username 传递值，<where>标签也能自动去掉 or 关键字而成功执行单独查询 age 年龄的 SQL 语句。

2.7.7 <choose>标签的使用

<choose>标签的作用是在众多的条件中选择出一个条件，有些类似于 Java 语言中的 switch 语句的作用。

在映射文件 userinfoMapping.xml 中添加如下配置代码。

```xml
<select id="selectUserinfo1" parameterType="orm.Userinfo"
    resultType="map">
    select * from userinfo where 1=1
    <choose>
        <when test="username!=null">and username like '%'||#{username}||'%'</when>
        <when test="password!=null">and password like '%'||#{password}||'%'</when>
        <otherwise>and age=200</otherwise>
    </choose>
</select>
```

配置代码<when test="username!=null">中的 username 是指传入的参数 Userinfo 实体类中的 username 属性值是否为 null。

创建名称为 test3 的 Servlet 对象，核心代码如下。

```java
public class test3 extends HttpServlet {

    public void doGet(HttpServletRequest request, HttpServletResponse response)
            throws ServletException, IOException {

        Userinfo userinfo1 = new Userinfo();
        userinfo1.setUsername("英");

        Userinfo userinfo2 = new Userinfo();
        userinfo2.setPassword("法");

        Userinfo userinfo3 = new Userinfo();

        SqlSession sqlSessionRef = GetSqlSession.getSqlSession();
        List<Map> listMap1 = sqlSessionRef.selectList("selectUserinfo1",
                userinfo1);
        for (int i = 0; i < listMap1.size(); i++) {
            Map eachMap = listMap1.get(i);
            System.out.println("listMap1 中的内容： " + eachMap.get("ID") + " "
                    + eachMap.get("USERNAME") + " " + eachMap.get("PASSWORD")
                    + " " + eachMap.get("AGE") + " "
                    + eachMap.get("INSERTDATE"));
        }

        List<Map> listMap2 = sqlSessionRef.selectList("selectUserinfo1",
                userinfo2);
        for (int i = 0; i < listMap2.size(); i++) {
            Map eachMap = listMap2.get(i);
            System.out.println("listMap2 中的内容： " + eachMap.get("ID") + " "
                    + eachMap.get("USERNAME") + " " + eachMap.get("PASSWORD")
                    + " " + eachMap.get("AGE") + " "
                    + eachMap.get("INSERTDATE"));
        }
```

```
        List<Map> listMap3 = sqlSessionRef.selectList("selectUserinfo1",
                userinfo3);
        for (int i = 0; i < listMap3.size(); i++) {
            Map eachMap = listMap3.get(i);
            System.out.println("listMap3 中的内容:    " + eachMap.get("ID") + " "
                    + eachMap.get("USERNAME") + " " + eachMap.get("PASSWORD")
                    + " " + eachMap.get("AGE") + " "
                    + eachMap.get("INSERTDATE"));
        }
        sqlSessionRef.commit();
        sqlSessionRef.close();

    }
}
```

数据表 userinfo 中的内容如图 2-13 所示。

ID	USERNAME	PASSWORD	AGE	INSERTDATE
1	a1	aa	100	2010/12/19 14:17:30
2	a2	aa	100	2010/12/19 14:17:30
3	a3	aa	100	2010/12/19 14:17:30
4	a4	aa	100	2010/12/19 14:17:30
5	a5	aa	100	2010/12/19 14:17:30
6	a6	aa	100	2010/12/19 14:17:30
7	a7	aa	100	2010/12/19 14:17:30
8	a8	aa	100	2010/12/19 14:17:30
198	英国		200	2010/12/26 9:10:56
199	法国	法国人	200	2010/12/26 9:10:56

图 2-13 userinfo 数据表内容

程序运行后，在控制台输出如图 2-14 所示的信息。

```
listMap1中的内容:  198 英国null 200 2010-12-26 09:10:56.0
listMap2中的内容:  199 法国法国人200 2010-12-26 09:10:56.0
listMap3中的内容:  198 英国null 200 2010-12-26 09:10:56.0
listMap3中的内容:  199 法国法国人200 2010-12-26 09:10:56.0
```

图 2-14 运行结果

标签<choose>起到了和 switch 结合 break 语句一样的作用。

2.7.8 <set>标签的使用

<set>标签可以用在 update 语句中，作用是动态指定要更新的列。

在映射文件 userinfoMapping.xml 中添加如下映射代码。

```
<update id="updateUserinfo" parameterType="orm.Userinfo">
    update userinfo
    <set>
        <if test="username!=null">username=#{username},</if>
        <if test="password!=null">password=#{password},</if>
        <if test="age!=null">age=#{age},</if>
        <if test="insertdate!=null">insertdate=#{insertdate}</if>
```

```
        </set>
        <if test="id!=null">where id=#{id}</if>
</update>
```

创建名称为 test4 的 Servlet 对象,核心代码如下。

```java
public class test4 extends HttpServlet {

    public void doGet(HttpServletRequest request, HttpServletResponse response)
            throws ServletException, IOException {

        Userinfo userinfo = new Userinfo();
        userinfo.setId(199L);
        userinfo.setUsername(null);
        userinfo.setPassword("新密码");
        userinfo.setAge(1000L);
        userinfo.setInsertdate(new Date());

        SqlSession sqlSessionRef = GetSqlSession.getSqlSession();
        sqlSessionRef.update("updateUserinfo", userinfo);
        sqlSessionRef.commit();
        sqlSessionRef.close();
    }

}
```

程序运行结果如图 2-15 所示。

| 199 法国 | 新密码 | 1000 | 2010/12/26 10:00:25 |

图 2-15 成功更新

2.7.9 <foreach>标签的使用

<foreach>标签有循环的功能,可以用来生成有规律的 SQL 语句。

<foreach>标签主要的属性有 item、index、collection、open、separator 和 close。

item 表示集合中每一个元素进行迭代时的别名,index 指定一个名字,用于表示在迭代过程中每次迭代到的位置,open 表示该语句以什么开始,separator 表示在每次迭代之间以什么符号作为分隔符,close 表示该语句以什么结束。

创建新的 Web 项目 foreachTest。

在映射文件 userinfoMapping.xml 中添加如下配置代码。

```xml
<select id="selectUserinfo2" parameterType="list" resultType="map">
    select * from userinfo where id in
    <foreach collection="list" item="eachId" index="currentIndex"
        open="(" separator="," close=")">
        #{eachId}
    </foreach>
</select>

<select id="selectUserinfo3" parameterType="orm.QueryUserinfo"
    resultType="map">
    select * from userinfo where id in
```

```xml
            <foreach collection="idList" item="eachId" index="currentIndex"
                open="(" separator="," close=")">
                #{eachId}
            </foreach>
            and username like '%'||#{username}||'%'
    </select>

    <select id="selectUserinfo4" parameterType="map" resultType="map">
        select * from userinfo where id in
            <foreach collection="idList" item="eachId" index="currentIndex"
                open="(" separator="," close=")">
                #{eachId}
            </foreach>
            and username like '%'||#{username}||'%'
    </select>
```

在 Oracle 数据库中，字符串的拼接要使用||符号。

创建名称为 test5 的 Servlet，核心代码如下。

```java
public class test5 extends HttpServlet {

    public void doGet(HttpServletRequest request, HttpServletResponse response)
            throws ServletException, IOException {

        // id 的值来自于 List
        List list = new ArrayList();
        list.add(1);
        list.add(3);
        list.add(5);

        SqlSession sqlSessionRef = GetSqlSession.getSqlSession();
        List<Map> listMap1 = sqlSessionRef.selectList("selectUserinfo2", list);
        for (int i = 0; i < listMap1.size(); i++) {
            Map eachMap = listMap1.get(i);
            System.out.println("list1 中的内容：   " + eachMap.get("ID") + " "
                    + eachMap.get("USERNAME") + " " + eachMap.get("PASSWORD")
                    + " " + eachMap.get("AGE") + " "
                    + eachMap.get("INSERTDATE"));
        }

        // id 的值来自于 QueryUserinfo 实体中的 List
        QueryUserinfo queryUserinfo = new QueryUserinfo();
        queryUserinfo.setUsername("法");
        queryUserinfo.getIdList().add(198);
        queryUserinfo.getIdList().add(199);

        listMap1 = sqlSessionRef.selectList("selectUserinfo3", queryUserinfo);
        for (int i = 0; i < listMap1.size(); i++) {
            Map eachMap = listMap1.get(i);
            System.out.println("list2 中的内容：   " + eachMap.get("ID") + " "
                    + eachMap.get("USERNAME") + " " + eachMap.get("PASSWORD")
                    + " " + eachMap.get("AGE") + " "
                    + eachMap.get("INSERTDATE"));
        }

        // id 的值来自于 Map 中的 List
        Map paramMap = new HashMap();
        paramMap.put("username", "5");
```

```
            paramMap.put("idList", list);

            listMap1 = sqlSessionRef.selectList("selectUserinfo4", paramMap);
            for (int i = 0; i < listMap1.size(); i++) {
                Map eachMap = listMap1.get(i);
                System.out.println("list3 中的内容:     " + eachMap.get("ID") + " "
                        + eachMap.get("USERNAME") + " " + eachMap.get("PASSWORD")
                        + " " + eachMap.get("AGE") + " "
                        + eachMap.get("INSERTDATE"));
            }

            sqlSessionRef.commit();
            sqlSessionRef.close();

    }

}
```

程序运行后，控制台输出结果如图 2-16 所示。

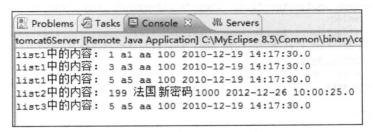

图 2-16　控制台输出结果

如何实现批量 insert 插入操作呢？由于批量插入的 SQL 语句在每种数据库中都不一样，比如 Oracle 是使用如下格式的语句。

```
INSERT INTO userinfo (id,
                  username,
                  password,
                  age,
                  insertdate)
    SELECT idauto.NEXTVAL,
        username,
        password,
        age,
        insertdate
      FROM (SELECT 'a' username,
              'aa' password,
              1 age,
              TO_DATE ('2000-1-1', 'yyyy-MM-dd') insertdate
          FROM DUAL
         UNION ALL
         SELECT 'b' username,
              'bb' password,
              1 age,
              TO_DATE ('2000-1-1', 'yyyy-MM-dd') insertdate
          FROM DUAL)
```

而 MySQL 是使用如下格式的 SQL 语句来实现批量 insert 插入。

```
insert into userinfo(username,password,age,insertdate)
values('a','aa',1,'2000-1-1'),
('b','aa',1,'2000-1-1'),
('c','aa',1,'2000-1-1'),
('d','aa',1,'2000-1-1'),
('e','aa',1,'2000-1-1')
```

由于批量 insert 插入的 SQL 语句在每种数据库中都不一样,就导致在 MyBatis 中 SQL 映射文件的写法也不一样。

下面来看看针对 Oracle 数据库的 SQL 映射文件的代码,具体如下。

```xml
<insert id="insertOracle" parameterType="list">
    INSERT INTO userinfo (id,
    username,
    password,
    age,
    insertdate)
    select
    idauto.nextval,username,password,age,insertdate from (
    <foreach collection="list" item="eachUserinfo" separator="union all">
        select
        #{eachUserinfo.username} username,#{eachUserinfo.password}
        password,#{eachUserinfo.age} age,#{eachUserinfo.insertdate}
        insertdate
        from dual
    </foreach>
    )
</insert>
```

Java 代码如下。

```java
public class Insert1 {

    public static void main(String[] args) throws IOException {

        List list = new ArrayList();

        for (int i = 0; i < 5; i++) {
            Userinfo userinfo = new Userinfo();
            userinfo.setUsername("中国" + (i + 1));
            userinfo.setPassword("中国人" + (i + 1));
            userinfo.setAge(100L);
            userinfo.setInsertdate(new Date());
            list.add(userinfo);
        }

        SqlSession session = GetSqlSession.getSqlSession();
        session.insert("insertOracle", list);
        session.commit();
        session.close();
    }

}
```

下面来看看针对 MySQL 数据库的 SQL 映射文件的代码,具体如下。

```xml
<insert id="insertMySQL" parameterType="list">
```

```xml
        INSERT INTO userinfo (
    username,
    password,
    age,
    insertdate)
    values
    <foreach collection="list" item="eachUserinfo" separator=",">
        (
        #{eachUserinfo.username},#{eachUserinfo.password},
        #{eachUserinfo.age},#{eachUserinfo.insertdate}
        )
    </foreach>
</insert>
```

Java 代码如下。

```java
public class Insert1 {

    public static void main(String[] args) throws IOException {

        List list = new ArrayList();

        for (int i = 0; i < 5; i++) {
            Userinfo userinfo = new Userinfo();
            userinfo.setUsername("中国MYSQL" + (i + 1));
            userinfo.setPassword("中国人MYSQL" + (i + 1));
            userinfo.setAge(100L);
            userinfo.setInsertdate(new Date());
            list.add(userinfo);
        }

        SqlSession session = GetSqlSession.getSqlSession();
        session.insert("insertMySQL", list);
        session.commit();
        session.close();

    }

}
```

2.7.10 使用<bind>标签对 like 语句进行适配

在 SQL 语句中使用 like 查询时，MySQL 在拼接字符串时使用 concat()方法，而 Oracle 却使用||运算符，两者的 SQL 语句并不相同，这就要创建 2 个 SQL 映射语句。下面创建实验用的项目 mysql_oracle_2_sql，SQL 映射文件示例代码如下。

```xml
<mapper namespace="userinfo">
    <select id="selectOracle" parameterType="string" resultType="entity.Userinfo">
        select * from userinfo where username like '%'||#{username}||'%'
    </select>
    <select id="selectMySQL" parameterType="string" resultType="entity.Userinfo">
        select * from userinfo where username like concat('%',#{username},'%')
    </select>
</mapper>
```

通过在运行代码中更改 selectList() 方法的第 1 个参数值来切换、执行不同的 SQL 语句，示例代码如下。

```
List<Userinfo> listUserinfo = session.selectList("selectOracle", "a");
List<Userinfo> listUserinfo = session.selectList("selectMySQL", "a");
```

这样会更改软件的源代码，不利于软件的稳定性，这时可以使用<bind>标签在多个数据库之间进行适配。

创建名称为 bindTAGNew 的项目，SQL 映射文件代码如下。

```xml
<?xml version="1.0" encoding="UTF-8" ?>
<!DOCTYPE mapper
PUBLIC "-//mybatis.org//DTD Mapper 3.0//EN"
"http://mybatis.org/dtd/mybatis-3-mapper.dtd">
<mapper namespace="userinfo">
    <select id="select1" parameterType="string" resultType="entity.Userinfo">
        <bind name="myQuerySQL" value="'%'+_parameter+'%'" />
        select * from userinfo
        where username like #{myQuerySQL}
    </select>
    <select id="select2" parameterType="entity.Userinfo" resultType="entity.Userinfo">
        <bind name="myQuerySQL" value="'%'+_parameter.getUsername()+'%'" />
        select * from userinfo
        where username like #{myQuerySQL}
    </select>
    <select id="select3" parameterType="entity.Userinfo" resultType="entity.Userinfo">
        <bind name="myQuerySQL" value="'%'+username+'%'" />
        select * from userinfo
        where username like #{myQuerySQL}
    </select>
    <select id="select4" parameterType="entity.Userinfo" resultType="entity.Userinfo">
        <bind name="myQuerySQL" value="'%'+_parameter.username+'%'" />
        select * from userinfo
        where username like #{myQuerySQL}
    </select>
    <select id="select5" parameterType="entity.Userinfo" resultType="entity.Userinfo">
        <bind name="myQuerySQL" value="'%'+#root._parameter.username+'%'" />
        select * from userinfo
        where username like #{myQuerySQL}
    </select>
    <select id="select6" parameterType="map" resultType="entity.Userinfo">
        <bind name="myQuerySQL" value="'%'+username+'%'" />
        select * from userinfo
        where username like #{myQuerySQL}
    </select>
</mapper>
```

<bind>标签的 value 属性值中的运算是使用 OGNL 表达式，OGNL 表达式的底层是使用 Java 语言，所以在 Java 语言中可以直接使用+号进行字符串的拼接，并不像在 Oracle 或 MySQL 中要使用||或 concat()。

多个运行类代码如下。

```java
public class Select1 {
    public static void main(String[] args) throws IOException {
        String configFile = "mybatis-config.xml";
        InputStream configStream = Resources.getResourceAsStream(configFile);
```

```java
        SqlSessionFactoryBuilder builder = new SqlSessionFactoryBuilder();
        SqlSessionFactory factory = builder.build(configStream);
        SqlSession session = factory.openSession();
        List<Userinfo> listUserinfo = session.selectList("select1", "a");
        for (int i = 0; i < listUserinfo.size(); i++) {
            Userinfo userinfo = listUserinfo.get(i);
            System.out.println(userinfo.getId());

        }
        session.commit();
        session.close();
    }

}

public class Select2 {
    public static void main(String[] args) throws IOException {
        Userinfo queryUserinfo = new Userinfo();
        queryUserinfo.setUsername("a");

        String configFile = "mybatis-config.xml";
        InputStream configStream = Resources.getResourceAsStream(configFile);
        SqlSessionFactoryBuilder builder = new SqlSessionFactoryBuilder();
        SqlSessionFactory factory = builder.build(configStream);
        SqlSession session = factory.openSession();
        List<Userinfo> listUserinfo = session.selectList("select2", queryUserinfo);
        for (int i = 0; i < listUserinfo.size(); i++) {
            Userinfo userinfo = listUserinfo.get(i);
            System.out.println(userinfo.getId());

        }
        session.commit();
        session.close();
    }

}

public class Select3 {
    public static void main(String[] args) throws IOException {
        Userinfo queryUserinfo = new Userinfo();
        queryUserinfo.setUsername("a");

        String configFile = "mybatis-config.xml";
        InputStream configStream = Resources.getResourceAsStream(configFile);
        SqlSessionFactoryBuilder builder = new SqlSessionFactoryBuilder();
        SqlSessionFactory factory = builder.build(configStream);
        SqlSession session = factory.openSession();
        List<Userinfo> listUserinfo = session.selectList("select3", queryUserinfo);
        for (int i = 0; i < listUserinfo.size(); i++) {
            Userinfo userinfo = listUserinfo.get(i);
            System.out.println(userinfo.getId());

        }
        session.commit();
        session.close();
    }
```

```java
    }

public class Select4 {
    public static void main(String[] args) throws IOException {
        Userinfo queryUserinfo = new Userinfo();
        queryUserinfo.setUsername("a");

        String configFile = "mybatis-config.xml";
        InputStream configStream = Resources.getResourceAsStream(configFile);
        SqlSessionFactoryBuilder builder = new SqlSessionFactoryBuilder();
        SqlSessionFactory factory = builder.build(configStream);
        SqlSession session = factory.openSession();
        List<Userinfo> listUserinfo = session.selectList("select4", queryUserinfo);
        for (int i = 0; i < listUserinfo.size(); i++) {
            Userinfo userinfo = listUserinfo.get(i);
            System.out.println(userinfo.getId());

        }
        session.commit();
        session.close();
    }

}

public class Select5 {
    public static void main(String[] args) throws IOException {
        Userinfo queryUserinfo = new Userinfo();
        queryUserinfo.setUsername("a");

        String configFile = "mybatis-config.xml";
        InputStream configStream = Resources.getResourceAsStream(configFile);
        SqlSessionFactoryBuilder builder = new SqlSessionFactoryBuilder();
        SqlSessionFactory factory = builder.build(configStream);
        SqlSession session = factory.openSession();
        List<Userinfo> listUserinfo = session.selectList("select5", queryUserinfo);
        for (int i = 0; i < listUserinfo.size(); i++) {
            Userinfo userinfo = listUserinfo.get(i);
            System.out.println(userinfo.getId());

        }
        session.commit();
        session.close();
    }

}

public class Select6 {
    public static void main(String[] args) throws IOException {
        Map map = new HashMap();
        map.put("username", "a");

        String configFile = "mybatis-config.xml";
        InputStream configStream = Resources.getResourceAsStream(configFile);
        SqlSessionFactoryBuilder builder = new SqlSessionFactoryBuilder();
        SqlSessionFactory factory = builder.build(configStream);
        SqlSession session = factory.openSession();
```

2.8 插入超大的字符串 String 文本内容

```
            List<Userinfo> listUserinfo = session.selectList("select6", map);
            for (int i = 0; i < listUserinfo.size(); i++) {
                Userinfo userinfo = listUserinfo.get(i);
                System.out.println(userinfo.getId());
            }
            session.commit();
            session.close();
        }
    }
```

标签<bind>中的代码 value="'%'+username+'%'"含义是从 Map 中根据 key 名称为 username 找到对应的值，然后再拼接成%value%的形式。

2.8 插入超大的字符串 String 文本内容

MyBatis 框架也非常支持 Oracle 的 CLOB，不需要特别的环境配置即可完成 CLOB 字段的读写操作。

创建名为 bigCLOB 的 Web 项目，映射文件 bigclob.xml 的代码如下。

```xml
<?xml version="1.0" encoding="UTF-8" ?>
<!DOCTYPE mapper
PUBLIC "-//mybatis.org//DTD Mapper 3.0//EN"
"http://mybatis.org/dtd/mybatis-3-mapper.dtd">
<mapper namespace="clob">
    <select id="selectById1" resultType="map">
        select * from bigtext where
        id=1445484
    </select>
    <select id="selectById2" resultType="entity.Bigclob">
        select * from bigtext where
        id=1445484
    </select>
    <insert id="insert" parameterType="entity.Bigclob">
        <selectKey order="BEFORE" resultType="java.lang.Long"
            keyProperty="id">
            select idauto.nextval from dual
        </selectKey>
        insert into bigtext(id,bigtext) values(#{id},#{bigtext})
    </insert>
</mapper>
```

创建实体类代码如下。

```java
package entity;

public class Bigclob {

    private long id;
    private String bigtext;

    public Bigclob() {
    }
```

```java
    public Bigclob(long id, String bigtext) {
        super();
        this.id = id;
        this.bigtext = bigtext;
    }

    public long getId() {
        return id;
    }

    public void setId(long id) {
        this.id = id;
    }

    public String getBigtext() {
        return bigtext;
    }

    public void setBigtext(String bigtext) {
        this.bigtext = bigtext;
    }
}
```

默认的情况下 Oracle 数据表中有 1 条近 80 万行的 CLOB 记录,如图 2-17 所示。

```
798383  "zzzz"="结缔组织"
798384  "zzzz"="蛮",<"zz"="蛮">
798385  end!
```

图 2-17 默认有 80 万行的记录

进行查询操作,将数据封装进 Map 的代码如下。

```java
public class Select1 {
    public static void main(String[] args) throws IOException, SQLException {
        String configFile = "mybatis-config.xml";
        InputStream configStream = Resources.getResourceAsStream(configFile);
        SqlSessionFactoryBuilder builder = new SqlSessionFactoryBuilder();
        SqlSessionFactory factory = builder.build(configStream);
        SqlSession session = factory.openSession();
        Map map = session.selectOne("selectById1");
        oracle.sql.CLOB clob = (oracle.sql.CLOB) map.get("BIGTEXT");

        Reader reader = clob.getCharacterStream();
        char[] charArray = new char[10000];
        int readLength = reader.read(charArray);
        while (readLength != -1) {
            String newString = new String(charArray, 0, readLength);
            System.out.println(newString);
            readLength = reader.read(charArray);
        }
        reader.close();
        session.commit();
        session.close();
    }
```

程序运行后在控制台打印出了最后一行信息，如图 2-18 所示。

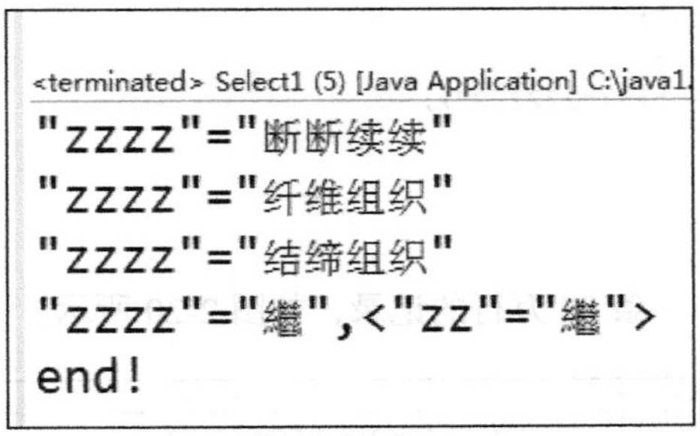

```
<terminated> Select1 (5) [Java Application] C:\java1.
"zzzz"="断断续续"
"zzzz"="纤维组织"
"zzzz"="结缔组织"
"zzzz"="繼",<"zz"="繼">
end!
```

图 2-18　控制台输出最后一条信息

进行查询操作，将数据封装进实体类 Bigclob.java 的代码如下。

```java
public class Select2 {
    public static void main(String[] args) throws IOException, SQLException {
        String configFile = "mybatis-config.xml";
        InputStream configStream = Resources.getResourceAsStream(configFile);
        SqlSessionFactoryBuilder builder = new SqlSessionFactoryBuilder();
        SqlSessionFactory factory = builder.build(configStream);
        SqlSession session = factory.openSession();
        Bigclob bigclob = session.selectOne("selectById2");
        System.out.println(bigclob.getBigtext());
        session.commit();
        session.close();
        System.out.println("封装进实体类运行结束！");
    }
}
```

程序运行后，在控制台输出最后一行信息，如图 2-19 所示。

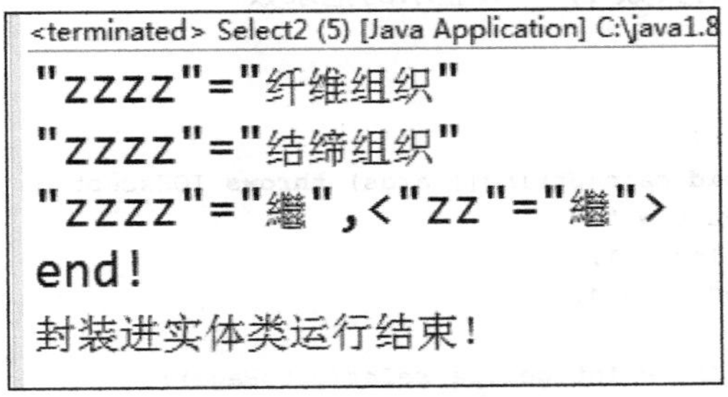

```
<terminated> Select2 (5) [Java Application] C:\java1.8
"zzzz"="纤维组织"
"zzzz"="结缔组织"
"zzzz"="繼",<"zz"="繼">
end!
封装进实体类运行结束！
```

图 2-19　控制台输出最后一条信息

插入新的 CLOB 记录的代码如下。

```java
public class Insert {
    public static void main(String[] args) throws IOException, SQLException {
        String configFile = "mybatis-config.xml";
        InputStream configStream = Resources.getResourceAsStream(configFile);
        SqlSessionFactoryBuilder builder = new SqlSessionFactoryBuilder();
        SqlSessionFactory factory = builder.build(configStream);
        SqlSession session = factory.openSession();
```

```
            Bigclob bigclob = session.selectOne("selectById2");

            session.insert("insert", bigclob);

            session.commit();
            session.close();
            System.out.println("222222");
        }

}
```

在数据表中插入了另外一条 80 万行的记录, 如图 2-20 所示。

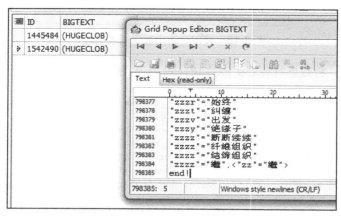

图 2-20　两条 80 万行的字段

2.9　对查询的数据进行分页

想要实现分页功能, 就要先算出起始位置, 起始位置的算法如下。
起始位置=(目标到达的页数-1)*一页显示的记录数

示例代码如下。

```
public class Select1 {
    public static void main(String[] args) throws IOException {
        String gotoPage = "a";
        int gotoPageInt = 1;
        int pageShowSize = 4;
        try {
            gotoPageInt = Integer.parseInt(gotoPage);
            if (gotoPageInt <= 0) {
                gotoPageInt = 1;
            }
        } catch (NumberFormatException e) {
            // insert db log
            gotoPageInt = 1;
        }

        int beginPosition = (gotoPageInt - 1) * pageShowSize;
        System.out.println(beginPosition);
    }
```

2.9 对查询的数据进行分页

MyBatis 还支持分页功能,创建名称为 pageTest 的 Web 项目。
映射文件配置代码如下。

```xml
<?xml version="1.0" encoding="UTF-8" ?>
<!DOCTYPE mapper
PUBLIC "-//mybatis.org//DTD Mapper 3.0//EN"
"http://mybatis.org/dtd/mybatis-3-mapper.dtd">
<mapper namespace="userinfo">
    <select id="selectAll" resultType="entity.Userinfo">
        select * from userinfo order by id asc
    </select>
</mapper>
```

使用 MyBatis 实现分页功能的代码如下。

```java
public class Select2 {
    public static void main(String[] args) throws IOException {

        String gotoPage = "5";
        int gotoPageInt = 1;
        int pageShowSize = 4;
        try {
            gotoPageInt = Integer.parseInt(gotoPage);
            if (gotoPageInt <= 0) {
                gotoPageInt = 1;
            }
        } catch (NumberFormatException e) {
            // insert db log
            gotoPageInt = 1;
        }

        int beginPosition = (gotoPageInt - 1) * pageShowSize;
        System.out.println(beginPosition);

        String configFile = "mybatis-config.xml";
        InputStream configStream = Resources.getResourceAsStream(configFile);
        SqlSessionFactoryBuilder builder = new SqlSessionFactoryBuilder();
        SqlSessionFactory factory = builder.build(configStream);
        SqlSession session = factory.openSession();
        List<Userinfo> list = session.selectList("selectAll", null, new RowBounds(beginPosition, pageShowSize));
        for (int i = 0; i < list.size(); i++) {
            Userinfo userinfo = list.get(i);
            System.out.println(userinfo.getId() + " " + userinfo.getUsername() + " " + userinfo.getPassword() + " "
                    + userinfo.getAge() + " " + userinfo.getInsertdate());
        }
        session.commit();
        session.close();
    }

}
```

但 MyBatis 提供的分页功能在执行效率上是比较差的,它的算法是先将数据表中符合条件的全部记录放入内存，然后在内存中进行分页，这样对内存占用率较高，所以可以参考第三方的 MyBatis 分页插件来优化分页执行的效率。由于第三方的插件不是官方提供的，请自行查询相关资料。

2.10 批处理 SQL 语句

创建名称为 BatchTest 的项目，示例代码如下。

```java
public class InsertUserinfo {

    public static void main(String[] args) {
        try {
            Userinfo userinfo = new Userinfo();
            userinfo.setUsername("高洪岩");
            userinfo.setPassword("大中国");
            userinfo.setAge(100L);

            String resource = "mybatis-config.xml";
            InputStream inputStream = Resources.getResourceAsStream(resource);
            SqlSessionFactory sqlSessionFactory = new SqlSessionFactoryBuilder().build(inputStream);
            SqlSession sqlSession = sqlSessionFactory.openSession();
            UserinfoMapper userinfoMapper = sqlSession.getMapper(UserinfoMapper.class);

            long beginTime = System.currentTimeMillis();
            for (int i = 0; i < 10000; i++) {
                userinfoMapper.insertUserinfo(userinfo);
            }
            long endTime = System.currentTimeMillis();

            System.out.println("非批量用时：" + (endTime - beginTime));

            //

            sqlSession = sqlSessionFactory.openSession(ExecutorType.BATCH);
            userinfoMapper = sqlSession.getMapper(UserinfoMapper.class);

            beginTime = System.currentTimeMillis();
            for (int i = 0; i < 10000; i++) {
                userinfoMapper.insertUserinfo(userinfo);
            }
            endTime = System.currentTimeMillis();

            System.out.println("批量用时：" + (endTime - beginTime));

            sqlSession.commit();
            sqlSession.close();

            System.out.println("createId=" + userinfo.getId());

        } catch (IOException e) {
            // TODO Auto-generated catch block
            e.printStackTrace();
        }
    }
```

    }
}

使用批量执行 SQL 语句的关键代码如下。

```
sqlSession = sqlSessionFactory.openSession(ExecutorType.BATCH);
```

批量与非批量的时间对比如图 2-21 所示。

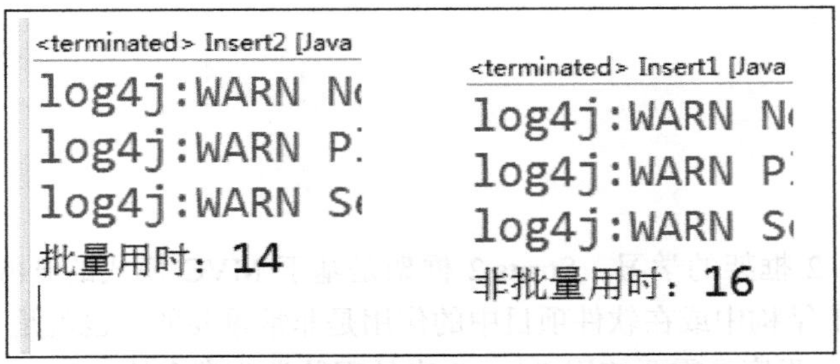

```
<terminated> Insert2 [Java      <terminated> Insert1 [Java
log4j:WARN N                    log4j:WARN N
log4j:WARN P                    log4j:WARN P
log4j:WARN S                    log4j:WARN S
批量用时：14                      非批量用时：16
```

图 2-21　批量与非批量执行的时间对比

如果在 Oracle 数据库中对序列进行池化，则两者的执行速度还会有进一步的大幅提升。如果使用固态硬盘，运行速度差距较小，如果使用机械硬盘则时间差距很大。

第 3 章　Struts 2 必备开发技能

本章进入 Struts 2 框架的学习，Struts 2 框架是基于 MVC 模式的开源 Java EE 框架，而 Struts 2 框架在 SSH 结构中或在软件项目中的作用是非常重要的。也就是，开发基于 Web 的 Java EE 软件项目时，程序员需要使用 Struts 2 来处理种类非常多的功能，比如，上传下载、数据验证等，所以本章的内容也较多，但读者应该着重掌握如下内容：

- 自定义 MVC 框架的使用；
- Struts 2 数据验证；
- Strust 2 国际化；
- 重定向与转发；
- 多模块处理；
- 紧耦合与松耦合开发。

3.1　使用 Struts 2 进行登录功能的开发

本章将用若干个常用案例学习 Struts 2 的开发，掌握这些案例是学习 Struts2 的必经之路。那么在本小节将用一个最经典的登录实例来学习 Struts2 框架，在此过程中可以掌握如何用 Struts2 框架搭建一个 Web 开发环境，Struts2 组件间的调用顺序及配置文件 struts.xml 的设计。

在开发 Struts2 框架之前先要下载相关的 jar 包，Struts 的官方网站截图如图 3-1 所示。

单击 Download 按钮开始下载，下载的 ZIP 文件内容有 Struts2 的示例，源代码及帮助文档。

图 3-1　Struts 的官网

3.1.1 为什么要使用 MVC

介绍到这，还有一个疑问，为什么要学习 Struts2 呢？其实解答这样的问题应该先知道什么是 MVC 模式，因为 Struts2 就是基于 MVC 模式的框架。MVC 模式是一种开发方式，它主要的用途就是对组件之间进行隔离、分层。M 代表业务逻辑层，也就是软件的功能；V 代表视图层，也就是用什么组件显示数据，常用的就是 HTML 和 JSP 等这些文件；而 C 代表控制层，代表软件大方向的执行流程以及用哪个视图对象将数据展示给客户。所以 MVC 就是将不同功能的组件进行隔离与分层，从而有利于代码的后期维护，所以使用 MVC 模式是有很多的优点。

那不使用 MVC 模式，对代码进行不分层的效果是什么样子的呢？好，来看下面的代码。

```jsp
<%@ page language="java" import="java.util.*,java.sql.*"
    pageEncoding="utf-8"%>
<!DOCTYPE HTML PUBLIC "-//W3C//DTD HTML 4.01 Transitional//EN">
<html>
    <body>
        <%
            boolean isLoginSuccess = false;//逻辑标记
            //准备数据库参数
            String url = "jdbc:sqlserver://localhost:1079;databaseName=y2106db";
            String driverName = "com.microsoft.sqlserver.jdbc.SQLServerDriver";
            String username = "sa";
            String password = "";

            Class.forName(driverName);
            //获得连接
            Connection connection = DriverManager.getConnection(url, username,
                    password);
            String sql = "select * from userinfo where username=? and password=?";
            PreparedStatement ps = connection.prepareStatement(sql);
            ps.setString(1, "a");
            ps.setString(2, "aa");
            ResultSet rs = ps.executeQuery();
            //逻辑判断
            while (rs.next()) {
                isLoginSuccess = true;
            }
            rs.close();
            ps.close();
            connection.close();

            //跳转控制
            if (isLoginSuccess == true) {
                response.sendRedirect("listUserinfo");
            } else {
                response.sendRedirect("wrong.jsp");
            }
        %>
    </body>
</html>
```

上面的代码是一个 JSP 文件，里面包含了很多种功能的代码。这样的设计虽然在软件开发的初期使项目的开发进度加快，但后期的维护量却是非常庞大的，所有的功能代码混杂在一起就像面条一样难以拆分。所以就要对代码进行分层，M 层用 JavaBean 服务组件，V 层用 JSP 文件只显示数据，C 层呢？使用 Servlet 来进行代替，分层后的 servletLogin 项目代码结构如图 3-2 所示。

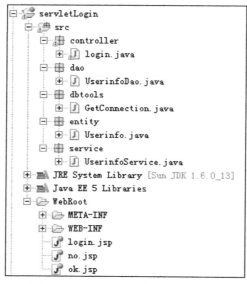

图 3-2 使用 Servlet 实现 MVC 模式

这些文件的代码细节如下所示。

视图 V 层文件 login.jsp 代码如下。

```jsp
<%@ page language="java" import="java.util.*" pageEncoding="utf-8"%>
<!DOCTYPE HTML PUBLIC "-//W3C//DTD HTML 4.01 Transitional//EN">
<html>
    <head>
    </head>
    <body>
        <form action="login" method="post">
            username:
            <input type="text" name="username">
            <br />
            password:
            <input type="text" name="password">
            <br />
            <input type="submit" value="登录">
        </form>
    </body>
</html>
```

此文件主要用来显示登录界面，在此 JSP 文件中并未出现<%%>标签，也就是，JSP 只负责显示数据，而不处理业务逻辑。

视图 V 层文件 ok.jsp 代码如下。

```jsp
<%@ page language="java" import="java.util.*" pageEncoding="utf-8"%>
<!DOCTYPE HTML PUBLIC "-//W3C//DTD HTML 4.01 Transitional//EN">
<html>
    <body>
```

```
            ok.jsp
        </body>
</html>
```

视图 V 层文件 no.jsp 代码如下。

```
<%@ page language="java" import="java.util.*" pageEncoding="utf-8"%>
<!DOCTYPE HTML PUBLIC "-//W3C//DTD HTML 4.01 Transitional//EN">
<html>
    <body>
        no.jsp
    </body>
</html>
```

上面 3 个 JSP 文件都是视图层，而文件 ok.jsp 和 no.jsp 负责显示登录成功或失败的提示，也是视图层组件。

再来看看业务逻辑层 UserinfoService.java 代码如下。

```
package service;

import java.sql.SQLException;

import dao.UserinfoDao;

public class UserinfoService {

    public boolean login(String username, String password) throws SQLException,
            ClassNotFoundException {
        UserinfoDao userinfoDao = new UserinfoDao();
        if (userinfoDao.findUserinfo(username, password) == null) {
            return false;
        } else {
            return true;
        }
    }

}
```

文件 UserinfoService.java 中的代码主要负责处理登录，而业务逻辑层需要调用 DAO 数据访问层 UserinfoDao.java 类进行数据库的查询操作，它的代码如下。

```
package dao;

import java.sql.Connection;
import java.sql.PreparedStatement;
import java.sql.ResultSet;
import java.sql.SQLException;

import dbtools.GetConnection;
import entity.Userinfo;

public class UserinfoDao {

    public Userinfo findUserinfo(String username, String password)
            throws SQLException, ClassNotFoundException {
        Userinfo userinfo = null;
        String sql = "select * from userinfo where username=? and password=?";
        Connection connection = GetConnection.getConnectionFromJDBC();
```

```
        PreparedStatement ps = connection.prepareStatement(sql);
        ps.setString(1, username);
        ps.setString(2, password);
        ResultSet rs = ps.executeQuery();
        while (rs.next()) {
            String iddb = rs.getString("id");
            String usernamedb = rs.getString("username");
            String passworddb = rs.getString("password");

            userinfo = new Userinfo();
            userinfo.setId(iddb);
            userinfo.setUsername(usernamedb);
            userinfo.setPassword(passworddb);
        }
        rs.close();
        ps.close();
        connection.close();

        return userinfo;
    }
}
```

在 UserinfoDao.java 类中将数据表中的记录封装进 Userinfo.java 类中,该类结构如图 3-3 所示。

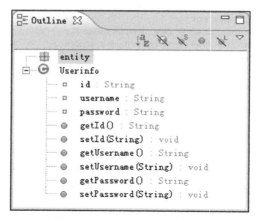

图 3-3 实体 Userinfo 类结构

在 UserinfoDao.java 数据访问层中调用的是 GetConnection.java 类获取 Connection 对象,它的代码如下。

```
package dbtools;

import java.sql.Connection;
import java.sql.DriverManager;
import java.sql.SQLException;

public class GetConnection {

    public static Connection getConnectionFromJDBC()
            throws ClassNotFoundException, SQLException {
```

```java
        String url = "jdbc:sqlserver://localhost:1079;databaseName=ghydb";
        String driverName = "com.microsoft.sqlserver.jdbc.SQLServerDriver";
        String username = "sa";
        String password = "";

        Class.forName(driverName);
        Connection connection = DriverManager.getConnection(url, username,
                password);
        return connection;
    }

}
```

以上是视图层 V，业务逻辑层 M 以及数据访问层 DAO 的代码，继续看看控制层 Controller 的代码，它使用的组件是 Servlet，核心代码如下。

```java
public class login extends HttpServlet {

    public void doPost(HttpServletRequest request, HttpServletResponse response)
            throws ServletException, IOException {

        try {
            String username = request.getParameter("username");
            String password = request.getParameter("password");

            UserinfoService usRef = new UserinfoService();
            if (usRef.login(username, password) == true) {
                response.sendRedirect("ok.jsp");
            } else {
                response.sendRedirect("no.jsp");
            }
        } catch (SQLException e) {
            // TODO Auto-generated catch block
            e.printStackTrace();
        } catch (ClassNotFoundException e) {
            // TODO Auto-generated catch block
            e.printStackTrace();
        }

    }

}
```

此项目运行时的调用顺序如下所示。

1. 在 IE 中调用 login.jsp 文件，输入用户名和密码（视图层 V）。
2. 单击提交按钮。
3. 进入 Servlet，取得 username 和 password 参数（控制层 C）。
4. 在 Servlet 中调用 UserinfoService.java 类（业务逻辑层 M）。
5. UserinfoService.java 类调用 UserinfoDao.java 类（DAO 数据访问层）进行数据库的查询操作。
6. UserinfoDao.java 类(DAO 数据访问层)将查询到的值返回给 UserinfoService.java 类（业务逻辑层 M）进行判断处理。

7. UserinfoService.java 类（业务逻辑层 M）再将登录的 true 或 false 值返回给 Servlet（控制层 C），Servlet 再对 true 或 false 值跳转到不同的 JSP 文件（视图层），以把登录的结果告诉给客户。

所以通过上面的代码，可以得知，使用 Servlet 就可以实现 MVC 设计模式，使组件之间进行分层、隔离，便于后期功能上的维护。但由于使用 Servlet 开发基于 MVC 模式的应用程序时还是接近于原生的 API 开发方式，因为大多数的功能还未得到封装，比如上传，国际化等还是需要由程序员自己写代码来处理，所以开发效率上并不算太高，这时就要学习并使用 Struts 2 框架了。

3.1.2 准备 jar 文件

什么是软件框架？软件框架就是软件功能的半成品，框架提供了针对某一个领域所写代码的基本模型。也就是大多数的功能已经得到封装实现，程序员只需要在这个软件功能的半成品上继续开发另外一半，这样可以增加程序员开发的效率，缩短软件设计的整体周期，而且还统一并规范了软件的整体架构。所有程序员只需要使用一种方式进行开发，有利于新职员快速加入开发进程中。

Struts 2 框架并不是 Oracle 公司官方发布的框架，它是在 WebWork 框架和 Struts 1 框架的基础上综合开发出来的产物，WebWork 和 Struts 1 都是实现 MVC 模式的 Web 框架，而 Struts2 框架的核心代码主要还是来自 WebWork 框架，所以使用 Strust 2 就得需要 jar 包文件，详见官网。

可以用几点来总结一下 Struts。

- 它是基于 MVC 模式的 Java EE 技术的 Web 开发框架，可以对使用 Java EE 技术的 Web 项目开发进行代码的分层，优点是有利于维护。
- Struts 2 框架的源代码主要来源于 WebWork 框架，是在 WebWork 框架基础上再与 Struts1 的优点进行整合而设计出新的 MVC 分层框架。
- Struts 2 的优点主要体现在解耦上，其他的附属技术也比 Struts 1 有所加强，使用上更加方便快捷，比如简化了配置文件的代码。软件代码的耦合性可以体现在类继承中，或在 A 类声明 B 类的属性，以及在 A 类中调用 B 类中的方法，这些情况都是耦合。类与类之间分隔不开，互相依赖，脱离不了，耦合造成了软件不利于扩展，绑定了指定的对象，所以就要尽可能地达到松耦合，也就是解耦，但世界上并没有真正的解耦情况，因为如果实现 100%的解耦合，造成的结果就是类之间没有任何的交互。
- 在官方网站上可以找到其全部相关资料，包括源代码、开发帮助文档、Java API 的使用帮助等。
- 如果以前使用过 Struts 1，那么阅读本教程将非常容易。如果没有接触过 Struts 1 框架或 MVC 框架，也要至少掌握如何使用 Servlet 开发基于 MVC 模式的应用，这样将有益于后面的学习。

在官方网站中下载当前最新版的 zip 文件 struts-all.zip，解压后的内容如图 3-4 所示。

其中 apps 是 Struts 2 的 demo 示例项目，docs 是开发文档及使用手册，lib 是 jar 包，src 则是 Strust 2 框架的源代码了。

进入 lib 文件夹看到很多 jar 包文件，如图 3-5 所示。

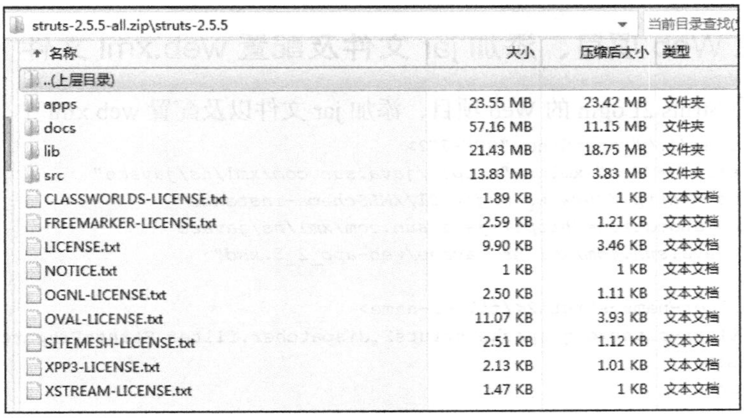

3.1 使用 Struts 2 进行登录功能的开发

图 3-4 解压内容

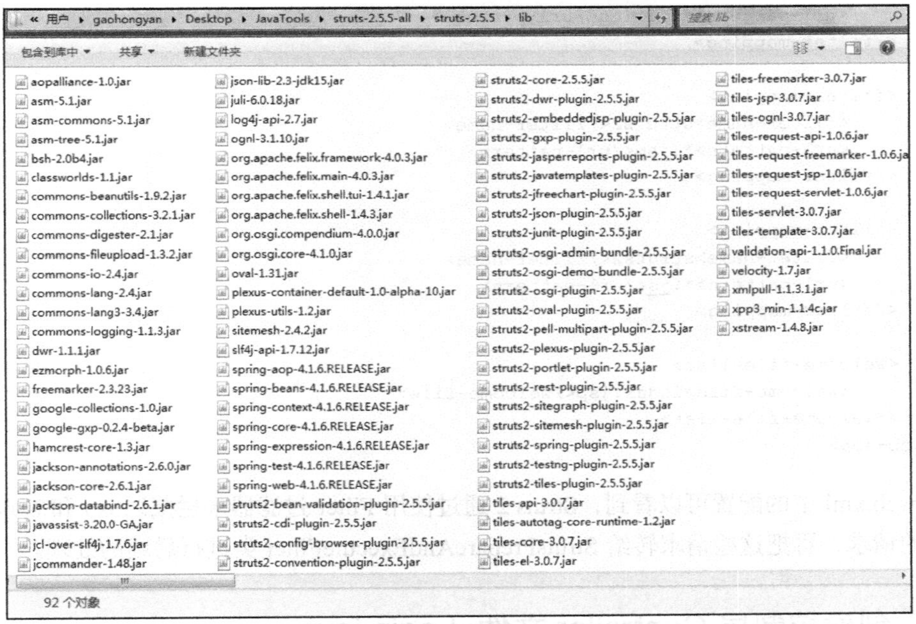

图 3-5 Struts 2 包中的 jar 文件

在我们开发登录功能时并不需要全部的这些 jar 文件，那需要哪几个呢？如图 3-6 所示。

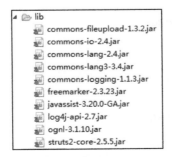

图 3-6 当前 Struts 2 版本必备的 jar 文件

3.1.3 创建 Web 项目、添加 jar 文件及配置 web.xml 文件

创建名称为 Struts 2Login 的 Web 项目,添加 jar 文件以及配置 web.xml 中,代码如下。

```xml
<?xml version="1.0" encoding="UTF-8"?>
<web-app version="2.5" xmlns="http://java.sun.com/xml/ns/javaee"
    xmlns:xsi="http://www.w3.org/2001/XMLSchema-instance"
    xsi:schemaLocation="http://java.sun.com/xml/ns/javaee
    http://java.sun.com/xml/ns/javaee/web-app_2_5.xsd">
    <filter>
        <filter-name>struts2</filter-name>
        <filter-class>org.apache.struts2.dispatcher.filter.StrutsPrepareAndExecuteFilter</filter-class>
    </filter>

    <filter-mapping>
        <filter-name>struts2</filter-name>
        <url-pattern>*.action</url-pattern>
    </filter-mapping>

    <filter-mapping>
        <filter-name>struts2</filter-name>
        <url-pattern>*.jsp</url-pattern>
    </filter-mapping>

    <filter-mapping>
        <filter-name>struts2</filter-name>
        <url-pattern>*.js</url-pattern>
    </filter-mapping>

    <welcome-file-list>
        <welcome-file>index.jsp</welcome-file>
    </welcome-file-list>
</web-app>
```

从 web.xml 中的配置可以看到,Struts 2 通过使用 Filter 过滤器来拦截以 jsp 和 js 或 action 为后缀的请求,再把这些请求转给 StrutsPrepareAndExecuteFilter 类进行后续的处理。

3.1.4 创建控制层 Controller 文件-Login.java

Login.java 核心代码如下。

```java
package controller;

import service.UserinfoService;

public class Login {

    private String username;
    private String password;

    public String getUsername() {
        return username;
    }

    public void setUsername(String username) {
```

```java
        this.username = username;
    }
    public String getPassword() {
        return password;
    }
    public void setPassword(String password) {
        this.password = password;
    }
    public String execute() {
        UserinfoService usRef = new UserinfoService();
        if (usRef.login(username, password)) {
            return "toOKJSP";
        } else {
            return "toNOJSP";
        }
    }
}
```

使用 Servlet 技术做为控制层时，自定义的 Servlet 类还需要继承自 HttpServlet 类，代码如下。

```java
public class login extends HttpServlet
```

而 Struts 2 并不需要继承自任何的类就可以实现 Controller 控制层的功能，这也是 Struts 2 框架一直倡导的"解耦合"。

方法 execute() 是固定的写法，也就是 Struts2 对这种写法是有声明格式规定的，必须是一个公开（public）的方法，该方法的返回数据类型是字符串（String），方法名称是 execute，该约定就像 public static void main(String args[ ])一样固定，它是 Struts 2 默认会调用的方法，返回值 String 类型代表一个逻辑的名称，通过这个逻辑的名称找到对应的物理 JSP 视图文件。

属性 username 和 password 并不是随便写的，它是与视图层中<input>标签中的 name 属性值进行对应。

3.1.5 创建业务逻辑层 Model 文件-UserinfoService.java

业务逻辑层 UserinfoService.java 文件代码如下。

```java
package service;

public class UserinfoService {
    public boolean login(String username, String password) {
        if (username.equals("a") && password.equals("aa")) {
            return true;
        } else {
            return false;
        }
    }
}
```

它是登录功能的核心业务，返回 true 或 false 做为登录成功或失败的结果值。

3.1.6 创建视图层 View 文件-login.jsp

视图层文件 login.jsp 代码如下。

```jsp
<%@ page language="java" import="java.util.*" pageEncoding="utf-8"%>
<!DOCTYPE HTML PUBLIC "-//W3C//DTD HTML 4.01 Transitional//EN">
<html>
    <head>
    </head>
    <body>
        <form action="login.action" method="post">
            username:
            <input type="text" name="username">
            <br />
            password:
            <input type="text" name="password">
            <br />
            <input type="submit" value="登录">
        </form>
    </body>
</html>
```

标签<form>的 action 属性值为 login.action,代表要把 username 和 password 提交给 Login.java 类,但在现在的情况下 Login.java 类仅仅是一个普通的 Java 类,并没有处理 HTTP 协议的功能,所以得在 src 中创建 struts.xml 配置文件。

3.1.7 添加核心配置文件 struts.xml 及解释

创建 struts.xml 配置文件,代码如下。

```xml
<?xml version="1.0" encoding="UTF-8"?>
<!DOCTYPE struts PUBLIC
    "-//Apache Software Foundation//DTD Struts Configuration 2.0//EN"
    "http://struts.apache.org/dtds/struts-2.0.dtd">
<struts>
    <constant name="struts.devMode" value="true" />
    <package name="struts2login" extends="struts-default">
        <action name="login" class="controller.Login">
            <result name="toOKJSP">/ok.jsp</result>
            <result name="toNOJSP">/no.jsp</result>
        </action>
    </package>
</struts>
```

配置代码<constant name="*struts.devMode*" value="*true*" />的作用是使 Struts 2 框架运行在开发模式,在此种模式下对异常的报错信息显示的比较完整,便于排错。

配置代码<package name="*struts2login*" extends="*struts-default*">的作用是定义一个 package 包。与 Java 中 package 包不同的是,Struts 2 中的 package 包可以继承,这也体现了配置代码也是可以复用的,属性 extends 值为 struts-default 的含义是自定义的包 struts2login 继承系统自带的包 struts-default,这样可以将系统自带的功能得到复用,比如上传下载等。

该包在 struts2-core.jar 文件中的 struts-default.xml 配置文件中进行定义，如图 3-7 所示。

```
176  <package name="struts-default" abstract="true" strict-method-invocation="true">
177    <result-types>
178      <result-type name="chain" class="com.opensymphony.xwork2.ActionChainResult"/
179      <result-type name="dispatcher" class="org.apache.struts2.result.ServletDispa
180      <result-type name="freemarker" class="org.apache.struts2.views.freemarker.Fr
181      <result-type name="httpheader" class="org.apache.struts2.result.HttpHeaderRe
182      <result-type name="redirect" class="org.apache.struts2.result.ServletRedirec
183      <result-type name="redirectAction" class="org.apache.struts2.result.ServletA
184      <result-type name="stream" class="org.apache.struts2.result.StreamResult"/>
```

图 3-7　struts-default 是系统自带的包 package

包 struts-default 的作用是提供一些默认的通用功能，比如上传、验证、下载等，在项目中只需要继承它，项目也就拥有了这些功能。

配置代码 <action name="*login*" class="*controller.Login*"> 的作用相当关键，controller 包中的 Login.java 类是一个普通的 JavaBean，但通过这行代码的配置，此类就可以处理 http 请求中的数据了，name 属性是访问的路径，也就是 <form> 标签中的 action 的值，但需要加 .action 后缀。

在 <action> 标签中有配置代码：

```
<result name="toOKJSP">/ok.jsp</result>
```

它的作用是用一个逻辑名称 toOKJSP 对应物理 ok.jsp 文件，它就是 execute() 方法的返回值。

3.1.8　添加 ok.jsp 和 no.jsp 登录结果文件

文件 ok.jsp 的代码如下。

```
<%@ page language="java" import="java.util.*" pageEncoding="utf-8"%>
<!DOCTYPE HTML PUBLIC "-//W3C//DTD HTML 4.01 Transitional//EN">
<html>
    <body>
        ok.jsp
    </body>
</html>
```

文件 no.jsp 的代码如下。

```
<%@ page language="java" import="java.util.*" pageEncoding="utf-8"%>
<!DOCTYPE HTML PUBLIC "-//W3C//DTD HTML 4.01 Transitional//EN">
<html>
    <body>
        no.jsp
    </body>
</html>
```

3.1.9　运行项目

设计完上面的步骤后，将项目部署到 Tomcat 中，在 IE 中输入网址，显示的界面如图 3-8 所示。

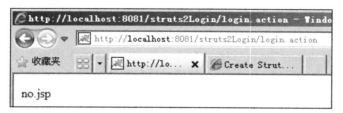

图 3-8　显示登录界面

什么都不输入，直接单击登录按钮，显示登录失败界面如图 3-9 所示。

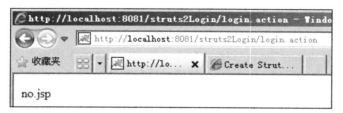

图 3-9　登录失败

输入 a 和 aa 显示登录成功界面如图 3-10 所示。

图 3-10　显示登录成功

下面来分析一下使用 Struts 2 实现 login 登录功能在执行时的主要过程。

1. 在 Web 容器 Tomcat 启动时首先要校验 struts.xml 配置文件是否正确，如果不正确则出现异常，如果正确则进入第 2 步。

2. 在浏览器上输入 login.jsp，然后输入 username 和 password 后再单击提交按钮开始访问 server 服务器端。

3. 进入到 server 服务器的请求被 web.xml 中配置的 Filter 过滤器进行拦截，前台<form>中的 action 值为 login.action，以 action 为后缀，和 web.xml 中拦截*.action 的策略一样，则将请求交给 StrutsPrepareAndExecuteFilter 类后进入到 Struts 2 框架内部开始运行。

4. struts 2 内部开始解析 struts.xml 文件，struts2 框架取得到 url 的网址 login.action，并且分离出 login 前缀，然后到 struts.xml 文件中进行匹配，尝试找到<action>标签的 name 值也为 login 的配置，如果没有找到则出现异常。

5. 如果找到则说明<action>标签的 name 值和 url 中的路径 login.action 的前缀相同，就把<action>标签对应的 class 属性值反射生成实例，再通过反射调用该实例的若干 set 和 get 方法，最后再调用 public String execute()方法。

6. 执行完 execute()方法之后，返回一个数据类型为 String 的 View 视图逻辑名称，Struts2 取得这个逻辑名称到 struts.xml 中。当前的<action>标签下的<result>标签找到有没有 name 同名的<result>逻辑名称配置，如果找到则转到物理 JSP 文件，如果没有找到，则出现异常。

3.1.10　Struts2 的拦截器

在实现登录的功能时，Action 中的属性会自动的进行设置值，这样的效果其实是拦截器在起作用，拦截器的作用是什么？Struts2 框架底层依赖的是 XWork 框架，XWork 框架是命令模式的实现，它也是 Struts2 框架的基础，它提供了 Action 的管理，Result 对象的处理以及最重要的组件"拦截器"都在它的内部进行默默的工作。在 Struts2 从接收请求（request）到完成响应（response）的过程中，Struts2 框架内部会有很多的类进行功能上的封装实现，比如上传、数据验证，以及登录功能中对 Action 属性的赋值，都是拦截器在起作用。有了拦截器可以将 Struts2 框架中的组件进行松耦合的开发，像七巧板一样根据业务的需要随意地拼装功能。

可以在项目中创建多个拦截器，但为了对拦截器有效地进行管理，可以对它们进行分组，形成拦截器栈，在 struts-default.xml 中就有一个名称为 defaultStack 拦截器栈，它也是默认的拦截器栈，Struts2 中的一些拦截器栈的示例代码如下。

```xml
<interceptor-stack name="basicStack">
</interceptor-stack>
<interceptor-stack name="validationWorkflowStack">
</interceptor-stack>
<interceptor-stack name="fileUploadStack">
</interceptor-stack>
<interceptor-stack name="modelDrivenStack">
</interceptor-stack>
<interceptor-stack name="chainStack">
</interceptor-stack>
<interceptor-stack name="i18nStack">
</interceptor-stack>
<interceptor-stack name="paramsPrepareParamsStack">
</interceptor-stack>
<interceptor-stack name="defaultStack">
</interceptor-stack>
<interceptor-stack name="completeStack">
</interceptor-stack>
<interceptor-stack name="executeAndWaitStack">
</interceptor-stack>
```

使用拦截器时必须要继承 AbstractInterceptor 类，然后重写 intercept()方法，在该方法中进行业务的处理，其中 ActionInvocation 的 invoke()方法的主要作用就是执行 Action 中的内容，返回的字符串就是 result 的逻辑名称。

Struts2 的核心组件是 Interceptor 拦截器，拦截器使 Struts2 功能之间进行解耦，便于分离，它非常类似于 Servlet 技术中的 Filter 过滤器，但是有本质上的不同，区别就是 Filter 过滤器在 Web 容器中，而 Interceptor 拦截器是在 Struts2 框架内部，它执行的过程如图 3-11 所示。

```
     Tomcat
        Porject
            Filter
            Struts2 framework
                Interceptor1
                Interceptor2
                        Action
                Interceptor3
```

图 3-11　执行流程

创建名称为 Struts2AbstractInterceptorTest 的 Web 项目，添加 Struts2 开发环境。创建控制层 PrintUsername.java 代码如下。

```java
package controller;

public class PrintUsername {

    public String execute() {
        System.out.println("PrintUsername execute()");
        return "toPrintUsernameJSP";
    }

}
```

配置文件 struts.xml 中的核心内容如下。

```xml
<package name="testtest" extends="struts-default">
    <action name="printUsername" class="controller.PrintUsername">
        <result name="toPrintUsernameJSP">
            /printUsername.jsp
        </result>
    </action>
</package>
```

程序运行后，在控制台准确打印出相关的信息，如图 3-12 所示。

图 3-12　运行结果 1

这是默认的结果，没有什么特别之处，下面开始添加拦截器，MyInterceptor.java 类的代码如下。

```java
package myinterceptor;

import com.opensymphony.xwork2.ActionInvocation;
import com.opensymphony.xwork2.interceptor.AbstractInterceptor;

public class MyInterceptor extends AbstractInterceptor {

    @Override
    public String intercept(ActionInvocation arg0) throws Exception {

        System.out.println("前");
        String resultValue = arg0.invoke();
        System.out.println("后");

        return resultValue;
    }

}
```

在配置文件 struts.xml 中注册拦截器并应用到 action 中，代码如下。

```xml
<?xml version="1.0" encoding="UTF-8" ?>
<!DOCTYPE struts PUBLIC "-//Apache Software Foundation//DTD Struts Configuration 2.1//EN"
 "http://struts.apache.org/dtds/struts-2.1.dtd">
<struts>
    <package name="testtest" extends="struts-default">

        <interceptors>
            <interceptor name="myinterceptor" class="myinterceptor.MyInterceptor"></interceptor>
        </interceptors>

        <action name="printUsername" class="controller.PrintUsername">
            <result name="toPrintUsernameJSP">
                /printUsername.jsp
            </result>
            <interceptor-ref name="myinterceptor"></interceptor-ref>
        </action>
    </package>
</struts>
```

程序运行后的结果如图 3-13 所示。

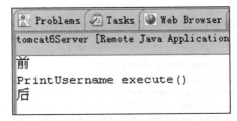

图 3-13 运行结果 2

通过上面的示例可以看到，在不更改任何 Action 代码的前题下，对功能进行了扩展，大大方便了组件式的开发。

下面继续更改 Action，代码更改如下。

```java
package controller;

public class PrintUsername {

    private String username;

    public String getUsername() {
        return username;
    }

    public void setUsername(String username) {
        this.username = username;
    }

    public String execute() {
        System.out.println("PrintUsername execute() username=" + username);
        return "toPrintUsernameJSP";
    }

}
```

输入 URL 网址：http://localhost:8081/Struts2AbstractInterceptorTest/printUsername.action?username=ghy，控制台输出的信息如图 3-14 所示。

图 3-14　运行结果 3

为什么取不到 username 了呢？因为现在使用的拦截器只有自己创建的 MyInterceptor.java，系统中的拦截器被屏蔽掉了，如何启用它呢？把它引入进来即可，struts.xml 配置文件更改代码如下。

```xml
<?xml version="1.0" encoding="UTF-8" ?>
<!DOCTYPE struts PUBLIC "-//Apache Software Foundation//DTD Struts Configuration 2.1//EN"
"http://struts.apache.org/dtds/struts-2.1.dtd">
<struts>
    <package name="testtest" extends="struts-default">

        <interceptors>
            <interceptor name="myinterceptor"
class="myinterceptor.MyInterceptor"></interceptor>
            <interceptor-stack name="myInterceptorStack">
                <interceptor-ref name="defaultStack"></interceptor-ref>
                <interceptor-ref name="myinterceptor"></interceptor-ref>
            </interceptor-stack>
        </interceptors>
```

```
            <action name="printUsername" class="controller.PrintUsername">
                <result name="toPrintUsernameJSP">
                    /printUsername.jsp
                </result>
                <interceptor-ref name="myInterceptorStack"></interceptor-ref>
            </action>
        </package>
</struts>
```

控制层 printUsername 引用了自定义的拦截器栈 myInterceptorStack，而 myInterceptorStack 中又包含了默认的拦截器栈 defaultStack，所以这样的设计就能正确输出 url 中的参数了，重启 Tomcat 重新执行 URL，控制台输出的信息如图 3-15 所示。

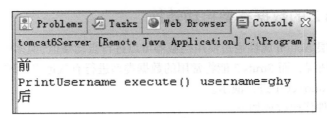

图 3-15　成功取得参数

如果想在每一个<action>标签中使用拦截器时，必须使用<interceptor-ref>子标签进行引用，有多少个<action>就需要引用多少次<interceptor-ref>。所以在这种情况下，可以创建 1 个默认的拦截器栈，配置代码如下。

```
<?xml version="1.0" encoding="UTF-8" ?>
<!DOCTYPE struts PUBLIC "-//Apache Software Foundation//DTD Struts Configuration 2.1//EN"
"http://struts.apache.org/dtds/struts-2.1.dtd">
<struts>
    <package name="testtest" extends="struts-default">

        <interceptors>
            <interceptor name="myinterceptor"
class="myinterceptor.MyInterceptor"></interceptor>
            <interceptor-stack name="myInterceptorStack">
                <interceptor-ref name="defaultStack"></interceptor-ref>
                <interceptor-ref name="myinterceptor"></interceptor-ref>
            </interceptor-stack>
        </interceptors>

        <default-interceptor-ref name="myInterceptorStack"></default-interceptor-ref>

        <action name="printUsername" class="controller.PrintUsername">
            <result name="toPrintUsernameJSP">
                /printUsername.jsp
            </result>
        </action>

    </package>
</struts>
```

在浏览器中输入网址，控制台输出如图 3-16 所示。

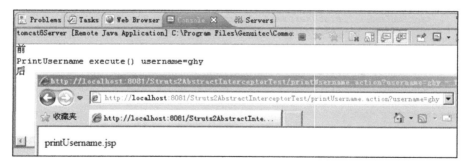

图 3-16 使用默认拦截器栈

3.1.11 Struts 2 的数据类型自动转换

Struts2 的控制层 Action 属性通过使用拦截器可以进行自动赋值,但多个属性的数据类型有时会出现不一样的情况,而 Struts 2 能将常用的数据类型进行有效地自动转换,下面在 Web 项目 Struts2DataTypeAutoSet 中进行演示。

将 index.jsp 文件代码改成如下。

```jsp
<%@ page language="java" import="java.util.*" pageEncoding="utf-8"%>
<%@ taglib uri="http://java.sun.com/jsp/jstl/core" prefix="c"%>
<%@ taglib uri="/struts-tags" prefix="s"%>
<!DOCTYPE HTML PUBLIC "-//W3C//DTD HTML 4.01 Transitional//EN">
<html>
    <body>
        <s:form action="register.action" method="post">
            <s:textfield name="username" label="username"></s:textfield>
            <s:textfield name="age" label="age"></s:textfield>
            <s:textfield name="insertdate" label="insertdate"></s:textfield>
            <s:textfield name="point" label="point"></s:textfield>
            <s:submit value="提交按钮"></s:submit>
        </s:form>
    </body>
</html>
```

创建实体类 Point.java 的代码如下。

```java
package entity;

public class Point {

    private String x;
    private String y;

    public String getX() {
        return x;
    }

    public void setX(String x) {
        this.x = x;
    }

    public String getY() {
        return y;
    }
```

3.1 使用 Struts 2 进行登录功能的开发

```java
    public void setY(String y) {
        this.y = y;
    }
}
```

创建 Register.java 控制层，代码如下。

```java
public class Register extends ActionSupport {

    private String username;
    private int age;
    private Date insertdate;
    private Point point;

    //省略 set get 方法

    public String execute() {
        System.out.println(username);
        System.out.println(age + 1);
        System.out.println(insertdate);
        System.out.println(point);

        StrutsTypeConverter converter;

        return null;
    }

}
```

运行 index.jsp 文件，输入的数据如图 3-17 所示。

单击"提交按钮"在控制台输出数据如图 3-18 所示。

图 3-17 在 index.jsp 中输入数据

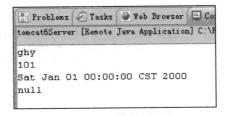

图 3-18 年龄被加 1 了

这说明类型被自动转换，再回到 index.jsp 输入信息如图 3-19 所示。
单击"提交按钮"在控制台输出信息如下。

```
abc
200
Sat Jan 01 00:00:00 CST 2000
null
```

说明字符串"100_200"不能转成 Point 数据类型，point 对象值为 null，也就是 Struts2 并不会将所有的数据类型都可以转换成功。遇到这种情况，可以创建 1 个自定义的数据类型转换器，创建自定义的类型转换器需要继承自 org.apache.struts2.util.StrutsTypeConverter 类，为了使用自定义的数据类型转换器，还需要在配置文件中进行注册，有两种方式。

图 3-19　point 有值了

- 应用于全局范围的类型转换器。在 src 目录创建 xwork-conversion.properties 属性文件，内容为：转换类全名=类型转换器类全名。
- 应用于特定类的类型转换器。特定类的相同目录下创建一个名为 ClassName-conversion.properties 的属性文件，内容为：特定类的属性名=类型转换器类全名。

继续解决上面 Point 类型转换异常的问题，创建一个自定义的类型转换器 PointConverter，代码如下。

```java
package extconverter;

import java.util.Map;

import org.apache.struts2.util.StrutsTypeConverter;

public class PointConverter extends StrutsTypeConverter {

    @Override
    public Object convertFromString(Map arg0, String[] arg1, Class arg2) {
        Object object = arg1;
        System.out.println(object);
        return null;
    }

    @Override
    public String convertToString(Map arg0, Object arg1) {
        return null;
    }

}
```

在 controller 类中创建 Register-conversion.properties 文件，其中 Register 和控制层 Java 文件同名，内容如图 3-20 所示。

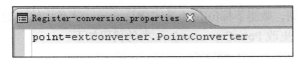

图 3-20 属性文件 1

上面代码主要定义的就是 point 这个对象数据类型转换器类的全路径。

部署项目，执行 URL，在控制台打印的信息如图 3-21 所示。

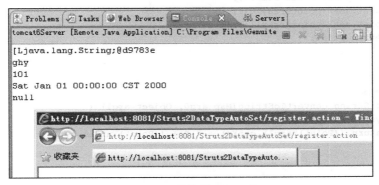

图 3-21 转换器已经工作

从打印结果来看，先调用的是 PointConverter.java 类中的方法。

```
@Override
public Object convertFromString(Map arg0, String[] arg1, Class arg2) {
    Object object = arg1;
    System.out.println(object);
    return null;
}
```

打印的是数组类型，数组中仅有 1 个值，继续更改代码。

```
@Override
public Object convertFromString(Map arg0, String[] arg1, Class arg2) {
    System.out.println(arg1[0]);
    return null;
}
```

再次运行项目，运行结果如图 3-22 所示。

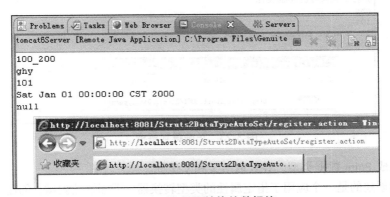

图 3-22 取得要转换的数据值

从图 3-22 可以看到，数组下标[0]的值其实就是<input>表单中输入的字符串。下面开始对这个字符串进行解析，更改 PointConverter.java 类的代码如下。

```java
public class PointConverter extends StrutsTypeConverter {

    @Override
    public Object convertFromString(Map arg0, String[] arg1, Class arg2) {
        System.out.println("执行了 convertFromString");
        String[] pointArray = arg1[0].split("\\_");
        Point point = new Point();
        point.setX(pointArray[0]);
        point.setY(pointArray[1]);
        return point;

    }

    @Override
    public String convertToString(Map arg0, Object arg1) {
        System.out.println("执行了 convertToString");
        return ((Point) arg1).getX() + " " + ((Point) arg1).getY();
    }

}
```

将控制层 Register.java 中的 execute()方法改成如下：

```java
public String execute() {
    System.out.println(username);
    System.out.println(age + 1);
    System.out.println(insertdate);
    System.out.println(point.getX() + " " + point.getY());

    StrutsTypeConverter converter;

    return null;
}
```

程序运行后的效果如图 3-23 所示。

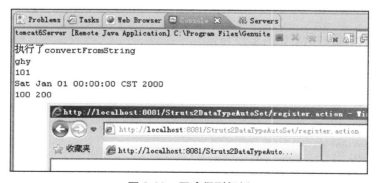

图 3-23　正确得到解析

成功转换成 Point 数据类型，并且取出 x 和 y 坐标。

上面示例创建的是一个应用于指定属性的类型转换器。下面继续测试，创建一个应用于全局范围的类型转换器。

创建名称为 Struts 2DateConverter 的 Web 项目，添加 Struts 2 开发环境。

创建数据类型转换器 DateConverter.java，代码如下。

```java
package extconverter;

import java.text.ParseException;
import java.text.SimpleDateFormat;
import java.util.Date;
import java.util.Map;

import javax.xml.bind.TypeConstraintException;

import org.apache.struts2.util.StrutsTypeConverter;

public class DateConverter extends StrutsTypeConverter {

    private static SimpleDateFormat[] formatArray = {
        new SimpleDateFormat("yyyy-MM-dd"),
        new SimpleDateFormat("yyyy年MM月dd日"),
        new SimpleDateFormat("yyyy/MM/dd") };

    @Override
    public Object convertFromString(Map arg0, String[] arg1, Class arg2) {
        System.out.println("执行了Date日期格式转换!");
        String dateString = arg1[0];
        for (int i = 0; i < formatArray.length; i++) {
            try {
                return formatArray[i].parse(dateString);
            } catch (ParseException e) {
                e.printStackTrace();
                continue;
            }
        }
        throw new TypeConstraintException("日期转换失败");
    }

    // convertToString()方法仅仅在使用Struts2标签输出时调用
    @Override
    public String convertToString(Map arg0, Object arg1) {
        System.out.println("DateConverter convertToString");
        return "在IE中输出："
            + new SimpleDateFormat("yyyy-MM-dd").format((Date) arg1);
    }

}
```

在新版的 Struts 2 框架中，不支持对日期格式为"2000 年 1 月 1 日"和"2000/1/1"进行数据类型转换，所以创建 Date 日期类型转换器来支持上面这两种写法。

在 src 下创建全局数据类型转换配置文件 xwork-conversion.properties，代码如图 3-24 所示。

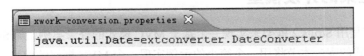

图 3-24 全局数据类型转换器配置文件

创建控制层 Register.java，代码如下。

```java
public class Register extends ActionSupport {

    private Date insertdate;

    public Date getInsertdate() {
        return insertdate;
    }

    public void setInsertdate(Date insertdate) {
        this.insertdate = insertdate;
    }

    public String execute() {
        System.out.println(insertdate);
        return "toIndexJSP";
    }

}
```

文件 index.jsp 代码如下。

```jsp
<%@ page language="java" import="java.util.*" pageEncoding="utf-8"%>
<%@ taglib uri="/struts-tags" prefix="s"%>
<!DOCTYPE HTML PUBLIC "-//W3C//DTD HTML 4.01 Transitional//EN">
<html>
    <body>
        <s:property value="insertdate" />
    </body>
</html>
```

重启 Tomcat，输入网址 http://localhost:8081/Struts2DateConverter/register.action?insertdate=2000-1-1，在控制台及 JSP 页面输出的信息如图 3-25 所示。

图 3-25　日期已被转换

3.2　MVC 框架的开发模型

前面的章节是用 Struts 2 框架来实现一个基于 MVC 模式的登录案例，那 Struts 2 到底是怎么实现 MVC 模式的呢？本节将用一个简小的示例实现一个 MVC 框架的模型，此模型的 MVC 功能和 Struts 2 一模一样。

3.2.1 基础知识准备 1——XML 文件的 CURD

在前面学习的过程中，一提到标记语言就想起 HTML，HTML 提供了很多标签来实现 Web 前端界面的设计，但 HTML 中的标签并不允许自定义，如果想定义一些自己独有的功能 HTML 就不再可行了，这时可以使用 XML 语言来进行实现。

XML 的全称是 eXtensible Markup Language，可扩展标记语言。它可以自定义标记名称与内容，与 HTML 相比它在灵活度上大幅改善，经常用在配置以及数据交互领域。

在使用 Struts 2 框架时，在 web.xml 和 struts.xml 中都有 xml 代码，xml 代码的主要作用就是配置，那在 Java 中如何读取 xml 中的内容呢？

创建名称为 struts 2_xml 的 java 项目，在项目中引入 dom4j-1.6.1.jar 文件，创建 struts.xml 文件，代码如下。

```xml
<mymvc>
    <actions>
        <action name="list" class="controller.List">
            <result name="toListJSP">
                /list.jsp
            </result>
            <result name="toShowUserinfoList" type="redirect">
                showUserinfoList.ghy
            </result>
        </action>

        <action name="showUserinfoList" class="controller.ShowUserinfoList">
            <result name="toShowUserinfoListJSP">
                /showUserinfoList.jsp
            </result>
        </action>
    </actions>
</mymvc>
```

创建 Reader.java 类，代码如下。

```java
package test;

import java.util.List;

import org.dom4j.Attribute;
import org.dom4j.Document;
import org.dom4j.DocumentException;
import org.dom4j.Element;
import org.dom4j.io.SAXReader;

public class Reader {

    public static void main(String[] args) {
        try {
            SAXReader reader = new SAXReader();
            Document document = reader.read(reader.getClass()
                    .getResourceAsStream("/struts.xml"));

            Element mymvcElement = document.getRootElement();
            System.out.println(mymvcElement.getName());
            Element actionsElement = mymvcElement.element("actions");
            System.out.println(actionsElement.getName());
```

```java
            System.out.println("");
            List<Element> actionList = actionsElement.elements("action");
            for (int i = 0; i < actionList.size(); i++) {
                Element actionElement = actionList.get(i);
                System.out.println(actionElement.getName());
                System.out.print("name="
                        + actionElement.attribute("name").getValue());
                System.out.println("action class="
                        + actionElement.attribute("class").getValue());

                List<Element> resultList = actionElement.elements("result");
                for (int j = 0; j < resultList.size(); j++) {
                    Element resultElement = resultList.get(j);
                    System.out.print("  result name="
                            + resultElement.attribute("name").getValue());
                    Attribute typeAttribute = resultElement.attribute("type");
                    if (typeAttribute != null) {
                        System.out.println(" type=" + typeAttribute.getValue());
                    } else {
                        System.out.println("");
                    }
                    System.out.println("   " + resultElement.getText().trim());
                    System.out.println("");
                }

                System.out.println("");
            }

        } catch (DocumentException e) {
            // TODO Auto-generated catch block
            e.printStackTrace();
        }
    }
}
```

程序运行后的结果如图 3-26 所示。

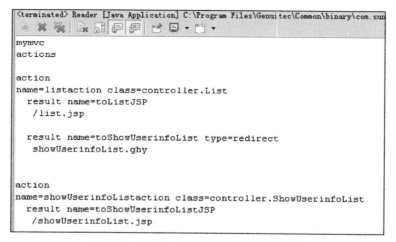

图 3-26　用 dom4j 解析的 XML 文件内容

上面的案例是读取解析 xml 文件,那如何创建 xml 文件呢?继续创建 1 个名称为 createXML

的 java 项目，创建 Java 类 Writer.java，代码如下。

```java
package test;

import java.io.FileWriter;
import java.io.IOException;

import org.dom4j.Document;
import org.dom4j.DocumentHelper;
import org.dom4j.Element;
import org.dom4j.io.OutputFormat;
import org.dom4j.io.XMLWriter;

public class Writer {

    public static void main(String[] args) {
        try {
            Document document = DocumentHelper.createDocument();
            Element mymvcElement = document.addElement("mymvc");
            Element actionsElement = mymvcElement.addElement("actions");
            // /
            Element listActionElement = actionsElement.addElement("action");
            listActionElement.addAttribute("name", "list");
            listActionElement.addAttribute("class", "controller.List");

            Element toListJSPResultElement = listActionElement
                    .addElement("result");
            toListJSPResultElement.addAttribute("name", "toListJSP");
            toListJSPResultElement.setText("/list.jsp");

            Element toShowUserinfoListResultElement = listActionElement
                    .addElement("result");
            toShowUserinfoListResultElement.addAttribute("name",
                    "toShowUserinfoList");
            toShowUserinfoListResultElement.addAttribute("type", "redirect");
            toShowUserinfoListResultElement.setText("showUserinfoList.ghy");
            // /

            Element showUserinfoListActionElement = actionsElement
                    .addElement("action");
            showUserinfoListActionElement.addAttribute("name",
                    "showUserinfoList");
            showUserinfoListActionElement.addAttribute("class",
                    "controller.ShowUserinfoList");

            Element toShowUserinfoListJSPResultElement = showUserinfoListActionElement
                    .addElement("result");
            toShowUserinfoListJSPResultElement.addAttribute("name",
                    "toShowUserinfoListJSP");
            toShowUserinfoListResultElement.setText("/showUserinfoList.jsp");
            // /

            OutputFormat format = OutputFormat.createPrettyPrint();
            XMLWriter writer = new XMLWriter(new FileWriter("ghy.xml"), format);
            writer.write(document);
            writer.close();

        } catch (IOException e) {
```

```
                // TODO Auto-generated catch block
                e.printStackTrace();
            }

        }

    }
```

程序运行后在当前项目中创建了名称为 ghy.xml，文件内容如图 3-27 所示。

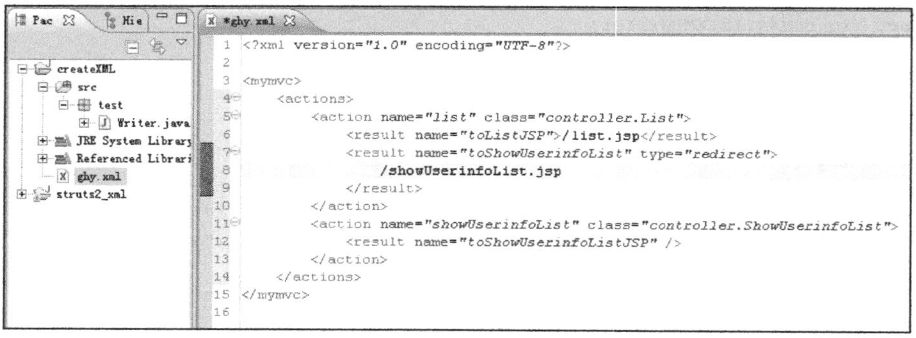

图 3-27 创建的 ghy.xml 中的内容

上面的案例是创建 xml 文件，那如何修改 xml 文件呢？继续创建 1 个名称为 Update.java 的 Java 类，代码如下。

```java
package dom4jTest;

import java.io.FileWriter;
import java.io.IOException;
import java.util.List;

import org.dom4j.Attribute;
import org.dom4j.Document;
import org.dom4j.DocumentException;
import org.dom4j.Element;
import org.dom4j.io.OutputFormat;
import org.dom4j.io.SAXReader;
import org.dom4j.io.XMLWriter;

public class Update {

    public static void main(String[] args) throws IOException {
        try {
            SAXReader reader = new SAXReader();
            Document document = reader.read(reader.getClass().getResourceAsStream("/test.xml"));
            Element mymvcElement = document.getRootElement();
            Element actionsElement = mymvcElement.element("actions");
            List<Element> actionList = actionsElement.elements("action");
            for (int i = 0; i < actionList.size(); i++) {
                Element actionElement = actionList.get(i);
                List<Element> resultList = actionElement.elements("result");
                for (int j = 0; j < resultList.size(); j++) {
                    Element resultElement = resultList.get(j);
                    String resultName = resultElement.attribute("name").getValue();
```

```java
                if (resultName.equals("toShowUserinfoList")) {
                    Attribute typeAttribute = resultElement.attribute("type");
                    if (typeAttribute != null) {
                        typeAttribute.setValue("zzzzzzzzzzzzzzzzzzzzz");
                        resultElement.setText("xxxxxxxxxxxxxxxxxxx");
                    }
                }
            }
        }
        OutputFormat format = OutputFormat.createPrettyPrint();
        XMLWriter writer = new XMLWriter(new FileWriter("src\\ghy.xml"), format);
        writer.write(document);
        writer.close();
    } catch (DocumentException e) {
        e.printStackTrace();
    }
  }
}
```

产生的 XML 文件内容如下所示。

```xml
<?xml version="1.0" encoding="UTF-8"?>

<mymvc>
  <actions>
    <action name="list" class="controller.List">
      <result name="toListJSP">/list.jsp</result>
      <result name="toShowUserinfoList" type="zzzzzzzzzzzzzzzzzzzzz">xxxxxxxxxxxxxxxxxxx</result>
    </action>
    <action name="showUserinfoList" class="controller.ShowUserinfoList">
      <result name="toShowUserinfoListJSP">/showUserinfoList.jsp</result>
    </action>
  </actions>
</mymvc>
```

成功更改 XML 文件中的属性值与文本内容。

那如何删除 XML 中的 Node 节点呢？创建 Delete.java 代码如下。

```java
package dom4jTest;

import java.io.FileWriter;
import java.io.IOException;
import java.util.List;

import org.dom4j.Document;
import org.dom4j.DocumentException;
import org.dom4j.Element;
import org.dom4j.io.OutputFormat;
import org.dom4j.io.SAXReader;
import org.dom4j.io.XMLWriter;

public class Delete {

    public static void main(String[] args) throws IOException {
        try {
            SAXReader reader = new SAXReader();
            Document document = reader.read(reader.getClass().getResourceAsStream("/test.xml"));
```

```java
            Element mymvcElement = document.getRootElement();
            Element actionsElement = mymvcElement.element("actions");
            List<Element> actionList = actionsElement.elements("action");
            for (int i = 0; i < actionList.size(); i++) {
                Element actionElement = actionList.get(i);
                List<Element> resultList = actionElement.elements("result");
                Element resultElement = null;
                boolean isFindNode = false;
                for (int j = 0; j < resultList.size(); j++) {
                    resultElement = resultList.get(j);
                    String resultName = resultElement.attribute("name").getValue();
                    if (resultName.equals("toShowUserinfoList")) {
                        isFindNode = true;
                        break;
                    }
                }
                if (isFindNode == true) {
                    actionElement.remove(resultElement);
                }
            }
            OutputFormat format = OutputFormat.createPrettyPrint();
            XMLWriter writer = new XMLWriter(new FileWriter("src\\ghy.xml"), format);
            writer.write(document);
            writer.close();
        } catch (DocumentException e) {
            e.printStackTrace();
        }
    }
}
```

产生的 xml 文件内容如下。

```xml
<?xml version="1.0" encoding="UTF-8"?>

<mymvc>
    <actions>
        <action name="list" class="controller.List">
            <result name="toListJSP">/list.jsp</result>
        </action>
        <action name="showUserinfoList" class="controller.ShowUserinfoList">
            <result name="toShowUserinfoListJSP">/showUserinfoList.jsp</result>
        </action>
    </actions>
</mymvc>
```

Node 节点成功删除。

那如何删除属性 Attr 呢？创建类 DeleteAttr.java 代码如下。

```java
package dom4jTest;

import java.io.FileWriter;
import java.io.IOException;
import java.util.List;

import org.dom4j.Attribute;
import org.dom4j.Document;
import org.dom4j.DocumentException;
```

```java
import org.dom4j.Element;
import org.dom4j.io.OutputFormat;
import org.dom4j.io.SAXReader;
import org.dom4j.io.XMLWriter;

public class DeleteAttr {

    public static void main(String[] args) throws IOException {
        try {
            SAXReader reader = new SAXReader();
            Document document = reader.read(reader.getClass().getResourceAsStream("/test.xml"));
            Element mymvcElement = document.getRootElement();
            Element actionsElement = mymvcElement.element("actions");
            List<Element> actionList = actionsElement.elements("action");
            for (int i = 0; i < actionList.size(); i++) {
                Element actionElement = actionList.get(i);
                List<Element> resultList = actionElement.elements("result");
                for (int j = 0; j < resultList.size(); j++) {
                    Element resultElement = resultList.get(j);
                    String resultName = resultElement.attribute("name").getValue();
                    if (resultName.equals("toShowUserinfoList")) {
                        Attribute typeAttribute = resultElement.attribute("type");
                        if (typeAttribute != null) {
                            resultElement.remove(typeAttribute);
                        }
                    }
                }
            }
            OutputFormat format = OutputFormat.createPrettyPrint();
            XMLWriter writer = new XMLWriter(new FileWriter("src\\ghy.xml"), format);
            writer.write(document);
            writer.close();
        } catch (DocumentException e) {
            e.printStackTrace();
        }
    }
}
```

产生的 XML 文件代码如下。

```xml
<?xml version="1.0" encoding="UTF-8"?>

<mymvc>
    <actions>
        <action name="list" class="controller.List">
            <result name="toListJSP">/list.jsp</result>
            <result name="toShowUserinfoList">showUserinfoList.ghy</result>
        </action>
        <action name="showUserinfoList" class="controller.ShowUserinfoList">
            <result name="toShowUserinfoListJSP">/showUserinfoList.jsp</result>
        </action>
    </actions>
</mymvc>
```

XML 文件中的 type 属性成功被删除。

3.2.2　基础知识准备 2——Java 的反射

反射技术可以对对象中的内容进行动态访问，在开发底层框架时经常使用到。

创建名为 reflect 的 Java 项目。

创建测试用的实体类 Userinfo.java，其代码如下。

```java
package entity;

import java.util.Date;

public class Userinfo {

    private int id;
    private String username;
    private String password;
    private int age;
    private Date insertDate;

    public Userinfo(int id, String username, String password, int age) {
        super();
        this.id = id;
        this.username = username;
        this.password = password;
        this.age = age;
    }

    public Userinfo() {
        super();
        id = 100;
        username = "中国";
        password = "中国人";
        age = 200;
        insertDate = new Date();
    }

    public Userinfo(int id, String username, String password, int age,
            Date insertDate) {
        super();
        this.id = id;
        this.username = username;
        this.password = password;
        this.age = age;
        this.insertDate = insertDate;
    }

    public String printDate() {
        System.out.println("调用了 printDate()无参方法!");
        return "将高洪岩返回! ";
    }

    //省略 get 和 set
}
```

创建运行类 Run.java，其代码如下。

```java
package test;
```

```java
import java.lang.reflect.Constructor;
import java.lang.reflect.InvocationTargetException;
import java.lang.reflect.Method;

import entity.Userinfo;

public class Run {
    public static void main(String[] args) {
        try {
            // /用3种方式取得Userinfo类的Class对象
            Class<?> classRef1 = Class.forName("entity.Userinfo");
            Class<?> classRef2 = new Userinfo().getClass();
            Class<?> classRef3 = Userinfo.class;
            System.out.println("证明获取的类名称一模一样: " + classRef1.getName() + "--"
                    + classRef2.getName() + "--" + classRef3.getName());

            // /使用Class类的newInstance()方法实例化Userinfo对象
            Class<?> classRef = Userinfo.class.getClassLoader().loadClass(
                    "entity.Userinfo");
            Userinfo userinfoRef = (Userinfo) classRef.newInstance();
            System.out.println("使用Class类的newInstance()方法调用无参的构造函数产生的默认值: id="
                    + userinfoRef.getId() + " username="
                    + userinfoRef.getUsername() + " password="
                    + userinfoRef.getPassword() + " age="
                    + userinfoRef.getAge() + " insertDate="
                    + userinfoRef.getInsertDate());

            // /使用Constructor类的newInstance()方法实例化Userinfo对象
            Constructor<?> constructorUserinfo1 = classRef
                    .getDeclaredConstructor(null);
            Userinfo userinfoRef1 = (Userinfo) constructorUserinfo1
                    .newInstance(null);
            System.out
                    .println("使用Constructor类的newInstance()方法调用无参的构造函数产生的默认值:id="
                            + userinfoRef1.getId()
                            + " username="
                            + userinfoRef1.getUsername()
                            + " password="
                            + userinfoRef1.getPassword()
                            + " age="
                            + userinfoRef1.getAge()
                            + " insertDate="
                            + userinfoRef1.getInsertDate());

            // /使用Constructor类的newInstance()方法实例化Userinfo对象
            Constructor<?> constructorUserinfo2 = classRef
                    .getDeclaredConstructor(int.class, String.class,
                            String.class, int.class);
            Userinfo userinfoRef2 = (Userinfo) constructorUserinfo2
                    .newInstance(100, "高洪岩账号", "高洪岩密码", 200);
            System.out
                    .println("使用Constructor类的newInstance()方法调用有参的构造函数产生的默认值:id="
                            + userinfoRef2.getId()
                            + " username="
                            + userinfoRef2.getUsername()
                            + " password="
                            + userinfoRef2.getPassword()
                            + " age="
```

```
                                    + userinfoRef2.getAge());
                System.out.println("打印通过反射对属性进行赋值的password: "
                                    + userinfoRef2.getPassword());

                // 对无参方法进行反射调用
                Method methodPrintDate = classRef.getDeclaredMethod("printDate");
                System.out.println(methodPrintDate.invoke(userinfoRef));
        } catch (ClassNotFoundException e) {
                // TODO Auto-generated catch block
                e.printStackTrace();
        } catch (InstantiationException e) {
                // TODO Auto-generated catch block
                e.printStackTrace();
        } catch (IllegalAccessException e) {
                // TODO Auto-generated catch block
                e.printStackTrace();
        } catch (SecurityException e) {
                // TODO Auto-generated catch block
                e.printStackTrace();
        } catch (NoSuchMethodException e) {
                // TODO Auto-generated catch block
                e.printStackTrace();
        } catch (IllegalArgumentException e) {
                // TODO Auto-generated catch block
                e.printStackTrace();
        } catch (InvocationTargetException e) {
                // TODO Auto-generated catch block
                e.printStackTrace();
        }
    }

}
```

程序运行结果如下所示。

证明获取的类名称一模一样：entity.Userinfo--entity.Userinfo--entity.Userinfo
使用 Class 类的 newInstance()方法调用无参的构造函数产生的默认值：id=100 username=中国 password=中国人 age=200 insertDate=Thu May 30 13:10:47 CST 2013
使用 Constructor 类的 newInstance()方法调用无参的构造函数产生的默认值：id=100 username=中国 password=中国人 age=200 insertDate=Thu May 30 13:10:47 CST 2013
使用 Constructor 类的 newInstance()方法调用有参的构造函数产生的默认值：id=100 username=高洪岩账号 password=高洪岩密码 age=200
打印通过反射对属性进行赋值的 password：高洪岩密码
调用了 printDate()无参方法！
将高洪岩返回！

3.2.3 实现 MVC 模型——自定义配置文件

创建 Web 项目 **newmymvc**。

此案例完整的项目结构如图 3-28 所示。

其中，**mymvc1.xml** 的代码如下。

```
<mymvc>
    <actions>
        <action name="listString" class="controller.ListString">
            <result name="toListJSP">
                /listString.jsp
```

```
            </result>
            <result name="toShowUserinfoList" type="redirect">
                showUserinfoList.ghy
            </result>
        </action>
    </actions>
</mymvc>
```

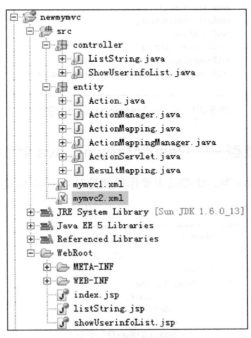

图 3-28 完整的项目结构

文件 mymvc2.xml 的代码如下。

```
<mymvc>
    <actions>
        <action name="showUserinfoList" class="controller.ShowUserinfoList">
            <result name="toShowUserinfoListJSP">
                /showUserinfoList.jsp
            </result>
        </action>
    </actions>
</mymvc>
```

上面的两个 xml 配置文件中各有一些<action>控制层的配置，也就是，可以将多个控制层<action>分散到不同的配置文件中，这样更有利于项目的模块化设计。

3.2.4 实现 MVC 模型——ActionMapping.java 封装<action>信息

创建 ActionMapping.java 类，此类的功能是对<action>中的信息进行封装，其代码结构如图 3-29 所示。

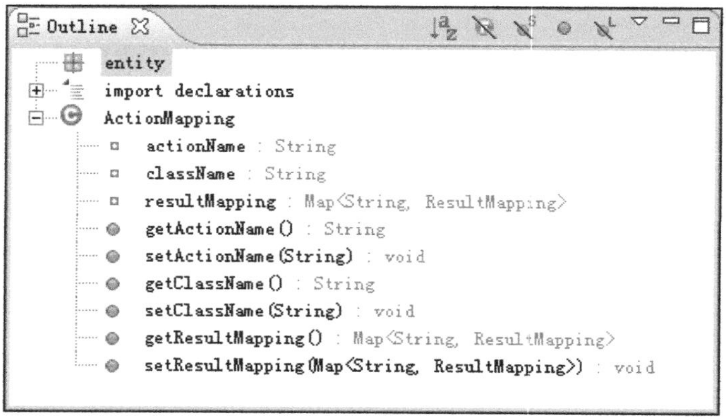

图 3-29　ActionMapping.java 代码结构

3.2.5　实现 MVC 模型——ResultMapping.java 封装<result>信息

创建 ResultMapping.java 类，此类的主要作用就是对<result>标签中的信息进行封装，代码结构如图 3-30 所示。

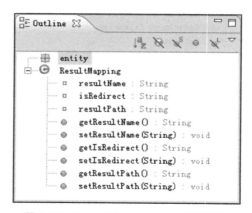

图 3-30　ResultMapping.java 代码结构

3.2.6　实现 MVC 模型——管理映射信息的 ActionMappingManager.java 对象

类 ActionMappingManager.java 主要的作用就是对 ActionMapping.java 进行管理，要将多个 ActionMapping.java 对象放入 Map<String, ActionMapping> actionMappingMap 容器中，因为在一个项目中有可能有多个<action>标签，它的代码如下。

```
package entity;

import java.util.HashMap;
import java.util.List;
import java.util.Map;

import javax.servlet.http.HttpServletRequest;

import org.dom4j.Document;
import org.dom4j.DocumentException;
```

3.2 MVC框架的开发模型

```java
import org.dom4j.Element;
import org.dom4j.io.SAXReader;

public class ActionMappingManager {

    private Map<String, ActionMapping> actionMappingMap = new HashMap<String,
ActionMapping>();

    public ActionMappingManager(String[] configFileArray) {
        createActionMapping(configFileArray);
    }

    private void createActionMapping(String[] configFileArray) {
        try {
            for (int i = 0; i < configFileArray.length; i++) {
                String configFile = configFileArray[i];

                SAXReader reader = new SAXReader();
                Document document = reader.read(this.getClass()
                        .getResourceAsStream("/" + configFile));

                List<Element> actionElementList = document.getRootElement()
                        .element("actions").elements("action");
                for (int j = 0; j < actionElementList.size(); j++) {
                    ActionMapping actionMapping = new ActionMapping();
                    Element actionElement = actionElementList.get(j);
                    String actionValue = actionElement.attributeValue("name");
                    String classValue = actionElement.attributeValue("class");

                    actionMapping.setActionName(actionValue);
                    actionMapping.setClassName(classValue);

                    List<Element> resultElementList = actionElement
                            .elements("result");
                    if (resultElementList.size() > 0) {
                        actionMapping.setResultMapping(new HashMap());

                        for (int y = 0; y < resultElementList.size(); y++) {
                            ResultMapping resultMappping = new ResultMapping();

                            Element resultElement = resultElementList.get(y);
                            String resultName = resultElement
                                    .attributeValue("name");
                            String resultPath = resultElement.getText().trim();

                            resultMappping.setResultName(resultName);
                            resultMappping.setResultPath(resultPath);

                            if (resultElement.attribute("type") != null) {
                                String typeValue = resultElement
                                        .attributeValue("type");
                                resultMappping.setIsRedirect("true");
                            }
                            actionMapping.getResultMapping().put(resultName,
                                    resultMappping);
                        }
                    }
```

```
                }
                actionMappingMap.put(actionValue, actionMapping);
            }

        }
    } catch (DocumentException e) {
        // TODO Auto-generated catch block
        e.printStackTrace();
    }
    String abc = "";
}

public ActionMapping getActionMapping(HttpServletRequest request) {
    String uri = request.getRequestURI();
    String contextPath = request.getContextPath();
    String actionPath = uri.substring(contextPath.length() + 1);
    String actionName = actionPath.substring(0, actionPath.indexOf("."));
    return actionMappingMap.get(actionName);
}
```

对象 Map<String, ActionMapping> actionMappingMap 存储实体类的结构如图 3-31 所示。

对象 actionMappingMap 中的 key 就是 actionName，也就是 action 的映射路径的值，actionMappingMap 中的 key 对应的值是 ActionMapping 实体类对象。

在 ActionMapping 对象中有 1 个 Map ResultMappingMap 属性，Map 的 key 就是 resultName，Map 的 key 对应的值就是 ResultMapping 实体类。

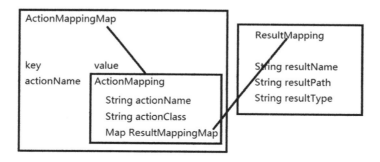

图 3-31　实体类之间的存储关系

3.2.7　实现 MVC 模型——创建反射 Action 的 ActionManager.java 对象

类 ActionManager.java 主要的作用就是创建 Action 对象，代码如下。

```
package entity;

public class ActionManager {

    public static Action getAction(String className) {
        Action action = null;
        try {
            action = (Action) Class.forName(className).newInstance();
        } catch (ClassNotFoundException e) {
```

```
            // TODO Auto-generated catch block
            e.printStackTrace();
        } catch (InstantiationException e) {
            // TODO Auto-generated catch block
            e.printStackTrace();
        } catch (IllegalAccessException e) {
            // TODO Auto-generated catch block
            e.printStackTrace();
        }
        return action;
    }

}
```

3.2.8　实现 MVC 模型——创建核心控制器 ActionServlet.java

ActionServlet.java 类主要的作用就是读取 web.xml 中的配置，用 Servlet API 可以读取 web.xml 中的一些配置信息，在本示例中使用两种方式来演示配置信息的读取，代码如下。

```xml
<servlet>
    <servlet-name>test1</servlet-name>
    <servlet-class>controller.test1</servlet-class>
    <init-param>
        <param-name>test1Key</param-name>
        <param-value>test1Value</param-value>
    </init-param>
</servlet>
<servlet>
    <servlet-name>test2</servlet-name>
    <servlet-class>controller.test2</servlet-class>
</servlet>

<servlet-mapping>
    <servlet-name>test1</servlet-name>
    <url-pattern>/test1</url-pattern>
</servlet-mapping>

<servlet-mapping>
    <servlet-name>test2</servlet-name>
    <url-pattern>/test2</url-pattern>
</servlet-mapping>

<context-param>
    <param-name>servletKey</param-name>
    <param-value>servletValue</param-value>
</context-param>
```

在上面的代码中创建了两个 Servlet，分别是 test1 和 test2，test1 的核心代码如下。

```java
public class test1 extends HttpServlet {

    public void doGet(HttpServletRequest request, HttpServletResponse response)
            throws ServletException, IOException {

        System.out.println(this.getInitParameter("test1Key"));
        System.out
                .println(this.getServletConfig().getInitParameter("test1Key"));
        System.out.println(this.getServletContext().getInitParameter(
```

```
            "servletKey"));
    }
}
```

程序运行后的效果如图 3-32 所示。

在图 3-32 中可以看到，test1 既可以取得<init-param>的参数配置，也可以取得<context-param>中的参数配置。

继续执行 test2，运行效果如图 3-33 所示。

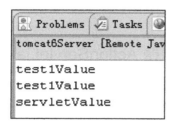

图 3-32　test1 运行效果

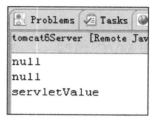

图 3-33　test2 运行效果

从图 3-32 中可以发现，<init-param>标签中的参数配置必须是当前的 Servlet 对象才可以取得，而<context-param>中的参数配置所有的 Servlet 都可以获取。

核心控制器 ActionServlet.java 也有从 web.xml 中读取配置参数的功能，但主要作用就是将 request 请求和自定义 MVC 框架的相关 Java 组件进行关联，代码如下。

```java
package entity;

import java.io.IOException;

import javax.servlet.ServletException;
import javax.servlet.http.HttpServlet;
import javax.servlet.http.HttpServletRequest;
import javax.servlet.http.HttpServletResponse;

public class ActionServlet extends HttpServlet {

    private ActionMappingManager actionMappingManager;

    public void doGet(HttpServletRequest request, HttpServletResponse response)
            throws ServletException, IOException {
        ActionMapping actionMapping = actionMappingManager
                .getActionMapping(request);
        Action action = ActionManager.getAction(actionMapping.getClassName());
        String resultName = action.execute(request, response);
        ResultMapping resultMapping = actionMapping.getResultMapping().get(
                resultName);
        if (resultMapping.getIsRedirect().equals("true")) {
            response.sendRedirect(resultMapping.getResultPath());
        } else {
            request.getRequestDispatcher(resultMapping.getResultPath().trim())
                    .forward(request, response);
        }
    }
}
```

```java
    public void doPost(HttpServletRequest request, HttpServletResponse response)
            throws ServletException, IOException {
        this.doGet(request, response);
    }

    public void init() throws ServletException {
        System.out.println("ActionServlet init()无参");
        String configFileValue = this.getInitParameter("configFile");
        actionMappingManager = new ActionMappingManager(configFileValue
                .split(","));
    }

}
```

3.2.9 实现 MVC 模型——创建 Action 接口及控制层 Controller 实现类

接口 Action 主要的作用就是定义 Controller 控制层的接口规范，代码如下。

```java
package entity;

import javax.servlet.http.HttpServletRequest;
import javax.servlet.http.HttpServletResponse;

public interface Action {

    public String execute(HttpServletRequest request,
            HttpServletResponse response);

}
```

创建控制层 ListString.java 的代码如下。

```java
package controller;

import java.util.ArrayList;
import java.util.List;

import javax.servlet.http.HttpServletRequest;
import javax.servlet.http.HttpServletResponse;

import entity.Action;

public class ListString implements Action {

    public String execute(HttpServletRequest request,
            HttpServletResponse response) {

        System.out.println("执行了 ListString");

        List listString = new ArrayList();
        for (int i = 0; i < 10; i++) {
            listString.add("高洪岩" + (i + 1));
        }
        request.setAttribute("listString", listString);

        return "toShowUserinfoList";
    }
```

创建控制层 ShowUserinfoList.java 的代码如下。

```java
package controller;

import java.util.ArrayList;
import java.util.List;

import javax.servlet.http.HttpServletRequest;
import javax.servlet.http.HttpServletResponse;

import entity.Action;

public class ShowUserinfoList implements Action {
    public String execute(HttpServletRequest request,
            HttpServletResponse response) {

        System.out.println("执行了 ShowUserinfoList");

        List userinfoList = new ArrayList();
        for (int i = 0; i < 5; i++) {
            userinfoList.add("userinfo" + (i + 1));
        }
        request.setAttribute("userinfoList", userinfoList);

        return "toShowUserinfoListJSP";
    }
}
```

3.2.10 实现 MVC 模型——创建视图层 V 对应的 JSP 文件

创建 JSP 文件 listString.jsp，代码如下。

```jsp
<%@ page language="java" import="java.util.*" pageEncoding="utf-8"%>
<!DOCTYPE HTML PUBLIC "-//W3C//DTD HTML 4.01 Transitional//EN">
<html>
    <body>
        listString.jsp!
        <br />
        ${listString}
    </body>
</html>
```

创建 JSP 文件 showUserinfoList.jsp，代码如下。

```jsp
<%@ page language="java" import="java.util.*" pageEncoding="utf-8"%>
<!DOCTYPE HTML PUBLIC "-//W3C//DTD HTML 4.01 Transitional//EN">
<html>
    <body>
        showUserinfoList.jsp!
        <br />
        ${userinfoList}
    </body>
</html>
```

3.2.11　实现 MVC 模型——在 web.xml 中配置核心控制器

文件 web.xml 中的代码如下。

```xml
<?xml version="1.0" encoding="UTF-8"?>
<web-app version="2.5" xmlns="http://java.sun.com/xml/ns/javaee"
    xmlns:xsi="http://www.w3.org/2001/XMLSchema-instance"
    xsi:schemaLocation="http://java.sun.com/xml/ns/javaee
    http://java.sun.com/xml/ns/javaee/web-app_2_5.xsd">
    <servlet>
        <servlet-name>ActionServlet</servlet-name>
        <servlet-class>entity.ActionServlet</servlet-class>
        <init-param>
            <param-name>configFile</param-name>
            <param-value>mymvc1.xml,mymvc2.xml</param-value>
        </init-param>
        <load-on-startup>0</load-on-startup>
    </servlet>

    <servlet-mapping>
        <servlet-name>ActionServlet</servlet-name>
        <url-pattern>*.ghy</url-pattern>
    </servlet-mapping>
    <welcome-file-list>
        <welcome-file>index.jsp</welcome-file>
    </welcome-file-list>
</web-app>
```

3.2.12　实现 MVC 模型——运行效果

在 IE 中输入网址 http://localhost:8081/newmymvc/listString.ghy，运行效果如图 3-34 所示。

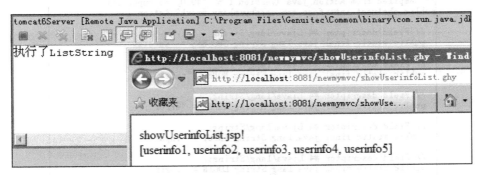

图 3-34　运行结果

至此，已经用基于 Servlet 的核心控制器实现了一个自定义的 MVC 框架模型。

3.3　Struts 2 的刷新验证功能

前面示例中的登录功能是通过跳转到不同的 JSP 文件来显示登录成功或失败的结果，在 Struts 2 中可以实现在当前页面中显示出错信息的功能，具有刷新浏览器的功能。

3.3.1 Action 接口

在 Struts 2 中验证功能使用 ActionSupport.java 类进行实现,使用的方式就是从控制层继承它,它的类继承关系如下。

```
public class com.opensymphony.xwork2.ActionSupport
implements
com.opensymphony.xwork2.Action,
com.opensymphony.xwork2.Validateable,
com.opensymphony.xwork2.ValidationAware,
com.opensymphony.xwork2.TextProvider,
com.opensymphony.xwork2.LocaleProvider,
java.io.Serializable
```

在上面的结构中可以看到 ActionSupport 实现了 6 个接口,其中 Action 接口的主要作用就是提供一个 execute()访问接口。也就是,所有的控制层必须要实现此方法,此方法在前面的知识点也应用过。

Action 接口的声明如图 3-35 所示。

在 Action 接口中声明了 5 个常量,这 5 个常量就是 execute()方法返回的字符串可以使用的 result 逻辑名称,较常用的是 success(成功)和 error(出错)。通过实现 Action 接口就可以将 result 对象的名称进行标准化,项目组中的程序员可以将常用的 result 名称进行统一,常量对应的常量值如图 3-36 所示。

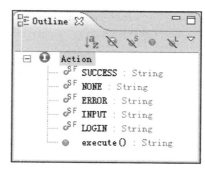

图 3-35 Action 接口的声明

```
// Compiled from Action.java (version 1.5 : 49.0, no super bit)
public abstract interface com.opensymphony.xwork2.Action {

  // Field descriptor #4 Ljava/lang/String;
  public static final java.lang.String SUCCESS = "success";

  // Field descriptor #4 Ljava/lang/String;
  public static final java.lang.String NONE = "none";

  // Field descriptor #4 Ljava/lang/String;
  public static final java.lang.String ERROR = "error";

  // Field descriptor #4 Ljava/lang/String;
  public static final java.lang.String INPUT = "input";

  // Field descriptor #4 Ljava/lang/String;
  public static final java.lang.String LOGIN = "login";

  // Method descriptor #16 ()Ljava/lang/String;
  public abstract java.lang.String execute() throws java.lang.Exception;
}
```

图 3-36 5 个常量值

应用常量的示例代码在项目 strutsActionTest 中,核心代码如下。

```
public class Login extends ActionSupport {
    public String execute() {
        if (1 == 1) {
            return Action.SUCCESS;
```

```
        } else {
            return Action.ERROR;
        }
    }
}
```

在 struts.xml 中的 action 配置代码如下。

```
<action name="login" class="controller.Login">
    <result name="success">/ok.jsp</result>
    <result name="error">/no.jsp</result>
</action>
```

但在实际的软件项目中这样的使用非常少，因为如果<result>的逻辑名称报错，也会很容易地进行排查。

当找不到 result 的逻辑名称 toOKJSP1 时，报如下错误。

```
No result defined for action controller.Login and result toOKJSP1
```

当找不到 action 路径 login1 时，报如下错误。

```
There is no Action mapped for namespace [/] and action name [login1] associated with context path [/struts2Login]. - [unknown location]
```

3.3.2 Validateable 和 ValidationAware 接口

ActionSupport 类实现了 Validateable 接口，接口的声明如图 3-37 所示。

Validateable 接口只有 1 个方法 validate()，ActionSupport 的子类必须覆盖该方法来实现自定义的验证功能。

图 3-37　Validateable 接口的声明

ValidationAware 接口的声明如图 3-38 所示。

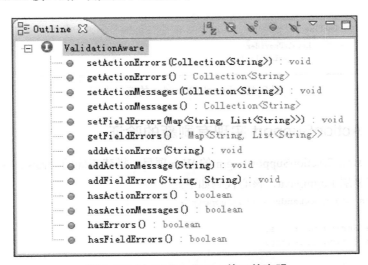

图 3-38　ValidationAware 接口的声明

Validateable 接口中 validate()方法的功能是把自定义的验证代码放在里面执行，那么如何才能对出错信息进行进一步的处理呢？此时就需要使用到 ValidationAware 接口，ValidationAware 接口规定了如果验证出错应该用哪些方法来对出错信息进行处理。所以 ActionSupport 类实现 Validateable 接口既取得了验证功能，而又实现了 ValidationAware 接口，从而还可以对出错信息进行处理。

3.3.3 TextProvider 和 LocaleProvider 接口

接口 TextProvider 的作用是从 properties 属性文件中取得资源文本，它的声明内容如图 3-39 所示。

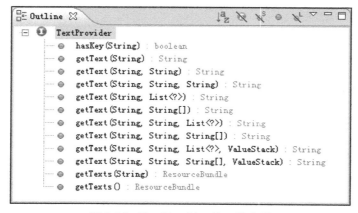

图 3-39　TextProvider 接口的声明

而接口 LocaleProvider 的主要作用就是访问国际化的环境，根据不同的国家语言环境就可以从不同语言的 properties 文件中取出对应的文本资源，LocaleProvider 的接口声明如图 3-40 所示。

图 3-40　LocaleProvider 的接口声明

3.3.4 使用 ActionSupport 实现有刷新的验证

创建名称为 struts2ActionSupportLogin 的 Web 项目，搭建正确的 Struts 2 开发环境。
（1）创建控制层 Login.java，核心代码如下。

```
public class Login extends ActionSupport {

    private String username;
    private String password;

    //set get 方法省略！一定要在 Login.java 中生成 get 和 set 方法

    @Override
    public void validate() {
```

```java
        super.validate();

        if ("".equals(this.getUsername())) {
            this.addFieldError("username", "对不起,用户名不可以为空!");
            // 通过使用 addFieldError 方法可以对出错信息进行处理
            // 再将这些出错信息文本在 JSP 页面上显示
        }
        if ("".equals(this.getPassword())) {
            this.addFieldError("password", "对不起,密码不可以为空!");
        }

    }

    public String execute() {
        UserinfoService usRef = new UserinfoService();
        if (usRef.login(username, password)) {
            return "toOKJSP";
        } else {
            return "toNOJSP";
        }
    }
}
```

重写的方法 public void validate()中基本都是格式型的数据验证代码,比如账号的输入值、身份证号的格式,邮箱的格式等。真正的业务型验证还是得在 execute()中进行处理。

(2) Login.java 中的 username 和 password 的属性值来自于 JSP,所以继续创建 login.jsp 文件,代码如下。

```jsp
<%@ page language="java" import="java.util.*" pageEncoding="utf-8"%>
<%@ taglib uri="/struts-tags" prefix="s"%>
<!DOCTYPE HTML PUBLIC "-//W3C//DTD HTML 4.01 Transitional//EN">
<html>
    <body>
        <s:form action="login" method="post">
            <s:textfield name="username" label="账号"></s:textfield>
            <br />
            <s:textfield name="password" label="密码"></s:textfield>
            <br />
            <s:submit value="提交"></s:submit>
        </s:form>
    </body>
</html>
```

在这里需要注意的是 JSP 文件中使用的是 Struts 2 的标签,目的是可以自动从作用域中获取出错信息并显示出来。

使用 s:textfield 标签的目的是在浏览器中生成一个单行文本域,而它的 name 属性的值一定要和 Login.java 类中的属性名称相同,这样才可以实现在 JSP 页面中单击"提交"按钮将 Login.java 类中的属性值进行自动填充。label 属性是标签内的文本提示内容。

(3) 更改 struts.xml 的代码如下。

```xml
<?xml version="1.0" encoding="UTF-8"?>
<!DOCTYPE struts PUBLIC
    "-//Apache Software Foundation//DTD Struts Configuration 2.0//EN"
    "http://struts.apache.org/dtds/struts-2.0.dtd">
<struts>
```

```xml
<constant name="struts.devMode" value="true" />
<package name="struts2login" extends="struts-default">
    <action name="login" class="controller.Login">
        <result name="success">/true.jsp</result>
        <result name="error">/false.jsp</result>
        <result name="input">/login.jsp</result>
    </action>
</package>
</struts>
```

配置代码 <result name="*input*">/login.jsp</result> 的作用是当 Login.java 类中的 this.addFieldError()函数被调用时，说明验证没有通过，则转到哪个 JSP 页面进行报错信息的显示。在 Struts 2 中使用 ActionSupport 类进行有刷新验证，则必须在 struts.xml 中配置名称为 input 的<result>，不然会出现异常。

```
No result defined for action controller.Login and result input
```

（4）运行项目。

输入 URL：http://localhost:8081/struts2ActionSupportLogin/login.jsp 后，按回车出现登录界面如图 3-41 所示。

图 3-41　Struts 2 标签自动转成 HTML 标签

什么都不输入，单击"提交"按钮出现界面如图 3-42 所示。

图 3-42　自动显示出错信息

至此，在 Struts 2 中使用 ActionSupport 类进行有刷新的登录验证的实验结束。

3.4　对 Struts 2 有刷新验证的示例进行升级

为什么升级？一定是有弊端，所以才会升级！有什么弊端呢？来看一下前面章节中使用 Struts2 的标签<%@ taglib uri="/struts-tags" prefix="s"%>显示登录界面的 HTML 代码，如图 3-43 所示。

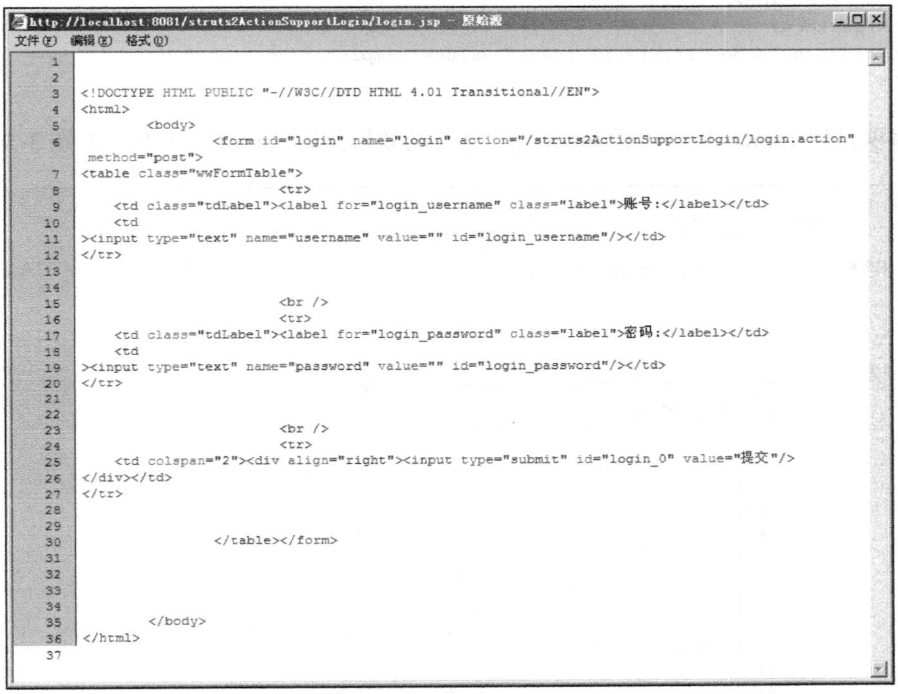

```
http://localhost:8081/struts2ActionSupportLogin/login.jsp - 原始源
文件(F) 编辑(E) 格式(O)


<!DOCTYPE HTML PUBLIC "-//W3C//DTD HTML 4.01 Transitional//EN">
<html>
        <body>
                <form id="login" name="login" action="/struts2ActionSupportLogin/login.action"
 method="post">
<table class="wwFormTable">
                        <tr>
    <td class="tdLabel"><label for="login_username" class="label">账号:</label></td>
    <td
><input type="text" name="username" value="" id="login_username"/></td>
</tr>


                        <br />
                        <tr>
    <td class="tdLabel"><label for="login_password" class="label">密码:</label></td>
    <td
><input type="text" name="password" value="" id="login_password"/></td>
</tr>


                        <br />
                        <tr>
    <td colspan="2"><div align="right"><input type="submit" id="login_0" value="提交"/>
</div></td>
</tr>


                </table></form>




        </body>
</html>
```

图 3-43　大量自动生成的 table/tr/td 代码存在

在查看 HTML 源代码时可以发现，使用 Struts2 的<s>标签自动加入了大量的 table、tr、td 等标签，这样在前、后台程序员联合开发时，会给前台程序员带来大量重定义标签的 CSS 样式代码。实际上，可以通过配置来将 Struts 2 自动生成的 HTML 标签进行屏蔽，并且配置的方式非常简单。

3.4.1　加入 xml 配置来屏蔽自动生成的 table/tr/td 代码

去掉 Struts 2 标签自动生成的 HTML 代码很简单，只需要在配置文件 struts.xml 中添加如下代码即可。

```
<constant name="struts.ui.theme" value="simple"></constant>
```

创建名为 struts2ActionSupportLogin_2 的 Web 项目，将前面名为 struts2ActionSupportLogin 的项目中的代码复制到 struts2ActionSupportLogin_2 项目中，更改后的 struts.xml 代码如下。

```
<?xml version="1.0" encoding="UTF-8"?>
<!DOCTYPE struts PUBLIC
```

```
    "-//Apache Software Foundation//DTD Struts Configuration 2.0//EN"
    "http://struts.apache.org/dtds/struts-2.0.dtd">
<struts>
    <constant name="struts.devMode" value="true" />
    <package name="struts2login" extends="struts-default">
        <action name="login" class="controller.Login">
            <result name="success">/true.jsp</result>
            <result name="error">/false.jsp</result>
            <result name="input">/login.jsp</result>
        </action>
    </package>
    <constant name="struts.ui.theme" value="simple"></constant>
</struts>
```

部署项目到 Tomcat 中，运行 login.jsp，运行的效果及生成的 HTML 代码如图 3-44 所示。

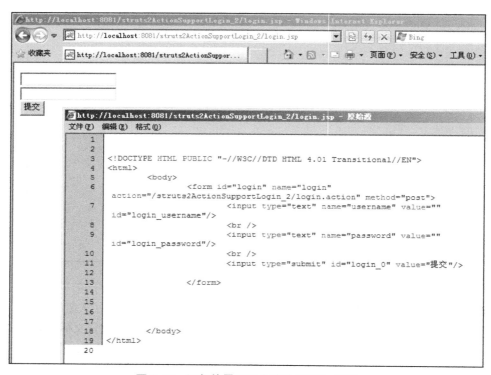

图 3-44 运行的界面及生成的 HTML 代码

生成的 HTML 代码并没有出现 table/tr/td 等标签，但在 IE 中并没有将 s:textfield 标签的 label 属性显示出来。

```
<s:textfield name="username" label="账号">
```

这正是代码 <constant name="struts.ui.theme" value="simple"></constant> 的执行结果，但是可以改由普通的文本标题来进行设计，也就是弃用 label 属性。重新设计的 login.jsp 代码如下。

```
<s:form action="login" method="post">
    账号：<s:textfield name="username"></s:textfield>
    <br />
    密码：<s:textfield name="password"></s:textfield>
    <br />
```

```
        <s:submit value="提交"></s:submit>
    </s:form>
```

在 IE 中重新运行的结果如图 3-45 所示。

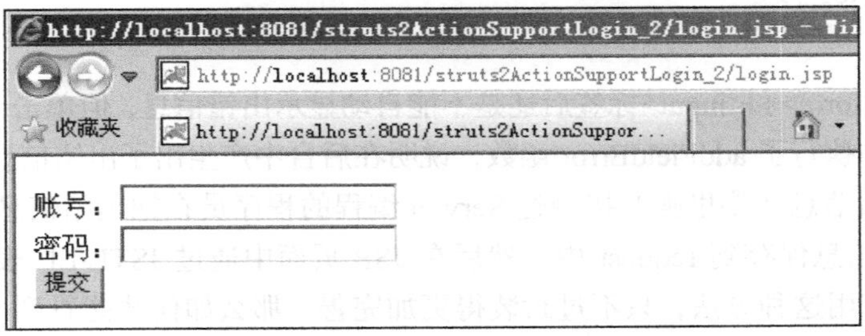

图 3-45 用文本代替 label 标签

3.4.2 解决"出错信息不能自动显示"的问题

什么也不输入，直接单击"提交"按钮进行登录，运行效果如图 3-46 所示。

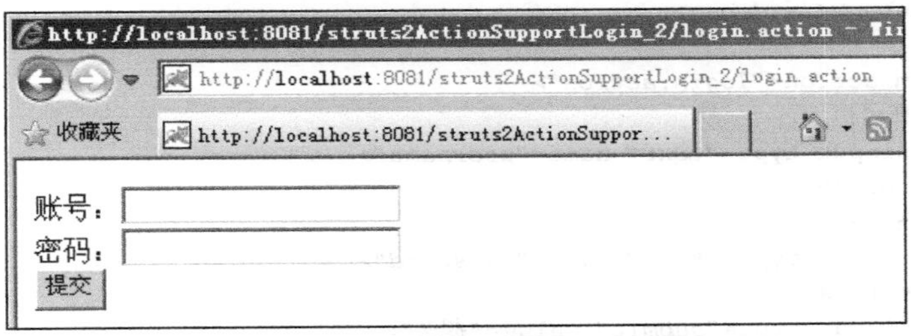

图 3-46 执行了 action 但出错信息并未显示

出现这样的效果还是由于如下的配置代码。

```
<constant name="struts.ui.theme" value="simple"></constant>
```

如果加入配置代码<constant name="struts.ui.theme" value="simple"></constant>就不会自动添加多余的 table/tr/td 等 HTML 标签，但又显示不了出错信息，如何解决这样的问题呢？

先别着急，仔细思考一下现在的情况：

- 使用 Struts 2 的标签的目的就是能自动显示出错信息，又增加了多余的 HTML 代码；
- 使用<constant name="struts.ui.theme">配置屏蔽了使用 Struts 2 标签后自动生成的多余 HTML 代码，但又显示不了出错信息；
- 所以现在的情况是没有必要再使用 Strust 2 标签了，理由是虽然现在不生成多余的 HTML 代码，但却不自动显示出错信息了；
- 所以弃用 Struts 2 的 HTML 标签改用普通的 HTML 代码来进行表单的设计。

更改后的 login.jsp 代码如下。

```
        <form action="login.action">
            账号：
            <input type="text" name="username">
```

```
                <br />
            密码：
            <input type="text" name="password">
            <br />
            <input type="submit" value="提交">
            <br />
        </form>
```

改用普通的<form>和<input>标签后还是不能自动显示出错信息，但单击"提交"按钮后却又在 Login.java 中执行了 addFieldError 函数，说明在后台中产生出了出错信息，仅仅取出的方法不得当，那出错信息在哪里呢？接触过 Servlet 编程的程序员在把出错信息显示到 JSP 页面时，都会将出错信息保存到 request 中，然后在 JSP 页面中通过 JSTL/EL 表达式显示出来，而 Struts 2 也是使用这种方法，只不过封装得更加完善。那么如何才能看到 Struts 2 的出错信息的封装格式呢？在 login.jsp 文件中加入<s:debug></s:debug>标签即可，完整的 login.jsp 代码如下。

```
<%@ page language="java" import="java.util.*" pageEncoding="utf-8"%>
<%@ taglib uri="/struts-tags" prefix="s"%>
<!DOCTYPE HTML PUBLIC "-//W3C//DTD HTML 4.01 Transitional//EN">
<html>
    <body>
        <s:debug></s:debug>
        <br />
        <form action="login.action">
            账号：
            <input type="text" name="username">
            <br />
            密码：
            <input type="text" name="password">
            <br />
            <input type="submit" value="提交">
            <br />
        </form>
    </body>
</html>
```

在 IE 中运行 login.jsp 的结果如图 3-47 所示。

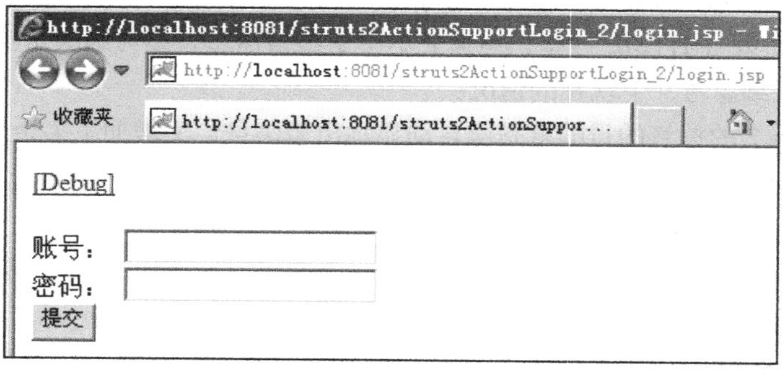

图 3-47　加入 s:debug 标签的运行结果

什么都不输入，继续单击"提交"按钮进行登录，再次跳转到 login.jsp 文件，显示结果如图 3-48 所示。

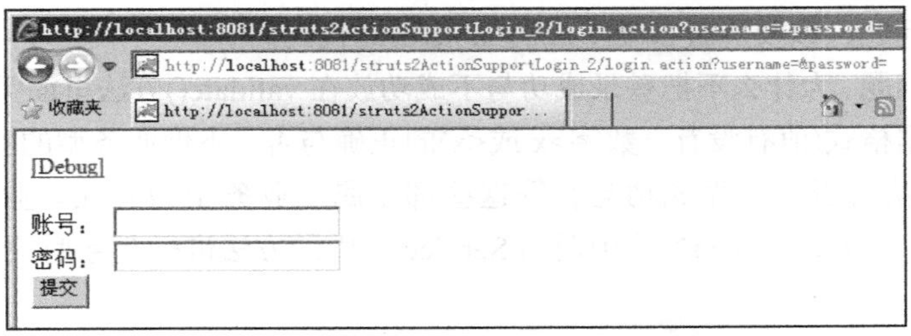

图 3-48 经过控制层的验证再返回 login.jsp 文件

单击"Debug"超级链接，出现的界面如图 3-49 所示。

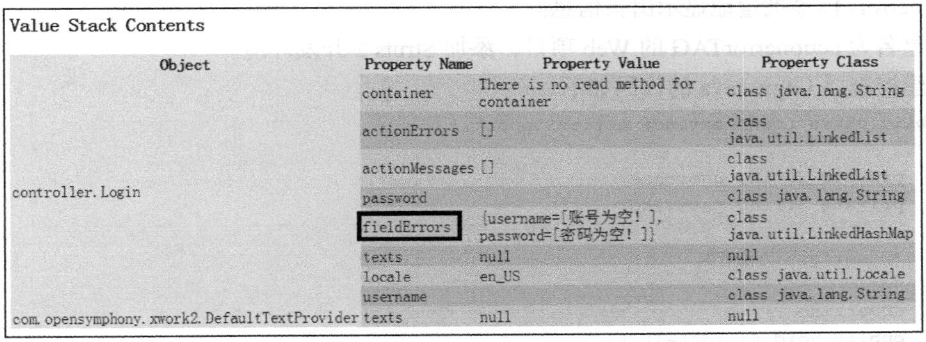

图 3-49 出错信息在 ValueStack 对象中

如何获取出错信息呢？设计 login.jsp 的代码如下。

```
<form action="login.action">
    账号：
    <input type="text" name="username">
    ${fieldErrors.username[0] }
    <br />
    密码：
    <input type="text" name="password">
    ${fieldErrors.password[0] }
    <br />
    <input type="submit" value="提交">
    <br />
</form>
```

重新运行 login.jsp，直接单击"提交"按钮，运行结果如图 3-50 所示。

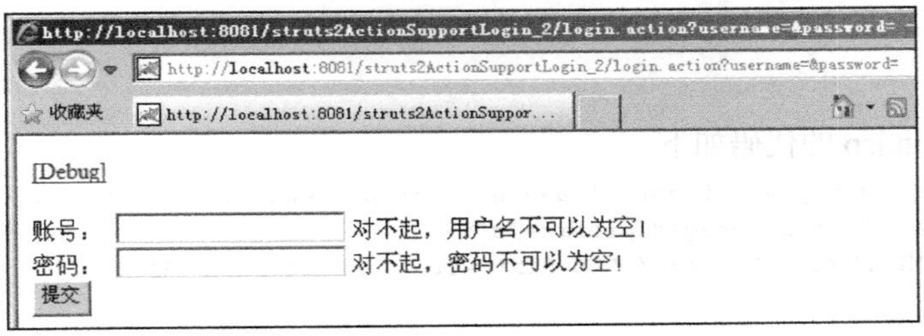

图 3-50 成功获取出错信息

到此，实验已经实现在 Struts 2 框架中使用 ActionSupport 类进行有刷新验证，解决了如何获取出错信息的问题，以及屏蔽自动生成的 table/tr/td 代码。

现在有一个问题，为什么不把登录成功与不成功放在 validate()方法里面呢？因为 validate()方法只是验证数据格式(值有没有，数字>x 或<x)的正确与否，不做业务型的验证，比如登录成功与否，转账成功与否，入库成功与否等这些都是属于业务型的验证。业务型验证还要在 execute()方法中写，在 execute()方法中调用 Service 层中的方法进行业务型验证。

3.5 用<s:actionerror>标签显示全部出错信息

前面的章节用${}EL 表达式来从作用域中一条一条地取出出错信息，Struts 2 中还可以使用<s:actionerror>标签批量地显示出错信息。

创建名为 actionerrorTAG 的 Web 项目，添加 Struts 2 开发环境。

创建控制层 Login.java 的代码如下。

```java
public class Login extends ActionSupport {

    private String username;
    private String password;

    // set get 方法省略

    @Override
    public void validate() {
        super.validate();

        if ("".equals(this.getUsername())) {
            this.addActionError("对不起,用户名不可以为空！");
        }
        if ("".equals(this.getPassword())) {
            this.addActionError("对不起,密码不可以为空！");
        }

    }

    public String execute() {
        UserinfoService usRef = new UserinfoService();
        if (usRef.login(username, password)) {
            return "toOKJSP";
        } else {
            return "toNOJSP";
        }
    }
}
```

登录界面 login.jsp 的代码如下。

```jsp
<%@ page language="java" import="java.util.*" pageEncoding="utf-8"%>
<%@ taglib uri="/struts-tags" prefix="s"%>
<!DOCTYPE HTML PUBLIC "-//W3C//DTD HTML 4.01 Transitional//EN">
<html>
    <body>
        <s:actionerror />
        <br />
        <s:debug></s:debug>
        <br />
```

3.5 用<s:actionerror>标签显示全部出错信息

```
<form action="login.action">
    账号：
    <input type="text" name="username">
    <br />
    密码：
    <input type="text" name="password">
    <br />
    <input type="submit" value="提交">
    <br />
</form>
</body>
</html>
```

运行结果如图 3-51 所示。

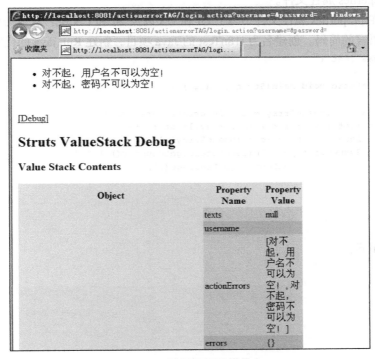

图 3-51 批量显示出错信息

在 JSP 页面中可以不使用 Struts 2 的<s:actionerror>标签，使用 JSTL+EL 技术也能取出异常信息，代码如下。

```
<c:forEach var="eachMessage" items="${actionErrors}">
    ${eachMessage}
    <br/>
</c:forEach>
```

对比前面的 ValueStack 中的内容可以发现，addFieldError 是使用 Map 来进行存储出错信息的，而 addActionError 是使用 List 进行存储的。掌握显示这两种出错信息的技能是掌握 Struts 2 验证功能的必要条件。

3.6 出错信息进行传参及国际化

前面章节中的出错文本信息是直接写在 Java 文件中的，不便于维护及国际化，在本节中将使用国际化技术来针对不同语言的浏览器客户端显示指定语言的文本信息，还可以对出错信息进行传参。

3.6.1 创建 info_en_US.properties 和 info_zh_CN.properties 属性文件

国际化的文本信息要放在 properties 文件中，所以要创建 info_en_US.properties 和 info_zh_CN.properties 属性文件。

为什么要在 properties 属性文件中写上 en 和 US 以及 zh 和 CN 呢？这就需要取得 Java 支持所有国家和语言的简码。创建名为 i18nTest 的 Java 项目，新建 Run.java 的代码如下。

```java
public class Test {

    public static void main(String[] args) {
        Locale[] localeArray = Locale.getAvailableLocales();
        for (int i = 0; i < localeArray.length; i++) {
            Locale eachLocale = localeArray[i];
            System.out.println(eachLocale.getCountry() + " "
                    + eachLocale.getLanguage());
        }

    }

}
```

运行结果如下。

```
JP ja
PE es
   en
JP ja
PA es
BA sr
   mk
GT es
AE ar
NO no
AL sq
   bg
IQ ar
YE ar
   hu
PT pt
CY el
QA ar
MK mk
   sv
CH de
US en
FI fi
   is
```

3.6 出错信息进行传参及国际化

```
      cs
MT en
SI sl
SK sk
      it
TR tr
      zh
      th
SA ar
      no
GB en
CS sr
      lt
      ro
NZ en
NO no
LT lt
NI es
      nl
IE ga
BE fr
ES es
LB ar
      ko
CA fr
EE et
KW ar
RS sr
US es
MX es
SD ar
ID in
      ru
      lv
UY es
LV lv
      iw
BR pt
SY ar
      hr
      et
DO es
CH fr
IN hi
VE es
BH ar
PH en
TN ar
      fi
AT de
      es
NL nl
EC es
TW zh
JO ar
      be
IS is
CO es
```

CR	es
CL	es
EG	ar
ZA	en
TH	th
GR	el
IT	it
	ca
HU	hu
	fr
IE	en
UA	uk
PL	pl
LU	fr
BE	nl
IN	en
ES	ca
MA	ar
BO	es
AU	en
	sr
SG	zh
	pt
	uk
SV	es
RU	ru
KR	ko
	vi
DZ	ar
VN	vi
ME	sr
	sq
LY	ar
	ar
CN	zh
BY	be
HK	zh
	ja
IL	iw
BG	bg
	in
MT	mt
PY	es
	sl
FR	fr
CZ	cs
CH	it
RO	ro
PR	es
CA	en
DE	de
	ga
LU	de
	de
AR	es
	sk
MY	ms
HR	hr

```
SG en
   da
   mt
   pl
OM ar
   tr
TH th
   el
   ms
SE sv
DK da
HN es
```

其中，在上面输出的内容中就有 en 和 US，以及 zh 和 CN，本实验要在项目中针对中文和英文的浏览器客户端显示指定的文本信息，所以要创建 info_en_US.properties 和 info_zh_CN.properties 文件，文件的命名格式在 Struts 2 中是固定的，而文件的前缀 info 就是此文件的标识。

3.6.2 在 JSP 文件中显示国际化的静态文本

创建名为 localeTest 的 Web 项目，创建 info_en_US.properties 和 info_zh_CN.properties 文件，两个属性文件的初始内容如图 3-52 所示。

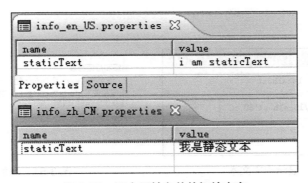

图 3-52 两个属性文件的初始内容

在 struts.xml 配置文件中配置属性文件，代码如下。

```
<?xml version="1.0" encoding="UTF-8"?>
<!DOCTYPE struts PUBLIC
    "-//Apache Software Foundation//DTD Struts Configuration 2.0//EN"
    "http://struts.apache.org/dtds/struts-2.0.dtd">
<struts>
    <constant name="struts.devMode" value="true" />
    <package name="localeTest" extends="struts-default">
    </package>
    <constant name="struts.custom.i18n.resources" value="info"></constant>
</struts>
```

起主要配置作用的就是如下代码。

```
<constant name="struts.custom.i18n.resources"
value="info"></constant>
```

属性 value 的值为 info 也就是 properties 属性文件的标识。

更改 index.jsp 文件内容的代码如下。

```
<%@ page language="java" import="java.util.*" pageEncoding="utf-8"%>
<%@ taglib uri="/struts-tags" prefix="s"%>
<!DOCTYPE HTML PUBLIC "-//W3C//DTD HTML 4.01 Transitional//EN">
<html>
    <body>
        <s:text name="staticText"></s:text>
        <br />
    </body>
</html>
```

部署项目，运行结果如图 3-53 所示。

那怎么显示英语文本呢？进入 IE 的 "Internet 选项" 菜单，单击 "语言" 菜单，添加英语，并将英语上移至顶端，效果如图 3-54 所示。

图 3-53　显示中文文本

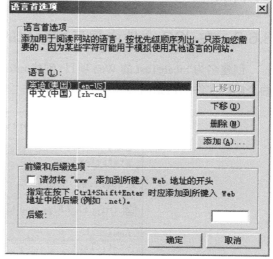

图 3-54　添加英语

"英语(美国)" 添加完成后单击 "确定" 按钮完成配置，再刷新 IE，则显示英语的提示文本，效果如图 3-55 所示。

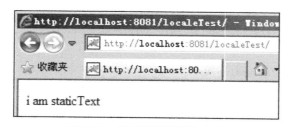

图 3-55　显示英语的提示文本

在 Struts 2 中根据如下代码来确定客户端浏览器的语言类型，根据不同的语言版本显示出指定语言的提示信息。

```
<%
    out.println(request.getLocale().getCountry());
%>
```

3.6.3 在 JSP 文件中显示国际化的静态文本时传递参数

更改属性文件内容，如图 3-56 所示。

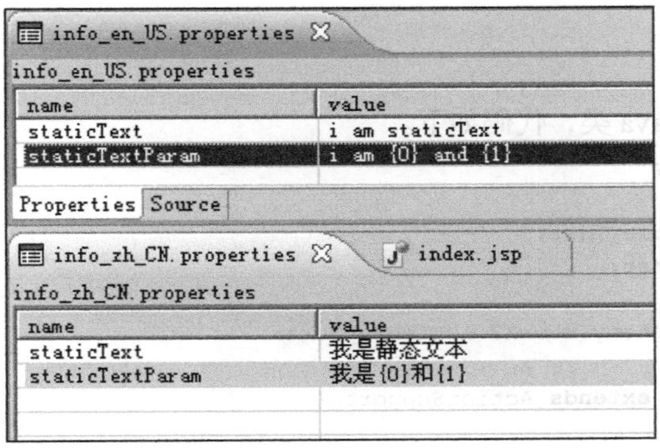

图 3-56 文本具有占位符

更改 index.jsp 的代码如下。

```
<body>
    <s:text name="staticText"></s:text>
    <br />
    <s:text name="staticTextParam">
        <s:param>高洪岩 1</s:param>
        <s:param>高洪岩 2</s:param>
    </s:text>
    <br />
</body>
```

重新运行 index.jsp，显示效果如图 3-57 所示。

图 3-57 静态文本具有参数值

3.6.4 在 Action 中使用国际化功能

在控制层的 Java 文件中也可以取得国际化属性文件中的文本。
创建 login.jsp 文件，代码如下。

```html
<body>
    ${errors.usernameKey1[0] }
    <br />
    ${errors.usernameKey2[0] }
    <br />
    <form action="login.action" method="post">
        username:
        <input type="text" name="username">
        <br />
        <input type="submit" value="登录">
    </form>
</body>
```

创建控制层 Login.java 类，代码如下。

```java
package controller;

import java.util.ArrayList;
import java.util.List;

import com.opensymphony.xwork2.ActionSupport;

public class Login extends ActionSupport {

    private String username;

    public String getUsername() {
        return username;
    }

    public void setUsername(String username) {
        this.username = username;
    }

    @Override
    public void validate() {
        List paramValue = new ArrayList();
        paramValue.add(" 我是参数值");

        if ("".equals(username)) {
            this.addFieldError("usernameKey1", this.getText("usernameNull1",
                paramValue));
        }
        if ("".equals(username)) {
            this.addFieldError("usernameKey2", this.getText("usernameNull2",
                paramValue));
        }
    }

    public String execute() {
        return null;
    }

}
```

更改属性文件，如图 3-58 所示。

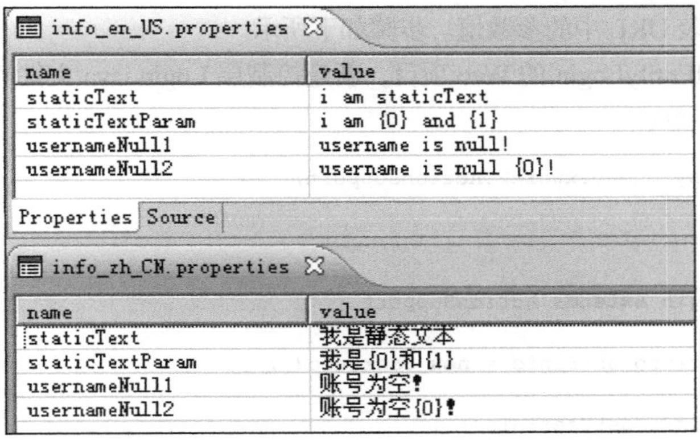

3.7 用实体类封装 URL 中的参数——登录功能的 URL 封装

图 3-58 更改属性文件

继续更改 struts.xml 中的代码，具体如下。

```xml
<?xml version="1.0" encoding="UTF-8"?>
<!DOCTYPE struts PUBLIC
    "-//Apache Software Foundation//DTD Struts Configuration 2.0//EN"
    "http://struts.apache.org/dtds/struts-2.0.dtd">
<struts>
    <constant name="struts.devMode" value="true" />
    <package name="localeTest" extends="struts-default">
        <action name="login" class="controller.Login">
            <result name="input">
                /login.jsp
            </result>
        </action>
    </package>
    <constant name="struts.custom.i18n.resources" value="info"></constant>
</struts>
```

部署项目并运行 URL：http://localhost:8081/localeTest/login.jsp，运行结果如图 3-59 所示。
单击"登录"按钮，界面显示效果如图 3-60 所示。

图 3-59 显示 login.jsp 运行结果

图 3-60 显示控制层中的国际化文本

3.7 用实体类封装 URL 中的参数——登录功能的 URL 封装

在登录 Login.java 类中有 username 和 password 属性，并且有相应的 get 和 set 方法。如果 JSP 的表单数量较大，那么 Action 类中的 set 和 get 方法代码将会非常多，不便于软件的代码维护，可以用一个实体类封装 URL 中的参数值。步骤如下所示。

（1）创建名为 urlEntityLogin 的 Web 项目，新建控制层 Login.java 文件的代码如下。

```java
package controller;

import com.opensymphony.xwork2.ActionSupport;

import entity.Userinfo;

public class Login extends ActionSupport {

    private Userinfo userinfo = new Userinfo();

    public Userinfo getUserinfo() {
        return userinfo;
    }

    public void setUserinfo(Userinfo userinfo) {
        this.userinfo = userinfo;
    }

    @Override
    public void validate() {
        if ("".equals(this.userinfo.getUsername())) {
            this.addFieldError("usernameNull", "账号为空！");
        }
        if ("".equals(this.userinfo.getPassword())) {
            this.addFieldError("passwordNull", "密码为空！");
        }
    }

    public String execute() {
        return null;
    }

}
```

（2）创建封装 URL 参数的实体，如图 3-61 所示。

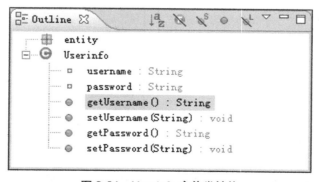

图 3-61　Userinfo 实体类结构

（3）login.jsp 的代码如下。

```jsp
<%@ page language="java" import="java.util.*" pageEncoding="utf-8"%>
<!DOCTYPE HTML PUBLIC "-//W3C//DTD HTML 4.01 Transitional//EN">
```

```html
<html>
    <head>
    </head>
    <body>
        ${fieldErrors.usernameNull[0] }
        <br />
        ${fieldErrors.passwordNull[0] }
        <br />
        <form action="login.action" method="post">
            username:
            <input type="text" name="userinfo.username">
            <br />
            password:
            <input type="text" name="userinfo.password">
            <br />
            <input type="submit" value="登录">
        </form>
    </body>
</html>
```

（4）登录失败时，出现的异常如图 3-62 所示。

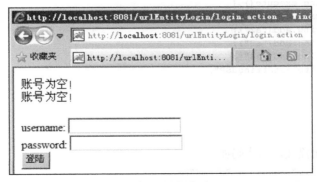

图 3-62　登录失败

将 URL 中的参数封装进实体类使用了 Struts 2 中的拦截器进行实现。

3.8　Struts 2 中的转发操作

在登录的章节中已经实现了转发的操作，登录成功转发到 ok.jsp，失败转发到 no.jsp。

3.8.1　Servlet 中的转发操作

转发操作是在服务器端的行为，也就是使用一个 request 对象，可以在 request 作用域中存入对象，最终在 JSP 页面显示。在 JSP 页面或 Servlet 中，使用如下语句实现转发操作。

```
request.getRequestDispatcher("jsp").forward(request, response);
```

3.8.2　Struts 2 中的转发操作

在本示例中将实现一个列表的功能，此示例也是一个转发操作。实现转发操作的步骤如下所示。

（1）创建名为 struts2Forward 的 Web 项目，创建控制层 ListString.java 的代码如下。

```java
package controller;

import java.util.ArrayList;
import java.util.List;

public class ListString {

    private List<String> listString = new ArrayList<String>();

    public List<String> getListString() {
        return listString;
    }

    public void setListString(List<String> listString) {
        this.listString = listString;
    }

    public String execute() {
        listString.add("a1");
        listString.add("a2");
        listString.add("a3");
        listString.add("a4");
        return "toListStringJSP";

    }

}
```

（2）struts.xml 配置文件的代码如下。

```xml
<?xml version="1.0" encoding="UTF-8"?>
<!DOCTYPE struts PUBLIC
    "-//Apache Software Foundation//DTD Struts Configuration 2.0//EN"
    "http://struts.apache.org/dtds/struts-2.0.dtd">
<struts>
    <constant name="struts.devMode" value="true" />
    <package name="struts2forward" extends="struts-default">
        <action name="listString" class="controller.ListString">
            <result name="toListStringJSP" type="dispatcher">
                /listString.jsp
            </result>
        </action>
    </package>
</struts>
```

（3）listString.jsp 的代码如下。

```jsp
<%@ page language="java" import=java.util.*" pageEncoding="utf-8"%>
<%@ taglib uri="http://java.sun.com/jsp/jstl/core" prefix="c"%>
<!DOCTYPE HTML PUBLIC "-//W3C//DTD HTML 4.01 Transitional//EN">
<html>
    <body>
```

```
            <c:forEach var="eachString" items="${listString}">
                ${eachString }<br />
            </c:forEach>
        </body>
</html>
```

（4）运行效果如图 3-63 所示。

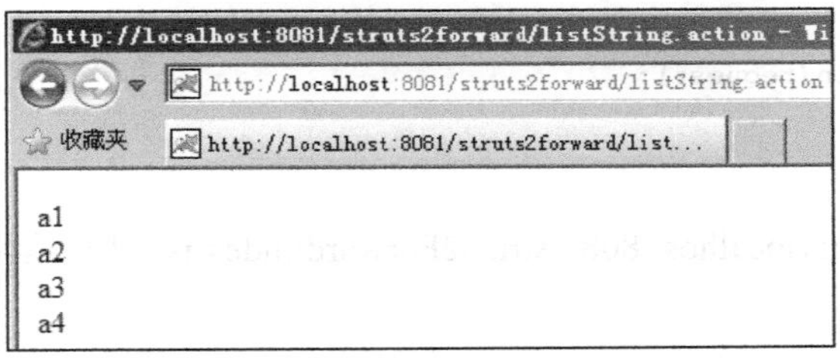

图 3-63　运行效果

Struts 2 中的转发操作主要是在 struts.xml 配置文件中使用 type="dispatcher"属性进行定义，而写不写 type="dispatcher"对转发操作毫无影响，默认的情况下不写 type 属性即是转发操作，因为在 struts-default.xml 配置文件中有默认的定义。查看 struts-default.xml 中的 result-types 节点下的相关配置，具体如下。

```
        <result-types>
            <result-type name="chain" class="com.opensymphony.xwork2.ActionChainResult"/>
            <result-type name="dispatcher" class="org.apache.struts2.result.ServletDispatcherResult" default="true"/>
            <result-type name="freemarker" class="org.apache.struts2.views.freemarker.FreemarkerResult"/>
            <result-type name="httpheader" class="org.apache.struts2.result.HttpHeaderResult"/>
            <result-type name="redirect" class="org.apache.struts2.result.ServletRedirectResult"/>
            <result-type name="redirectAction" class="org.apache.struts2.result.ServletActionRedirectResult"/>
            <result-type name="stream" class="org.apache.struts2.result.StreamResult"/>
            <result-type name="velocity" class="org.apache.struts2.result.VelocityResult"/>
            <result-type name="xslt" class="org.apache.struts2.views.xslt.XSLTResult"/>
            <result-type name="plainText" class="org.apache.struts2.result.PlainTextResult" />
            <result-type name="postback" class="org.apache.struts2.result.PostbackResult" />
        </result-types>
```

其中，result-types 即为可以返回的种类。dispatcher 有一个属性 default，且值为 true，含义是如果在 result 标记中不注明 type，则默认就是 dispatcher 转发操作。

上面列表中有 1 个 result 对象——redirectAction 在开发中也经常被使用到，它通常在重定向到 1 个 Action 时使用。

在上面的示例中，通过在 Action 控制层里声明 1 个 List 属性并添加 add()一些数据，然后就能在 JSP 页面中使用 JSTL+EL 取得 List 中的属性值，这是如何做到的呢？分步骤来解决这个问题。

创建测试用的项目 servletProject，index.jsp 文件的核心代码如下。

```
    <body>
        <%
            out.println(request);
        %>
    </body>
```

执行 URL 网址 http://localhost:8081/servletProject/index.jsp。

程序运行后，在 JSP 页面上输出的信息如下。

```
org.apache.catalina.connector.RequestFacade
```

然后在刚才实验的 struts2Forward 项目中的 index.jsp 中设计的代码如下。

```
<body>
    <%
        out.println(request);
    %>
</body>
```

执行 URL 网址 http://localhost:8081/struts2Forward/index.jsp，程序运行后在 JSP 页面上输出的信息如下。

```
org.apache.struts2.dispatcher.StrutsRequestWrapper
```

从打印的结果来看，是 2 个不同的 request 对象。

在 Struts 2 中使用 JSTL+EL 表达式，之所以能取得到 Action 中的 List 属性里面的值，就是因为使用了 org.apache.struts2.dispatcher.StrutsRequestWrapper，其关键代码如下。

```java
public Object getAttribute(String key) {
    if (key == null) {
        throw new NullPointerException("You must specify a key value");
    }

    if (disableRequestAttributeValueStackLookup || key.startsWith("javax.servlet")) {
        return super.getAttribute(key);
    }

    ActionContext ctx = ActionContext.getContext();
    Object attribute = super.getAttribute(key);//######尝试在 request 作用域中找这个 key
                                               //对应的值

    if (ctx != null && attribute == null) {
//######如果 request 作用域中没有这个值说明 attribute == null 条件成立
//######进入这个 if 语句
        boolean alreadyIn = isTrue((Boolean) ctx.get(REQUEST_WRAPPER_GET_ATTRIBUTE));

        if (!alreadyIn && !key.contains("#")) {
            try {
                // If not found, then try the ValueStack
                ctx.put(REQUEST_WRAPPER_GET_ATTRIBUTE, Boolean.TRUE);
                ValueStack stack = ctx.getValueStack();
                if (stack != null) {
                    //######如果 request 中没有这个值就尝试去 ValueStack 对象里找对应的值
                    //######Struts2 的 Action 中的属性对象都放入这个 ValueStack 值栈中
                    attribute = stack.findValue(key);
                }
            } finally {
                ctx.put(REQUEST_WRAPPER_GET_ATTRIBUTE, Boolean.FALSE);
            }
        }
    }
    return attribute;
```

}

上面的源代码中已经写了注释，说明 org.apache.struts2.dispatcher.StrutsRequestWrapper 对象要在两处对象里面尝试找到 Value 值。一处是 request 作用域中，另一处就是 ValueStack 值栈中，Struts 2 的 Action 的属性值自动放入 ValueStack 值栈中。

3.9 由 Action 重定向到 Action——无参数

在 Servlet 中使用代码 "response.sendRedirect("jsp");" 进行重定向的操作，它是客户端的行为。

3.9.1 什么样的情况下使用重定向

使用重定向的经典场景就是登录成功后显示全部的书籍列表，这是从一个登录的 Action 重定向到另一个列表的 Action。

3.9.2 新建起始控制层 Login.java

新建 Web 项目 strutsRedirectActionTest1，并且新建模拟登录验证类，代码如下。

```java
package controller;

public class Login {

    public String execute() {

        return "list";

    }

}
```

代码 "return "list";" 在本示例要达到的目标是一个 "重定向" 的操作，目的是改变 IE 的 URL 地址，执行 List.action 而列出图书列表。

重定向的操作是由 Login.java 重定向到 List.java。

3.9.3 新建目的控制层 List.java

设计 List.java 类，将所有书籍名称放入一个 ArrayList 中，并且在 JSP 文件显示出来，代码如下。

```java
package controller;

import java.util.ArrayList;

public class List {

    private ArrayList bookList = new ArrayList();

    public ArrayList getBookList() {
        return bookList;
```

```
    }
    public void setBookList(ArrayList bookList) {
        this.bookList = bookList;
    }

    public String execute() {

        bookList.add("book1");
        bookList.add("book2");
        bookList.add("book3");
        bookList.add("book4");

        return "bookListJsp";

    }

}
```

代码很简单，声明一个 ArrayList 变量 bookList，然后在 execute 中往这个 bookList 中存放书籍的名称，最后执行代码 "return "bookListJsp"" 转发到一个显示书籍列表的 JSP 文件。

3.9.4 在 struts.xml 文件中配置重定向的重点

struts.xml 配置文件的代码设计如下。

```
<?xml version="1.0" encoding="UTF-8" ?>
<!DOCTYPE struts PUBLIC
    "-//Apache Software Foundation//DTD Struts Configuration 2.0//EN"
    "http://struts.apache.org/dtds/struts-2.0.dtd">

<struts>

    <package name="ghyStruts1" extends="struts-default">

        <action name="login" class="controller.Login">
            <result name="list" type="redirectAction">
                <param name="actionName">List</param>
            </result>
        </action>

        <action name="List" class="controller.List">
            <result name="bookListJsp">/bookListJsp.jsp</result>
        </action>


    </package>

</struts>
```

下述标记的功能是从当前的 Action，即 login 重定向到 List 这个 Action，其中重定向的关键操作是 type 的属性值：redirectAction。

```
            <result name="list" type="redirectAction">
                <param name="actionName">List</param>
            </result>
```

3.9.5 新建显示列表的 JSP 文件

显示书籍列表的 bookListJsp.jsp 代码如下。

```
<%@ page language="java" import="java.util.*" pageEncoding="utf-8"%>
<%@ taglib uri="http://java.sun.com/jsp/jstl/core" prefix="c"%>
<%@ page isELIgnored="false"%>

<!DOCTYPE HTML PUBLIC "-//W3C//DTD HTML 4.01 Transitional//EN">
<html>
    <body>
        <c:forEach var="bookName" items="${bookList}">
            <c:out value="${bookName}"></c:out>
            <br />
        </c:forEach>
    </body>
</html>
```

在 IE 地址栏中输入地址 http://localhost:8081/strutsRedirectActionTest1/login.action 后，它立即变成地址 http://localhost:8081/strutsRedirectActionTest1/listUserinfo.action，IE 界面如图 3-64 所示。

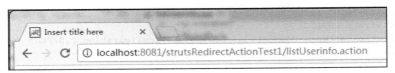

图 3-64　程序运行结果

并且在 JSP 页面上输出了 List 中的数据。

3.10　由 Action 重定向到 Action——有参数

3.10.1　什么样的情况下需要重定向传递参数

如果有一个需求，在登录之后，默认显示的图书列表是第 100 页，这样的需求不仅仅是登录 Action 与图书列表 Action 进行重定向的操作，还需要向这两个 Action 传递参数。

比如在登录成功之后显示图书列表第 100 页，本小节就实现这样的需求。

3.10.2　新建起始控制层 Login.java 文件

新建项目 struts 2.6，新建控制层 Login.java 类，代码如下。

```
package controller;

public class Login {

    private int id;
```

```
    public int getId() {
        return id;
    }

    public void setId(int id) {
        this.id = id;
    }

    public String execute() {

        this.setId(100);

        return "list";

    }

}
```

把 id 的值 100 要传递给 List.java 中，也就是图书列表的控制层。

3.10.3 更改 struts.xml 配置文件

struts.xml 配置文件如下。

```
<?xml version="1.0" encoding="UTF-8" ?>
<!DOCTYPE struts PUBLIC
    "-//Apache Software Foundation//DTD Struts Configuration 2.0//EN"
    "http://struts.apache.org/dtds/struts-2.0.dtd">

<struts>

    <package name="ghyStruts1" extends="struts-default">

        <action name="login" class="controller.Login">
            <result name="list" type="redirectAction">
                <param name="idInList">${id}</param>
                <param name="actionName">List</param>
            </result>
        </action>

        <action name="List" class="controller.List">
            <result name="bookListJsp">/bookListJsp.jsp</result>
        </action>
    </package>

</struts>
```

如何让 Login.java 中的 id 传递给 List.java 呢？下面是最关键的配置。

```
            <result name="list" type="redirectAction">
                <param name="idInList">${id}</param>
                <param name="actionName">List</param>
            </result>
```

在从 Login.java 控制层重定向到 List.java 这个 Action 时，还要传递参数${id}，这个值是从

Login.java 类的 id 属性中来，将 Login.java 类中的 id 属性传出去，传到哪里去？传到 List.action 中去。参数名称是什么？是 idInList。那么这段配置其实就是本小节的核心：揭示如何在 Action 与 Action 中传递参数，并且是重定向的操作。

type 属性为 redirectAction 的作用正是从一个 action 重定向到另外一个 action。

Action 之间重定向传递参数使用如下配置也能实现同样的效果。

```xml
<action name="login" class="controller.Login">
    <result name="toListUserinfoAction" type="redirectAction">
        listUserinfo.action?zzzzzzzzzzz=${gotoPage}
    </result>
</action>
```

3.10.4 新建目的控制层 List.java 文件

显示图书列表的控制层 List.java 的代码如下。

```java
package controller;

import java.util.ArrayList;

public class List {

    private ArrayList bookList = new ArrayList();

    private int idInList;

    public ArrayList getBookList() {
        return bookList;
    }

    public void setBookList(ArrayList bookList) {
        this.bookList = bookList;
    }

    public int getIdInList() {
        return idInList;
    }

    public void setIdInList(int idInList) {
        this.idInList = idInList;
    }

    public String execute() {

        bookList.add("book1");
        bookList.add("book2");
        bookList.add("book3");
        bookList.add("book4");

        return "bookListJsp";
    }

}
```

在 List.java 中可以看到，并没有发现针对 idInList 这个 int 类型变量赋值的操作，其实这个操作是由 Struts 2 底层来实现赋值的，只需要在 List.java 文件中声明一个 int 类型的 idInList 变量即可，框架底层自动将 url 中的同名参数赋给 idInList 同名属性。

3.10.5　用 JSTL 和 EL 在 JSP 文件中打印数据

显示图书列表的 bookListJsp.jsp 页面的代码如下。

```
<%@ page language="java" import="java.util.*" pageEncoding="utf-8"%>
<%@ taglib uri="http://java.sun.com/jsp/jstl/core" prefix="c"%>
<%@ page isELIgnored="false"%>

<!DOCTYPE HTML PUBLIC "-//W3C//DTD HTML 4.01 Transitional//EN">
<html>
    <body>
        显示第${param.idInList }页
        <br />
        <c:forEach var="bookName" items="${bookList}">
            <c:out value="${bookName}"></c:out>
            <br />
        </c:forEach>
    </body>
</html>
```

在 IE 地址栏输入地址 http://localhost:8081/struts2.6/login.action 来进行模拟的登录表单提交，提交之后，页面立即变化，并且地址为 http://localhost:8081/struts2.6/List.action?idInList=100。显示内容如图 3-65 所示。

图 3-65　IE 显示的内容

在使用重定向传递中文时需要使用如下代码进行发送与接收。

发送端代码如下。

```
public class Login {

    private String gotoPage = "中%%%%AD";

    public String getGotoPage() {
        return gotoPage;
```

```java
    }

    public void setGotoPage(String gotoPage) {
        this.gotoPage = gotoPage;
    }

    public String execute() throws UnsupportedEncodingException {
        System.out.println("Login");
        gotoPage = java.net.URLEncoder.encode(gotoPage, "utf-8");
        gotoPage = gotoPage.replaceAll("%", "_");
        return "toListUserinfoAction";
    }
}
```

接收端代码如下。

```java
public class ListUserinfo {

    private List list = new ArrayList();
    private String zzzzzzzzzzz;

    public String getZzzzzzzzzzz() {
        return zzzzzzzzzzz;
    }

    public void setZzzzzzzzzzz(String zzzzzzzzzzz) {
        this.zzzzzzzzzzz = zzzzzzzzzzz;
    }

    public List getList() {
        return list;
    }

    public void setList(List list) {
        this.list = list;
    }

    public String execute() throws UnsupportedEncodingException {
        zzzzzzzzzzz = zzzzzzzzzzz.replace("_", "%");
        zzzzzzzzzzz = java.net.URLDecoder.decode(zzzzzzzzzzz, "utf-8");
        System.out.println("ListUserinfo zzzzzzzzzzz=" + zzzzzzzzzzz);
        list.add("大中国1");
        list.add("大中国2");
        list.add("大中国3");
        list.add("大中国4");
        list.add("大中国5");
        return "toListUserinfoJSP";
    }

}
```

在发送时要将%替换成_（下划线），接收时进行反操作，将_（下划线）替换成%，这样做是为了防止出现在 Tomcat7 中的发送端的网址存在%而导致接收端接收到的值为 null 的情况，示例网址如下。

```
http://localhost:8081/servletTest2/test2?username=abc%
```

所以要将%替换成_（下划线）。本测试在项目 servletTest2 中有源代码。

学习到这，需要掌握 4 个知识点：
- 转发；
- 重定向无传参；
- 重定向有传参；
- 重定向传递中文参数。

3.11 让 Struts 2 支持多模块多配置文件开发

Struts 2 支持多模块多配置文件开发，并且使用起来比 Struts 1 更加方便，下面就模拟项目里面有"仓库""前台""财务"3 个模块，并且把这 3 个模块整合到一起的实验。

3.11.1 新建 4 个模块的控制层

新建项目 Struts 2.7，创建 4 个控制层 Action 类，分别是"财务""仓库""前台"和"从财务模块的财务重定向到仓库模块的财务"，代码如下。

```java
package controller;

public class Caiwu {//财务类

    public String execute() {

        System.out.println("执行了财务统计模块");

        return "toCaiwu";

    }
}
```

```java
package controller;

public class Qiantai {//前台类

    public String execute() {

        System.out.println("执行了前台模块");

        return "toQiantai";

    }
}
```

```java
package controller;

public class Cangku {//仓库
```

```java
    public String execute() {

        System.out.println("执行了仓库模块");

        return "toCangku";

    }
}

package controller;

public class Fromcaiwu_caiwu_tocangku_caiwu {//不同模块之间的重定向的操作

    private int id;//属性 id 的作用是在重定向的过程中传递参数

    public int getId() {
        return id;
    }

    public void setId(int id) {
        this.id = id;
    }

    public String execute() {

        id = 999;

        System.out.println("从财务模块中的财务转到了仓库中的财务");

        return "fromcaiwu_caiwu_tocangku_caiwu";

    }
}
```

3.11.2 新建 3 个模块的配置文件

创建 3 个配置文件，分别是"仓库模块 xml 文件""前台模块 xml 文件""财务模块 xml 文件"，3 个 xml 配置文件的代码如下。

caiwu.xml 配置文件的代码如下。

```xml
<?xml version="1.0" encoding="UTF-8" ?>
<!DOCTYPE struts PUBLIC
    "-//Apache Software Foundation//DTD Struts Configuration 2.0//EN"
    "http://struts.apache.org/dtds/struts-2.0.dtd">

<struts>

    <package name="caiwu" extends="struts-default" namespace="/caiwu">

        <action name="caiwu" class="controller.Caiwu">
            <result name="toCaiwu">/caiwu.jsp</result>
        </action>

        <action name="fromcaiwu_caiwu_tocangku_caiwu"
            class="controller.Fromcaiwu_caiwu_tocangku_caiwu">
```

```xml
            <result name="fromcaiwu_caiwu_tocangku_caiwu"
                type="redirectAction">
                <param name="namespace">/cangku</param>
                <param name="actionName">caiwu?id=${id}</param>
            </result>
        </action>

    </package>

</struts>
```

cangku.xml 配置文件的代码如下。

```xml
<?xml version="1.0" encoding="UTF-8" ?>
<!DOCTYPE struts PUBLIC
    "-//Apache Software Foundation//DTD Struts Configuration 2.0//EN"
    "http://struts.apache.org/dtds/struts-2.0.dtd">

<struts>

    <package name="cangku" extends="struts-default" namespace="/cangku">

        <action name="caiwu" class="controller.Cangku">
            <result name="toCangku">/cangku.jsp</result>
        </action>

    </package>

</struts>
```

前面这两个配置文件有一个共同的特点就是都有路径为 caiwu 的 Action，但它们在不同的 namespace 命名空间中，命名空间就像 Java 中的包一样，将相同的类名放在不同的包中是允许的。那么在 Struts 2 中进行多模块开发时，如果请求的 action 路径相同，是根据命名空间的路径来划分。也就是说，在请求 action 时，路径要加上命名空间，由于命名空间不允许重复，所以即使 Action 路径重复也没有问题。

注意： 在使用新版本 Struts 2 时，如果想正确实现本示例的目的，就要在每个 struts.xml 中设置 package 包名不要重复，namespace 命名空间在配置文件之间也不要重复。

qiantai.xml 配置文件的代码如下。

```xml
<?xml version="1.0" encoding="UTF-8" ?>
<!DOCTYPE struts PUBLIC
    "-//Apache Software Foundation//DTD Struts Configuration 2.0//EN"
    "http://struts.apache.org/dtds/struts-2.0.dtd">

<struts>

    <package name="qiantai" extends="struts-default" namespace="/qiantai">

        <action name="qiantai" class="controller.Qiantai">
            <result name="toQiantai">/qiantai.jsp</result>
        </action>

    </package>

</struts>
```

3.11.3 使用 include 标记导入多个配置文件

将各子模块整合的 struts.xml 配置文件的代码如下。

```xml
<?xml version="1.0" encoding="UTF-8" ?>
<!DOCTYPE struts PUBLIC
    "-//Apache Software Foundation//DTD Struts Configuration 2.0//EN"
    "http://struts.apache.org/dtds/struts-2.0.dtd">

<struts>

    <include file="caiwu.xml"></include>
    <include file="cangku.xml"></include>
    <include file="qiantai.xml"></include>

</struts>
```

前 3 个配置文件中的代码与普通配置文件配置 Action 信息别无两样，但 struts.xml 却和以前做过的示例代码完全不同。由于本示例模拟 3 个模块的整合，所以在 struts.xml 配置文件中要将这 3 个模块进行集中配置，主要使用的就是 include 标记，file 属性是其他模块配置文件所在的路径。

3.11.4 创建各模块使用的 JSP 文件

各模块显示数据的 JSP 文件代码如下。
Caiwu.jsp 文件的代码如下。

```jsp
<%@ page language="java" import="java.util.*" pageEncoding="utf-8"%>

<!DOCTYPE HTML PUBLIC "-//W3C//DTD HTML 4.01 Transitional//EN">
<html>
    <body>
        显示财务模块
    </body>
</html>
```

cangku.jsp 文件的代码如下。

```jsp
<%@ page language="java" import="java.util.*" pageEncoding="utf-8"%>

<!DOCTYPE HTML PUBLIC "-//W3C//DTD HTML 4.01 Transitional//EN">
<html>
    <body>
        显示仓库模块
    </body>
</html>
```

qiantai.jsp 文件的代码如下。

```jsp
<%@ page language="java" import="java.util.*" pageEncoding="utf-8"%>

<!DOCTYPE HTML PUBLIC "-//W3C//DTD HTML 4.01 Transitional//EN">
<html>
    <body>
```

显示前台模块
    </body>
</html>
```

## 3.11.5 运行各模块的效果

程序运行结果如图 3-66 所示。

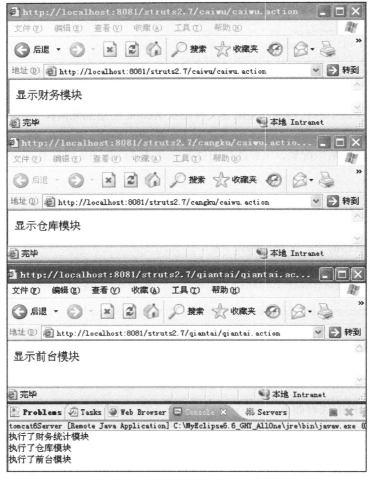

图 3-66　多模块程序示例运行结果

打开 3 个 IE 窗口，输入各模块 action 的 url 地址，一定要把 namespace 命名空间的路径加进去，然后在控制台输出执行各模块功能代码的字符串。

还要执行最重要的步骤，也就是从一个模块重定向到另外一个模块的操作。打开 IE，输入地址 http://localhost:8081/struts2.7/caiwu/fromcaiwu_caiwu_tocangku_caiwu.action，然后按下回车键，出现如图 3-67 所示的界面效果。
```

3.12 在 Action 中有多个业务方法时的处理

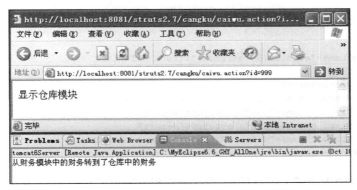

图 3-67 从一个模块重定向到另外一个模块

3.12 在 Action 中有多个业务方法时的处理

在 Action 中可以声明多个业务方法，调用不同的业务方法需要在 struts.xml 的<action>标记中加入 method 属性。

新建 Struts 2.8.1 项目，添加 Struts 2 的 jar 包及必要的 xml 配置文件，新建控制层 Action 类 List.java 的代码如下。

```java
package controller;

public class List {

    public String listUser() {
        System.out.println("列出员工信息");
        return "user";
    }

    public String listSalarySum() {
        System.out.println("列出员工总工资信息");
        return "sum";
    }

    public String listSalaryAvg() {
        System.out.println("列出员工平均工资信息");
        return "avg";
    }

    public String execute() {
        System.out.println("执行了 execute 方法");
        return "default";
    }

}
```

新建 struts.xml 配置文件，代码如下。

```xml
<?xml version="1.0" encoding="UTF-8" ?>
<!DOCTYPE struts PUBLIC
    "-//Apache Software Foundation//DTD Struts Configuration 2.0//EN"
    "http://struts.apache.org/dtds/struts-2.0.dtd">
```

```xml
<struts>

    <package name="struts2.8" extends="struts-default">

        <global-results>
            <result name="user">/user.jsp</result>
            <result name="sum">/sum.jsp</result>
            <result name="avg">/avg.jsp</result>
            <result name="default">/default.jsp</result>
        </global-results>

        <action name="listUser" class="controller.List"
            method="listUser">
        </action>
        <action name="listSalarySum" class="controller.List"
            method="listSalarySum">
        </action>
        <action name="listSalaryAvg" class="controller.List"
            method="listSalaryAvg">
        </action>
        <action name="list" class="controller.List"></action>
    </package>

</struts>
```

在本示例的配置文件中定义了全局的 result 转发，而在定义 action 标记时加入了 method 属性，这个属性就是执行 action 里所对应的方法，也就是 Action 控制层中的方法名。

在 IE 中依次输入从屏幕上方到下方的网址，在控制台中查看运行后的结果，如图 3-68 所示。

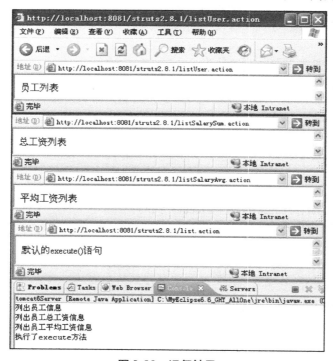

图 3-68　运行结果

从结果中可以看到，直接输入 url 后就可以运行指定 action 中的自定义方法了，主要的实现技术点为 struts.xml 文件中 action 节点中的 method 属性。

3.13 自定义全局 result

看下面的配置定义。

```xml
<action name="List" class="controller.List">
    <result name="bookListJsp">/bookListJsp.jsp</result>
</action>
```

Action 的路径 List 下有一个名为 bookListJsp 的 result 对象，但这个 bookListJsp 对象只能在当前的 List 的 Action 路径下使用，如果有其他的 Action 也想转到这个 bookListJsp.jsp 页面，必须在其他的 Action 下还要注册一个 result。

```xml
<result name="bookListJsp">/bookListJsp.jsp</result>
```

这样造成配置的冗余。

完全可以配置一个全局的 result，使所有的 Action 都可以访问这个 result 就可以了。

3.13.1 新建全局 result 实例和控制层代码

新建 Web 项目 Struts 2.9。

新建两个 Action，代码如下。

```java
package controller;

import com.opensymphony.xwork2.Action;

public class List1 implements Action {

    public String execute() {

        return "toOneResult";

    }
}
```

```java
package controller;

import com.opensymphony.xwork2.Action;

public class List2 implements Action {

    public String execute() {

        return "toOneResult";

    }
}
```

3.13.2 声明全局的 result 对象

在 struts.xml 配置文件中声明全局的 result，配置代码如下。

```xml
<?xml version="1.0" encoding="UTF-8" ?>
<!DOCTYPE struts PUBLIC
    "-//Apache Software Foundation//DTD Struts Configuration 2.0//EN"
    "http://struts.apache.org/dtds/struts-2.0.dtd">

<struts>

    <package name="ghyStruts1" extends="struts-default">

        <global-results>
            <result name="toOneResult">/toOneResult.jsp</result>
        </global-results>

        <action name="list1" class="controller.List1"></action>

        <action name="list2" class="controller.List2"></action>

    </package>

</struts>
```

使用 global-results 标记创建一个全局的 result。

JSP 代码如下。

```jsp
<%@ page language="java" import="java.util.*" pageEncoding="utf-8"%>

<!DOCTYPE HTML PUBLIC "-//W3C//DTD HTML 4.01 Transitional//EN">

<html>
    <body>
        转到同一个 JSP 文件
    </body>
</html>
```

3.13.3 部属项目并运行

部属项目，启动 Tomcat。

在 IE 地址栏中输入下面的 2 个地址：

http://localhost:8081/struts2.9/list1.action
http://localhost:8081/struts2.9/list2.action

发现同时显示了一个页面，如图 3-69 所示。

图 3-69　全局 result 的运行效果

注意：如果局部的 result 名称和全局的名称一样，那么局部的优先级比全局的要高！

3.14　在 Action 中使用 servlet 的 API（紧耦版）

由于在 Struts 2 中采用的是松耦合的方式进行设计，所以在 Struts 2 的 Action 类中就可以不用 Servlet 的 API 来实现功能,但在控制层 Action 中想使用 Servlet 的 API 时,该如何操作呢？使用 ServletActionContext 类。

3.14.1　将数据放到不同的作用域中

新建 Struts 2.10 项目，新建控制层 List.java 文件，代码如下。

```
package controller;

import org.apache.struts2.ServletActionContext;

public class List {

    private String usernameRequest;
    private String usernameSession;
    private String usernameApplication;

    public String getUsernameApplication() {
        return usernameApplication;
    }

    public String getUsernameRequest() {
        return usernameRequest;
    }
```

```java
    public void setUsernameRequest(String usernameRequest) {
        this.usernameRequest = usernameRequest;
    }

    public String getUsernameSession() {
        return usernameSession;
    }

    public void setUsernameSession(String usernameSession) {
        this.usernameSession = usernameSession;
    }

    public void setUsernameApplication(String usernameApplication) {
        this.usernameApplication = usernameApplication;
    }

    public String execute() {
        usernameRequest = "usernameRequestValue";
        usernameSession = "usernameSessionValue";
        usernameApplication = "usernameApplicationValue";

        ServletActionContext.getRequest().setAttribute("usernameRequest",
                usernameRequest);
        ServletActionContext.getRequest().getSession().setAttribute(
                "usernameSession", usernameSession);
        ServletActionContext.getServletContext().setAttribute(
                "usernameApplication", usernameApplication);

        return "default";
    }
}
```

想使用 Servlet 的 API，必须得通过 ServletActionContext 类。

新建 struts.xml 配置文件，代码如下。

```xml
<?xml version="1.0" encoding="UTF-8" ?>
<!DOCTYPE struts PUBLIC
    "-//Apache Software Foundation//DTD Struts Configuration 2.0//EN"
    "http://struts.apache.org/dtds/struts-2.0.dtd">

<struts>

    <package name="struts2.8" extends="struts-default">
        <action name="list" class="controller.List">
            <result name="default">/default.jsp</result>
        </action>

    </package>

</struts>
```

3.14.2 从不同作用域中取值

新建 default.jsp 文件，代码如下。

```
<%@ page language="java" import="java.util.*" pageEncoding="utf-8"%>
```

```
<%@ page isELIgnored="false"%>
<%@ taglib uri="/struts-tags" prefix="s"%>

<!DOCTYPE HTML PUBLIC "-//W3C//DTD HTML 4.01 Transitional//EN">
<html>
    <body>
        request: ${requestScope.usernameRequest}
        <br />
        session: ${sessionScope.usernameSession}
        <br />
        application: ${applicationScope.usernameApplication}
        <br />
    </body>
</html>
```

在 IE 中的运行效果如图 3-70 所示。

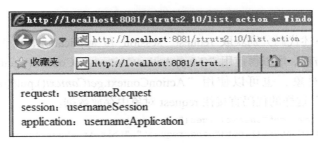

图 3-70　运行效果

在 IE 中输入 action 的路径后，JSP 显示出在不同作用域中的值。

3.15　在 Action 中使用 Servlet 的 API（松耦版）

本小节就使用 ActionContext 类来对 request、session 和 application 对象进行操作，不需要再使用 ServletActionContext 对象了。

新建 Web 项目 Struts 2.11。

添加必要的 Struts 2 支持环境和 jar 文件。

3.15.1　新建控制层

新建控制层 ShowScopeValue.java 文件，代码如下。

```java
package controller;

import java.util.Map;

import com.opensymphony.xwork2.ActionContext;

public class ShowScopeValue {

    public String execute() {

        // 第一种方法往 request 对象中放数据
```

```java
        Map request = (Map) ActionContext.getContext().get("request");
        request.put("requestValue", "this is reqeustValue");

        // 第二种方法往 request 对象中放数据
        ActionContext.getContext().put("otherrequest",
                "this is otherrequest value");

        Map session = (Map) ActionContext.getContext().getSession();
        session.put("sessionValue", "this is sessionValue");

        Map application = (Map) ActionContext.getContext().getApplication();
        application.put("applicationValue", "this is applicationValue");

        return "showscopevalue";
    }
}
```

在程序中使用下述代码分别获取 request、session 和 application，由于 Struts 2 没有类似 getRequest()这样的方法，所以得使用"(Map) ActionContext.getContext().get("request");"这样的代码获得 request 对象，也可以使用"ActionContext.getContext().put("otherrequest","this is otherrequest value");"这样的代码直接往 request 对象中存放数据。

```java
(Map) ActionContext.getContext().get("request");
ActionContext.getContext().put("otherrequest","this is otherrequest value");
(Map) ActionContext.getContext().getSession();
(Map) ActionContext.getContext().getApplication();
```

3.15.2 新建 JSP 视图

新建 showscopevalue.jsp 的代码如下。

```jsp
<%@ page language="java" import="java.util.*" pageEncoding="utf-8"%>
<%@ taglib uri="/struts-tags" prefix="s"%>

<!DOCTYPE HTML PUBLIC "-//W3C//DTD HTML 4.01 Transitional//EN">
<html>
    <body>
        request1:${requestValue}
        <br />
        request1:${otherrequest}
        <br />
        session:${sessionValue}
        <br />
        application:${applicationValue}
    </body>
</html>
```

struts.xml 配置文件的代码如下。

```xml
<?xml version="1.0" encoding="UTF-8" ?>
<!DOCTYPE struts PUBLIC
    "-//Apache Software Foundation//DTD Struts Configuration 2.0//EN"
    "http://struts.apache.org/dtds/struts-2.0.dtd">
```

3.15 在 Action 中使用 Servlet 的 API（松耦版）

```
<struts>

    <package name="strutsTest" extends="struts-default">

        <action name="showscopevalue"
            class="controller.ShowScopeValue">
            <result name="showscopevalue">/showscopevalue.jsp</result>
        </action>

    </package>

</struts>
```

程序运行效果如图 3-71 所示。

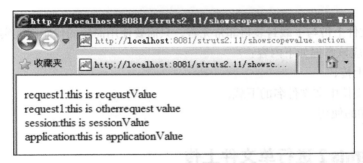

图 3-71 运行效果

此案例的知识点总结。

1. 写法：

ActionContext.getContext().put("requestKey1", "requestValue1");

向值栈中存放值

2. 写法：

Map map = (Map) ActionContext.getContext().get("request");

map.put("requestKey2", "requestValue2");

向 request 作用域存放值

3. 写法：

ActionContext.getContext().getSession().put("sessionKey", "sessionValue");

向 session 作用域存放值

4. 写法：

ActionContext.getContext().getApplication().put("applicationKey", "applicationValue");

向 application 作用域存放值

第 4 章　Struts 2 文件的上传与下载

Struts 2 对文件上传与下载封装得非常完善，只需要使用简短的代码就能实现对应的功能，所以读者在本章应该掌握如下内容：
- Struts 2 上传；
- Struts 2 支持中文文件名的下载；
- 属性驱动的使用。

4.1 使用 Struts 2 进行单文件上传

4.1.1 Struts 2 上传功能的底层依赖

文件上传的功能在 Struts 2 中也得到了非常好的封装，主要使用的是 FileUpload 上传组件。本节就来实现一个单文件上传的示例。

4.1.2 新建上传文件的 JSP 文件

新建一个 Web 项目 struts3.1。
新建具有上传表单的 JSP 文件 register.jsp，代码如下。

```jsp
<%@ page language="java" import="java.util.*" pageEncoding="utf-8"%>
<%@ page isELIgnored="false"%>
<%@ taglib uri="/struts-tags" prefix="s"%>

<!DOCTYPE HTML PUBLIC "-//W3C//DTD HTML 4.01 Transitional//EN">
<html>
    <head>
    </head>
    <body>
        <s:form action="register" method="post" enctype="multipart/form-data">
            username:<s:textfield name="username"></s:textfield>
            <br />
            <s:file name="uploadFile"></s:file>
            <br />
            <s:submit value="提交"></s:submit>
        </s:form>
    </body>
```

```
</html>
```

从 JSP 代码中可以看到，在 form 标签加入了属性：enctype="multipart/form-data"，而且还加入了 s:file 文件域上传标签，标签的 name 属性值为 uploadFile。

4.1.3　新建上传文件的控制层 Register.java 文件

新建控制层 Register.java 文件，代码如下。

```java
package controller;

import java.io.File;
import java.io.IOException;

import org.apache.commons.io.FileUtils;
import org.apache.struts2.ServletActionContext;

import com.opensymphony.xwork2.ActionSupport;

public class Register {

    private String username;

    private File uploadFile;

    private String uploadFileFileName;

    public String getUsername() {
        return username;
    }

    public void setUsername(String username) {
        this.username = username;
    }

    public File getUploadFile() {
        return uploadFile;
    }

    public void setUploadFile(File uploadFile) {
        this.uploadFile = uploadFile;
    }

    public String getUploadFileFileName() {
        return uploadFileFileName;
    }

    public void setUploadFileFileName(String uploadFileFileName) {
        this.uploadFileFileName = uploadFileFileName;
    }

    public String execute() throws IOException {
        //打印 username 的值
        System.out.println("username 的值是：" + username);
        //取得上传后文件要存放的路径
        String targetDirectory = ServletActionContext.getRequest().getRealPath(
                "/upload");
```

```
        //生成上传的File对象
        File target = new File(targetDirectory, uploadFileFileName);
        //复制File对象,从而实现上传文件
        FileUtils.copyFile(uploadFile, target);

        return "register";
    }
}
```

4.1.4 Action 中 File 实例的命名规则

在这个控制层 Register.java 文件中,有 3 个属性。

```
    private String username;
    private File uploadFile;
    private String uploadFileFileName;
```

其中,数据类型为 File 的变量 uploadFile 和 JSP 的 s:file 表单的 name 属性值要一致对应,这样就可以使用 Struts 2 的拦截器进行属性值的自动封装。

而数据类型是 String 的 username 则是一个普通的 s:textfield 单行文本域表单数据值。

但其中最为重要的是类型为 String 的变量 uploadFileFileName,这个变量存储的是上传文件的文件名。这个变量的命名是有规则的,必须在类型为 File 的变量名后面加上 FileName,比如,File 的变量名是 abc,则这个上传文件的文件名的变量就得取名为 abcFileName,最后再生成这 3 个属性的 get 和 set 方法。

4.1.5 设置上传文件的大小

新建 struts.properties 属性文件,代码如下。

```
struts.multipart.maxSize=2048000000
#上面的意义是设置最大的上传文件大小
struts.multipart.saveDir=/tempUploadFile
#上面的意义是设置上传文件的临时目录
```

注意:一定要在项目中创建一个名称为 upload 的目录!

4.1.6 设计 struts.xml 配置文件

struts.xml 配置文件的代码如下。

```
<?xml version="1.0" encoding="UTF-8" ?>
<!DOCTYPE struts PUBLIC
    "-//Apache Software Foundation//DTD Struts Configuration 2.0//EN"
    "http://struts.apache.org/dtds/struts-2.0.dtd">

<struts>
    <package name="struts3.1" extends="struts-default">
        <action name="register" class="controller.Register">
            <result name="register">/showregister.jsp</result>
            <result name="input">/register.jsp</result>
        </action>
```

```
</package>

<constant name="struts.ui.theme" value="simple"></constant>
</struts>
```

4.1.7　成功上传单个文件

部属项目，运行程序，效果如图 4-1 所示。

图 4-1　初始运行效果

在 username 的表单输入字符串，然后在 File 表单中单击"浏览"选择一个待上传的文件，单击"提交"按钮后，控制台打印出了 username 的值，如图 4-2 所示。

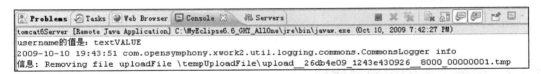

图 4-2　控制台中的效果

从控制台中可以看到，在上传的同时可以取得普通文本域 username 的值，最后又删除了上传的临时文件，并且可以在 Tomcat 目录中看到上传的文件，如图 4-3 所示。

图 4-3　上传后的文件

4.2　使用 Struts 2 进行多文件上传

Struts 2 也支持多文件批量上传，只是要将 File 类型改成 File[]数组类型，然后通过循环遍历的方法进行多文件的上传。

4.2.1 新建上传多个文件的 JSP

新建 struts 3.2 的 Web 项目。
在 register.jsp 文件中加入多个同名的 File 表单，代码如下。

```jsp
<%@ page language="java" import="java.util.*" pageEncoding="utf-8"%>
<%@ page isELIgnored="false"%>
<%@ taglib uri="/struts-tags" prefix="s"%>

<!DOCTYPE HTML PUBLIC "-//W3C//DTD HTML 4.01 Transitional//EN">
<html>
    <head>
    </head>
    <body>
        <s:form action="register" method="post" enctype="multipart/form-data">
            username:<s:textfield name="username"></s:textfield>
            <br />
            <s:file name="uploadFile"></s:file>
            <br />
            <s:file name="uploadFile"></s:file>
            <br />
            <s:file name="uploadFile"></s:file>
            <br />
            <s:submit value="提交"></s:submit>
        </s:form>
    </body>
</html>
```

4.2.2 设计上传的控制层代码

新建 Action 控制层 Register.java 文件，代码如下。

```java
package controller;

import java.io.File;
import java.io.IOException;
import java.text.SimpleDateFormat;
import java.util.Date;

import org.apache.commons.io.FileUtils;
import org.apache.struts2.ServletActionContext;

import com.opensymphony.xwork2.ActionSupport;

public class Register extends ActionSupport {

    private String username;

    private File uploadFile[];// 是 File 的数组类型

    private String uploadFileFileName[];// 由于上传多文件，所以文件名也是数组类型

    @Override
```

```java
public void validate() {

}

public String getUsername() {
    return username;
}

public void setUsername(String username) {
    this.username = username;
}

public File[] getUploadFile() {
    return uploadFile;
}

public void setUploadFile(File[] uploadFile) {
    this.uploadFile = uploadFile;
}

public String[] getUploadFileFileName() {
    return uploadFileFileName;
}

public void setUploadFileFileName(String[] uploadFileFileName) {
    this.uploadFileFileName = uploadFileFileName;
}

public String execute() throws IOException {
    System.out.println("username 的值是: " + username);
    // 取得上传后文件所存放的路径
    String targetDirectory = ServletActionContext.getRequest().getRealPath(
            "/upload");
    // 循环得到 File[]数组中的每一个对象,然后分别进行复制 File 对象,实现上传文件
    for (int i = 0; i < uploadFile.length; i++) {
        // 在文件名上加上时间来标识每一个上传的文件
        File target = new File(targetDirectory, new SimpleDateFormat(
                "yyyy_MM_dd_HH_mm_ss").format(new Date()).toString()
                + System.nanoTime() + uploadFileFileName[i]);

        FileUtils.copyFile(uploadFile[i], target);
    }

    return "register";
}

}
```

File 类型的变量 uploadFile 改成了数组类型,然后通过 for 循环再依次取出其中的每 1 个 File 对象,再通过 io 框架包中的 copy 方法进行文件的二进制数据上传。

但由于上传的文件可能与以前上传的文件名重名,这里笔者将文件名包含上传时间来作为文件名的唯一标识,精确到了纳秒级。而代码 ServletActionContext.getRequest().getRealPath("/upload")的功能是取得上传的路径,上传的文件要放到这个目录中。

注意:属性的命名规则如下所示。

```java
private File uploadFile[];
```

```
private String uploadFileFileName[];
```

和单文件上传时的命名规则一样，这里大家要注意。

还要新建一个 struts.properties 文件，内容如图 4-4 所示。

```
1 struts.multipart.maxSize=2048000000
2
3 struts.multipart.saveDir=/tempUploadFile
```

图 4-4 struts.properties 文件内容

注意：一定要在项目中创建一个名称为 upload 的目录！

配置文件 struts.xml 的代码如下。

```xml
<?xml version="1.0" encoding="UTF-8" ?>
<!DOCTYPE struts PUBLIC
    "-//Apache Software Foundation//DTD Struts Configuration 2.0//EN"
    "http://struts.apache.org/dtds/struts-2.0.dtd">

<struts>
    <package name="struts3.2" extends="struts-default">
        <action name="register" class="controller.Register">
            <result name="register">/showregister.jsp</result>
            <result name="input">/register.jsp</result>
        </action>
    </package>

    <constant name="struts.ui.theme" value="simple"></constant>
</struts>
```

4.2.3 成功上传多个文件

部属项目，运行程序，效果如图 4-5 所示。

图 4-5 初始运行效果

4.3 使用属性驱动形式的文件上传

在3个file表单域中依次单击"浏览"按钮，上传3个文件，单击"提交"按钮后的控制台效果如图4-6所示。

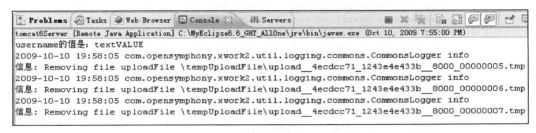

图4-6 控制台运行效果

上传多个文件的同时也取得了单行文本域中的值，并且删除了3个上传临时文件。

在Tomcat目录中也可以看到上传的3个同名文件用上传的时间作为标识以防止重名，效果如图4-7所示。

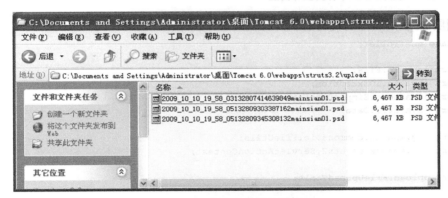

图4-7 上传多个文件的效果

4.3 使用属性驱动形式的文件上传

在前面的两小节中，类型为File的变量都是直接声明在Action中。如果变量较多，则Action中的get和set方法非常臃肿，那么就可以使用属性驱动的形式来将属性用JavaBean进行封装，本节就来实现这样的例子。

4.3.1 创建上传多个文件的JSP

创建一个Web项目struts3.3。

新建register.jsp文件，代码如下。

```
<%@ page language="java" import="java.util.*" pageEncoding="utf-8"%>
<%@ page isELIgnored="false"%>
<%@ taglib uri="/struts-tags" prefix="s"%>

<!DOCTYPE HTML PUBLIC "-//W3C//DTD HTML 4.01 Transitional//EN">
<html>
```

```
<head>
</head>
<body>
    <s:form action="register" method="post" enctype="multipart/form-data">
        username:<s:textfield name="fileUploadTools.username"></s:textfield>
        <br />
        <s:file name="fileUploadTools.uploadFile"></s:file>
        <br />
        <s:file name="fileUploadTools.uploadFile"></s:file>
        <br />
        <s:file name="fileUploadTools.uploadFile"></s:file>
        <br />
        <s:submit value="提交"></s:submit>
    </s:form>
</body>
</html>
```

4.3.2　设计上传文件的控制层

控制层 Register.java 的代码如下。

```java
package controller;

import java.io.File;
import java.io.IOException;
import java.text.SimpleDateFormat;
import java.util.Date;

import org.apache.commons.io.FileUtils;
import org.apache.struts2.ServletActionContext;

import upload.FileUploadTools;

import com.opensymphony.xwork2.ActionSupport;

public class Register {
//声明封装了 File 上传的 FileUploadTools 类的实例
//FileUploadTools 类也封装了上传的属性及 get 和 set 方法
    private FileUploadTools fileUploadTools = new FileUploadTools();

    public FileUploadTools getFileUploadTools() {
        return fileUploadTools;
    }

    public void setFileUploadTools(FileUploadTools fileUploadTools) {
        this.fileUploadTools = fileUploadTools;
    }

    public String execute() throws IOException {

        fileUploadTools.beginUpload();

        return "register";
    }

}
```

4.3 使用属性驱动形式的文件上传

在控制层中加入一个属性，代码如下。

```
private FileUploadTools fileUploadTools = new FileUploadTools();
```

这个 FileUploadTools 类封装的就是上传的表单数据，这样 Action 的代码非常清爽，维护起来也非常方便，set 和 get 的代码量大大减少。下面一起来看一下 FileUploadTools.java 类的代码。

4.3.3 新建上传文件的封装类

FileUploadTools.java 的代码如下。

```java
package upload;

import java.io.File;
import java.io.IOException;
import java.text.SimpleDateFormat;
import java.util.Date;

import org.apache.commons.io.FileUtils;
import org.apache.struts2.ServletActionContext;

public class FileUploadTools {

    private String username;

    private File uploadFile[];// 上传的文件是数组类型

    private String uploadFileFileName[];// 文件名是数组类型

    private String uploadFileContentType[];

    // 上传的 ContentType 文件类型也是数组类型
    // 必须要加上对 ContentType 的声明，不然会报异常
    public String[] getUploadFileContentType() {
        return uploadFileContentType;
    }

    public void setUploadFileContentType(String[] uploadFileContentType) {
        this.uploadFileContentType = uploadFileContentType;
    }

    public String getUsername() {
        return username;
    }

    public void setUsername(String username) {
        this.username = username;
    }

    public File[] getUploadFile() {
        return uploadFile;
    }

    public void setUploadFile(File[] uploadFile) {
        this.uploadFile = uploadFile;
    }
```

```java
public String[] getUploadFileFileName() {
    return uploadFileFileName;
}
public void setUploadFileFileName(String[] uploadFileFileName) {
    this.uploadFileFileName = uploadFileFileName;
}
public void beginUpload() throws IOException {
    System.out.println("username 的值是： " + username);

    String targetDirectory = ServletActionContext.getRequest().getRealPath(
            "/upload");

    System.out.println("");
    for (int i = 0; i < uploadFile.length; i++) {

        File target = new File(targetDirectory, new SimpleDateFormat(
                "yyyy_MM_dd_HH_mm_ss").format(new Date()).toString()
                + System.nanoTime() + uploadFileFileName[i]);

        FileUtils.copyFile(uploadFile[i], target);
    }
    System.out.println("");
}
}
```

在 FileUploadTools.java 类中有一个 beginUpload()方法，就是用来上传文件及取得普通表单域的值的。

4.3.4 将 JSP 文件中 s:file 标签的 name 属性进行更改

既然在 Action 中声明了一个封装类，那么细心的朋友也会发现 JSP 文件中的表单 name 属性值也悄然发生了变化，变化后的代码如下。

```
username:<s:textfield name="fileUploadTools.username"></s:textfield>
<br />
<s:file name="fileUploadTools.uploadFile"></s:file>
<br />
<s:file name="fileUploadTools.uploadFile"></s:file>
<br />
<s:file name="fileUploadTools.uploadFile"></s:file>
<br />
```

主要的变化就是 name 的属性值加入了 fileUploadTools.，这个值要与 Action 的属性名一一对应，而且必须要一致，Action 中的属性代码如下。
```
private FileUploadTools fileUploadTools = new FileUploadTools();
```

所以要写成 fileUploadTools.uploadFile 这样的格式。
还要新建一个 struts.properties 文件，内容如图 4-8 所示。

4.3 使用属性驱动形式的文件上传

```
struts.properties ⊠
1 struts.multipart.maxSize=2048000000
2
3 struts.multipart.saveDir=/tempUploadFile
```

图 4-8 struts.properties 文件内容

注意：一定要在项目中创建一个名为 upload 的目录！

struts.xml 配置文件代码如下。

```xml
<?xml version="1.0" encoding="UTF-8" ?>
<!DOCTYPE struts PUBLIC
    "-//Apache Software Foundation//DTD Struts Configuration 2.0//EN"
    "http://struts.apache.org/dtds/struts-2.0.dtd">

<struts>
    <package name="struts3.1" extends="struts-default">
        <action name="register" class="controller.Register">
            <result name="register">/showregister.jsp</result>
            <result name="input">/register.jsp</result>
        </action>
    </package>

    <constant name="struts.ui.theme" value="simple"></constant>
</struts>
```

4.3.5 以属性驱动方式成功上传多个文件

部属项目，运行程序，效果如图 4-9 所示。

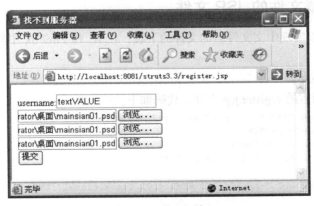

图 4-9 初始运行效果

单击"提交"按钮后的控制台变化如图 4-10 所示。

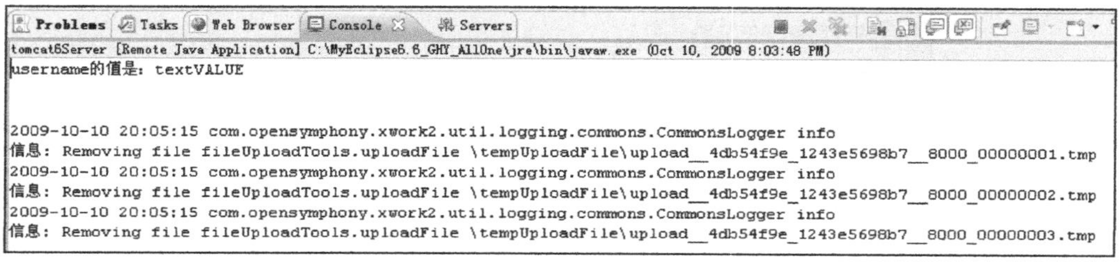

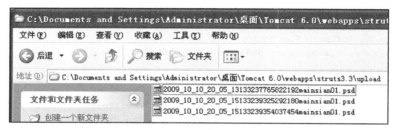

图 4-10　控制台的变化

在 Tomcat 上传的文件如图 4-11 所示。

图 4-11　上传后的文件

基于属性驱动的上传模型非常有利于代码的整洁及良好的后期维护性。另外建议大家在开发项目时，应该尽量多使用属性驱动形式，不管是普通的注册功能还是上传功能。

4.4　用 Struts 2 实现下载文件的功能（支持中文文件名与 IE 和 FireFix 兼容）

4.4.1　新建下载文件的 JSP 文件

Struts 2 完全可以实现中文文件名的文件的下载，在 IE 和 FireFox 中程序代码能实现兼容的功能。

新建 Web 项目 z36。

新建下载文件列表的 register.jsp 文件，代码如下。

```
<%@ page language="java" import="java.util.*,java.net.*"
    pageEncoding="utf-8"%>
<%@ page isELIgnored="false"%>
<%@ taglib uri="/struts-tags" prefix="s"%>

<!DOCTYPE HTML PUBLIC "-//W3C//DTD HTML 4.01 Transitional//EN">
<html>
    <head>
    </head>
    <body>
        <a
            href="download.action?fileName=<%=java.net.URLEncoder.encode("测试用的 RAR.rar", "utf-8")
                .toString().replace("%", "_")%>">测试用的 RAR.rar</a>
        <br />
```

```html
            <a href="download.action?fileName=abc.rar">abc.rar</a>
            <br />
        </body>
</html>
```

4.4.2 新建下载文件的控制层文件

新建 Action 控制层类 Download.java 文件，代码如下。

```java
package controller;

import java.io.File;
import java.io.FileInputStream;
import java.io.IOException;
import java.net.URLEncoder;

import javax.servlet.ServletOutputStream;
import javax.servlet.http.HttpServletRequest;
import javax.servlet.http.HttpServletResponse;

import org.apache.commons.io.IOUtils;
import org.apache.struts2.ServletActionContext;

public class Download {

    private String fileName = "";

    public String getFileName() {
        return fileName;
    }

    public void setFileName(String fileName) {
        this.fileName = fileName;
    }

    public String execute() throws IOException {
        String downPath = ServletActionContext.getRequest().getServletContext().getRealPath("/");
        fileName = fileName.replace("_", "%");
        fileName = java.net.URLDecoder.decode(fileName, "utf-8");
        HttpServletRequest request = ServletActionContext.getRequest();
        HttpServletResponse response = ServletActionContext.getResponse();
        String downfileName = "";
        if (request.getHeader("USER-AGENT").toLowerCase().indexOf("msie") > 0) {// IE
            fileName = URLEncoder.encode(fileName, "UTF-8");
            downfileName = fileName.replace("+", "%20");// 处理空格变"+"的问题
        } else {// FF
            downfileName = new String(fileName.getBytes("UTF-8"), "ISO-8859-1");
        }
        System.out.println(downPath + fileName);
        File downloadFile = new File(downPath + fileName);
        response.setContentType("application/octet-stream;");
        response.setHeader("Content-disposition", String.format("attachment; filename=\"%s\"", downfileName)); // 文件名外的双引号处理 firefox 的空格截断问题
        response.setHeader("Content-Length", String.valueOf(downloadFile.length()));
        FileInputStream fis = new FileInputStream(downloadFile);
        ServletOutputStream out = response.getOutputStream();
        IOUtils.copy(fis, out);
```

```
            return null;
        }
    }
```

4.4.3 更改 struts.xml 配置文件

struts.xml 文件的代码如下。

```xml
<?xml version="1.0" encoding="UTF-8" ?>
<!DOCTYPE struts PUBLIC
    "-//Apache Software Foundation//DTD Struts Configuration 2.0//EN"
    "http://struts.apache.org/dtds/struts-2.0.dtd">
<struts>
    <package name="struts3.5" extends="struts-default">
        <action name="download" class="controller.Download">
        </action>
    </package>
</struts>
```

4.4.4 成功下载中文文件名的文件

程序的初始运行效果如图 4-12 所示。

图 4-12 IE 初始运行效果

单击"测试用的 RAR.rar"超链接,弹出"文件下载"窗口,可以看到正确地解析了的中文文件名,如图 4-13 所示。

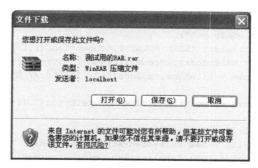

图 4-13 正确地解析了中文文件名

至此用 Struts 2 实现支持中文文件名下载的示例就介绍到这,超链接 abc.rar 也可以正确地下载文件。

第 5 章 JSON、Ajax 和 jQuery 与 Struts 2 联合使用

学习完 Struts 2 框架后可以结合 jQuery 框架中的 Ajax 模块来实现无刷新的效果,那么前台与后台进行数据传递时用什么技术呢？那就是 JSON,它是一种 JavaScript 对象,它可以非常方便地对数据进行封装,而且 Java 也支持对 JSON 代码的解析,所以本章应该掌握如下内容：
- 使用 JavaScript 解析 JSON 字符串；
- 如何将 JavaScript 对象转成 JSON 字符串；
- 如何将 JSON 字符串传递到 Struts 2 的 Action 中；
- 如何在 Struts 2 的 Action 中解析 JSON 字符串；
- 在 IE 中解析从 Action 返回的各种格式的 JSON 对象；
- 使用基于 jQuery+JSON+Struts 2 进行联合开发。

5.1 JSON 介绍

JavaScript 语言首要的目的是为 Web 浏览器提供一种页面脚本语言,用来控制 Web 浏览器中的 DOM 对象,虽然 JavaScript 发展多年,但仍被普遍认为是 Java 的一个子集,而事实并非如此。它是一种语法类似 C 语言和 Java,并且支持面向对象的 Scheme 语言。JavaScript 使用了 ECMAScript 语言规范第三版进行了标准化。所以 JavaScript 只是与 C 及 Java 语法相似。

什么是 JSON

JSON 是 JavaScript 面向对象语法的一个子集。由于 JSON 是 JavaScript 的一个子集,因此它可清晰地运用于此语言中。JSON（JavaScript Object Notation）是一个轻量级数据交换格式,程序员可以非常容易地写出符合 JSON 格式的字符串,而且在各种编程语言中都有相应的类库对 JSON 的对象进行解析和生成。JSON 是完全独立的语言,它使用标准的语法格式,来与其他各种编程语言进行数据交换。

JSON 主要创建两种数据对象：
- 由 JSON 格式字符串创建转换成 JavaScript 的 object 对象；
- 由 JSON 格式字符串创建转换成 JavaScript 的 List 或数组链表对象。

5.2 用 JSON 创建对象

5.2.1 JSON 创建对象的语法格式

用 JSON 创建对象的格式非常简单，格式如图 5-1 所示。

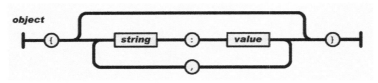

图 5-1　JSON 创建对象格式

从图 5-1 中可以看到，用 JSON 创建对象使用键值对的形式，中间用逗号进行分隔，但要在大括号里面进行定义。

5.2.2 在 JSP 中用 JSON 创建一个对象

下面的程序就是用这种格式来创建对象，并且显示出对象的属性。

```
<!DOCTYPE HTML PUBLIC "-//W3C//DTD HTML 4.01 Transitional//EN">
<html>
    <head>
        <meta http-equiv="Content-Type" content="text/html; charset=utf-8">
        <script>
            function test(){
                var myJSONObject = {
                    "name": "高洪岩",
                    "age": 10,
                    "address": "北京东城"
                };
                alert(myJSONObject.name + " " + (myJSONObject.age + 10) + " " + myJSONObject.address);
            }
        </script>
    </head>
    <body>
        <input type="button" value="test" onclick="test()"/>
    </body>
</html>
```

5.2.3 运行效果

单击 "button" 按钮，调用 test()函数，弹出对话框，效果如图 5-2 所示。

需要注意的是，在本段代码中，必须要加上代码<meta http-equiv="Content-Type" content="text/html; charset=utf-8">来定义编码格式。因为本示例属性值有中文，这种情况也适

图 5-2　显示对象属性

用于通过 Ajax 远程返回 JSON 字符串。

5.3 用 JSON 创建字符串的限制

JSON 是一种通用的字符串格式语言，它也有一些使用上的注意事项。JSON 规定除了字符 """ "\" "/" 和一些控制符（\b、\f、\n、\r、\t）等需要编码外，其他 Unicode 字符可以直接输出。

5.3.1 需要转义的特殊字符

需要转义的特殊字符如图 5-3 所示。

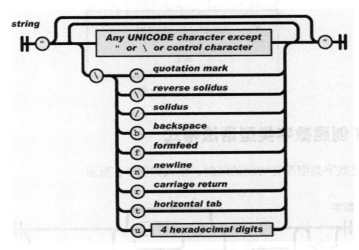

图 5-3　字符串需要值注意事项

如果字符串中有双引号、\、/、b、f、n、r、t 这些字符，那么就需要进行编码。

5.3.2 在 JSP 中对 JSON 特殊字符进行转义

下面实现一个编码转义的示例，代码如下。

```
<!DOCTYPE HTML PUBLIC "-//W3C//DTD HTML 4.01 Transitional//EN">
<html>
    <head>
        <meta http-equiv="Content-Type" content="text/html; charset=utf-8">
        <script>
            function test(){
                var myJSONObject = {
                    "name": "高洪岩\"\\\/",
                    "age": 10,
                    "address": "北京东城"
                };
                alert(myJSONObject.name + " " + (myJSONObject.age + 10) + " " + myJSONObject.address);
            }
        </script>
```

```
        </head>
    <body>
        <input type="button" value="test" onclick="test()"/>
    </body>
</html>
```

name 属性的值是：高洪岩\"\\\/，通过使用\符号对双引"，\和/符号进行转义，从而变成普通的字符串。

5.3.3 运行效果

运行效果如图 5-4 所示。

图 5-4 运行效果

5.4 用 JSON 创建数字类型语法格式

用 JSON 创建数字类型没有特别的限制，格式如图 5-5 所示。

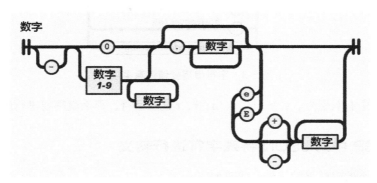

图 5-5 创建数字类型的格式

JSON 创建的数字支持整数,小数和科学数字的格式,例如 12(整数)、12.3(小数)和-23.3e10（科学数字格式）。

5.4.1 在 JSP 中用 JSON 创建数字类型

下面就用 JSON 创建数字，代码如下。
```
<!DOCTYPE HTML PUBLIC "-//W3C//DTD HTML 4.01 Transitional//EN">
<html>
    <head>
        <meta http-equiv="Content-Type" content="text/html; charset=utf-8">
```

```
            <script>
                function test(){
                    var myJSONObject = {
                        "type1": 12,
                        "type2": 10.4,
                        "type3": -12,
                        "type4": -10.4,
                        "type5": -30.4e2,
                        "type6": 30.4e2,
                        "type7": 30.4e-1
                    };
                    var resultValue = myJSONObject.type1 + "\n " + myJSONObject.type2 + "\n 
" + myJSONObject.type3 + "\n " + myJSONObject.type4 + "\n " + myJSONObject.type5 + "\n " + 
myJSONObject.type6 + "\n " + myJSONObject.type7;
                    alert(resultValue);
                }
            </script>
    </head>
    <body>
        <input type="button" value="test" onclick="test()"/>
    </body>
</html>
```

5.4.2 运行效果

单击"test"按钮后的运行效果如图 5-6 所示。

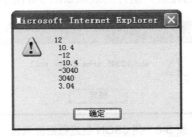

图 5-6　用 JSON 创建数字运行效果

5.5　用 JSON 创建数组对象的语法格式

用 JSON 创建数组对象的语法格式非常简单，格式如图 5-7 所示。

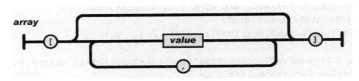

图 5-7　用 JSON 创建数组对象的语法格式

创建数组对象的格式是在方括号中写入每一个元素的值，如果是多个元素，就用逗号进行分隔。注意事项是不要忘写方括号[]。

如果是 Boolean 类型就用 true 或 false 表示。此外，JavaScript 中的 null 被表示为 null，注

意：true、false 和 null 都没有双引号，否则将被视为一个 String 类型。

5.5.1 JSON 创建一个数组对象

下面就用 JSON 创建一个数组对象，代码如下。

```
<!DOCTYPE HTML PUBLIC "-//W3C//DTD HTML 4.01 Transitional//EN">
<html>
    <head>
        <meta http-equiv="Content-Type" content="text/html; charset=utf-8">
        <script>
            function test(){
                var myJSONObject = ["abc", 12345, true, false, null];
                alert(myJSONObject[0] + " " + myJSONObject[1] + " " + myJSONObject[2] + " " + myJSONObject[3] + " " + myJSONObject[4]);
            }
        </script>
    </head>
    <body>
        <input type="button" value="test" onclick="test()"/>
    </body>
</html>
```

5.5.2 运行效果

通过数组的下标访问对应的元素，运行的效果如图 5-8 所示。

图 5-8　用 JSON 创建数组并取值

5.6 用 JSON 创建嵌套的对象类型

JSON 可以创建嵌套的对象类型，包括对象类型和数组类型。下面的示例就演示了这个功能，代码如下。

```
<!DOCTYPE HTML PUBLIC "-//W3C//DTD HTML 4.01 Transitional//EN"
"http://www.w3.org/TR/html4/loose.dtd">
<html xmlns="http://www.w3.org/1999/xhtml">
    <head>
        <meta http-equiv="Content-Type" content="text/html; charset=utf-8" />
        <title>New Web Project</title>
        <script>
            function testMethod(){
                var userinfo = {
                    "myArray": [{
                        "username1": "usernameValue11"
                    }, {
                        "username2": "usernameValue22"
```

5.6 用 JSON 创建嵌套的对象类型

```
            }, ["abc", 123, true, [123, 456]]],
        "myObject": {
            "username": "大中国"
        },
        "myObject1": {
            "address": [{
                "name": "name1"
            }, {
                "name": "name2"
            }]
        },
    };

    document.writeln(userinfo.myArray[0].username1 + "<br/>");
    document.writeln(userinfo.myArray[1].username2 + "<br/>");
    document.write(userinfo.myArray[2][0] + "<br/>");
    document.write(userinfo.myArray[2][1] + "<br/>");
    document.write(userinfo.myArray[2][2] + "<br/>");
    document.write(userinfo.myArray[2][3][0] + "<br/>");
    document.write(userinfo.myArray[2][3][1] + "<br/>");
    document.write(userinfo.myObject.username + "<br/>");
    document.write(userinfo.myObject1.address[0].name + "<br/>");
    document.write(userinfo.myObject1.address[1].name + "<br/>");
    }
    </script>
</head>
<body>
    <input type="button" onclick="javascript:testMethod()">
</body>
</html>
```

运行效果

单击 "test" 按钮后，运行效果如图 5-9 所示。

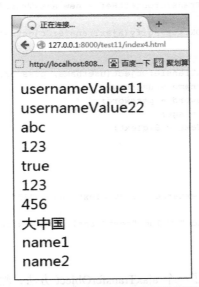

图 5-9　在嵌套的情况下显示对象中的属性及数组中元素值

在本示例中嵌套的情况为对象及数组，使用 JSON 创建对象和取得对象的值非常方便。这也就是现阶段的开发中，XML 逐步被 JSON 替换掉的原因。

5.7 将对象转成 JSON 字符串

5.7.1 什么情况下需要将对象转成 JSON 字符串

在什么样的情况下需要将 JavaScript 对象转成 JSON 字符串？在大多数的场景下是在联合应用 Ajax 技术时。

在常规的技术操作下，如果需要将多个来源的数据通过 Ajax 的方式传到服务器端，就需要在 Web 客户端将不同来源的数据使用 "+" 加号进行拼接，形成完整的字符串，然后再进行传递。这样拼接字符串的方式很明显不具有 OOP 特性。而且如果数据格式复杂，数据来源多，拼接的字符串也容易出现错误，这时就可以使用 json.org 网站提供的 json2.js 框架文件来将 javascript 的对象转换成 JSON 字符串，转换的方式非常简单方便

5.7.2 使用 stringify 方法将对象转成 JSON 字符串

在 JPS 中，可以把对象转换成 JSON 字符串。下面就是实现代码。

```html
<!DOCTYPE HTML PUBLIC "-//W3C//DTD HTML 4.01 Transitional//EN">
<html>
    <head>
        <meta http-equiv="Content-Type" content="text/html; charset=utf-8">
        <script type="text/javascript" src="json2.js">
        </script>
        <script>
            function ajaxTransferText(){
                var BigText = document.getElementById("BigText").value;
                var ajaxTransferObjectRef = new ajaxTransferObject("高洪岩", "高洪岩密码", 10, BigText);
                alert(JSON.stringify(ajaxTransferObjectRef));
            }

            function ajaxTransferObject(username, password, age, BigText){
                this.username = username;
                this.password = password;
                this.age = age;
                this.BigText = BigText;
            }
        </script>
    </head>
    <body>
        <textarea name="textarea" id="BigText" cols="45" rows="5"></textarea>
        <br>
        <input type="button" value="test" onclick="ajaxTransferText()"/>
    </body>
</html>
```

从程序中看到，首先声明了一个 ajaxTransferObject 方法，然后在 ajaxTransferText 方法中进行实例化，最后再通过 JSON 对象中的 stringify 方法将对象 ajaxTransferObject 转换成了字符

串。单击"test"按钮后弹出的警告对话框如图 5-10 所示。

图 5-10　alert 弹出的效果

在 alert 中的字符串显示 textarea 标签中的文本都自动通过 JSON 对象转义了。要想使用 JSON 对象，必须导入 json2.js 文件，代码如下。

```
<script type="text/javascript" src="json2.js">
```

以上示例代码在项目 5_test 中。

5.8　将对象转成 JSON 字符串提交到 Action 并解析（以 post 方式提交）

如果想在服务器端解析传递过来的 JSON 字符串，那么就需要针对 Java 语言的 JSON 解析包。这个功能由 json-lib-2.3-jdk15.jar 包提供，但这个包需要依赖其他的几个 jar 包文件，不然会出现莫名其妙的错误，比如简单的 new JSONObject()对象的代码都会出错。这几个 jar 文件列表如图 5-11 所示。

本节就来实现一个由 JSON+Ajax+Struts 2 联合开发解析由客户端传递过来的 JSON 字符串的示例，Ajax 提交方式是 POST。

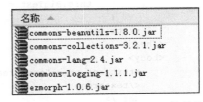

图 5-11　json-lib-2.3-jdk15.jar 依赖的 jar 文件列表

5.8.1　在 JSP 中创建 JSON 和 Ajax 对象

新建 Web 项目 struts 5.7。
新建 JSP 文件并且设计 JavaScript 的 Ajax 对象，代码如下。

```
<%@ page language="java" import="java.util.*" pageEncoding="utf-8"%>
<!DOCTYPE HTML PUBLIC "-//W3C//DTD HTML 4.01 Transitional//EN">
<html>
    <head>
        <meta http-equiv="Content-Type" content="text/html; charset=utf-8">
        <script type="text/javascript" src="json2.js">
        </script>
        <script type="text/javascript">
            var myAjaxObject;

            function ajaxTransferText(){
```

```
                var BigText = document.getElementById("BigText").value;
                var ajaxTransferObjectRef = new ajaxTransferObject("username",
"password", 10, BigText);
                var JSONString = JSON.stringify(ajaxTransferObjectRef);

                if (window.ActiveXObject) {
                    myAjaxObject = new ActiveXObject("Microsoft.XMLHTTP");
                }
                else {
                    myAjaxObject = new XMLHttpRequest();
                }

                var urlString = "jsonString=" + JSONString;

                alert(urlString);
                myAjaxObject.open("POST", "getJSON.action"+"?timestamp=" + new
Date().getTime(), true);
                myAjaxObject.setRequestHeader("Content-Type",
"application/x-www-form-urlencoded");

                myAjaxObject.send(urlString);
            }

            function ajaxTransferObject(username, password, age, BigText){
                this.username = username;
                this.password = password;
                this.age = age;
                this.BigText = BigText;
            }
        </script>
    </head>
    <body>
        <textarea name="textarea" id="BigText" cols="45" rows="5">
        </textarea>
        <br />
        <input type="button" value="test" onclick="ajaxTransferText()" />
    </body>
</html>
```

在程序中导入了 JavaScript 文件。

```
<script type="text/javascript" src="json2.js">
```

它的主要作用就是将 JavaScript 的对象转成 JSON 的字符串,很简单地使用 JSON.stringify() 方法就可以了。然后代码中又创建了 Ajax 对象,数据提交的方式是 post。需要说明的是,使用支持 ECMA262 规范的浏览器内置了 JSON 对象,不需要导入外部的 json2.js 文件。

上面使用的代码写法是原生 Ajax API,原生 Ajax API 只有这一种写法,它可以和任何后台编程技术进行数据交互,属于只写一种 Ajax API 代码,更改一下 Ajax 提交路径就可以将数据提交到不同语言的后台中。

5.8.2 用 Action 控制层接收通过 Ajax 传递过来的 JSON 字符串

设计取得 JSON 字符串转成对象的 Action 对象 GetJSON.java 文件,代码如下。

5.8 将对象转成 JSON 字符串提交到 Action 并解析（以 post 方式提交）

```java
package controller;

import java.io.IOException;

import net.sf.json.JSONObject;

import com.opensymphony.xwork2.ActionSupport;

public class GetJSON extends ActionSupport {

    private String jsonString;

    public String getJsonString() {
        return jsonString;
    }

    public void setJsonString(String jsonString) {
        this.jsonString = jsonString;
    }

    public String execute() {
        System.out.println(jsonString);
        JSONObject json = JSONObject.fromObject(jsonString);
        System.out.println("username=" + json.get("username"));
        System.out.println("username=" + json.get("password"));
        System.out.println("username=" + json.get("age"));
        System.out.println("username=" + json.get("BigText"));

        return null;
    }
}
```

从前台接收传过来的 JSON 字符串后，使用 JSONObject 对象将这个字符串转成对象格式，再通过 get 方法取出属性对应的值，并且将其输出。

5.8.3 运行效果

程序初始运行结果如图 5-12 所示。

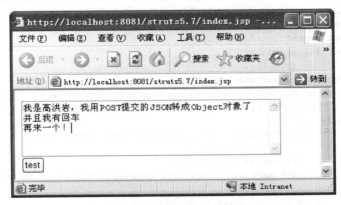

图 5-12　程序初始运行结果

单击"test"按钮就可以看到转成 JSON 字符串的样式,结果如图 5-13 所示。

图 5-13 转成 JSON 的字符串

5.8.4 在控制台输出的数据

单击"确定"按钮后继续提交到 Action,控制台输出解析后的属性值,结果如图 5-14 所示。

图 5-14 解析后的结果

从这个示例来看,由 JSON 转成对象的方法非常简单,前台需要使用 json2.js 文件,后台导入那几个有关联的 JSON 的 jar 包即可。

5.9 将对象转成 JSON 字符串提交到 Action 并解析(以 get 方式提交)

5.9.1 新建创建 JSON 字符串的 JSP 文件

新建 Web 项目 struts5.8,新建 JSP 文件,代码如下。

```
<%@ page language="java" import="java.util.*" pageEncoding="utf-8" %>
 <!DOCTYPE HTML PUBLIC "-//W3C//DTD HTML 4.01 Transitional//EN">
<html>
    <head>
        <meta http-equiv="Content-Type" content="text/html; charset=utf-8">
        <script type="text/javascript" src="json2.js">
        </script>
        <script type="text/javascript">
            var myAjaxObject;

            function ajaxTransferText(){
                var BigText = document.getElementById("BigText").value;
                var ajaxTransferObjectRef = new ajaxTransferObject("username", "password", 10, BigText);
                var JSONString = JSON.stringify(ajaxTransferObjectRef);

                if (window.ActiveXObject) {
                    myAjaxObject = new ActiveXObject("Microsoft.XMLHTTP");
                }
                else {
                    myAjaxObject = new XMLHttpRequest();
```

5.9 将对象转成 JSON 字符串提交到 Action 并解析(以 get 方式提交)

```
                    }
                    var urlString = "jsonString=" + JSONString + "&timestamp=" + new Date().getTime();
                    alert(urlString);
                    myAjaxObject.open("GET", "getJSON.action?"+urlString, true);
                    myAjaxObject.send(null);
                }

                function ajaxTransferObject(username, password, age, BigText){
                    this.username = username;
                    this.password = password;
                    this.age = age;
                    this.BigText = BigText;
                }
            </script>
    </head>
    <body>
        <textarea name="textarea" id="BigText" cols="45" rows="5">
        </textarea>
        <br/>
        <input type="button" value="test" onclick="ajaxTransferText()"/>
    </body>
</html>
```

在 JSP 文件中，只需要关注 Ajax 的 get 方式提交的代码和 post 提交的代码的区别，get 方式提交的数据要放到请求路径的后面，当作 URL 的参数来传递，而 post 提交则是放在 send() 方法中当作参数进行提交到服务器端。

5.9.2 新建接收 JSON 字符串的 Action 控制层

新建 GetJSON.java 文件，代码如下。

```java
package controller;

import java.io.IOException;
import java.io.UnsupportedEncodingException;

import net.sf.json.JSONObject;

import com.opensymphony.xwork2.ActionSupport;

public class GetJSON extends ActionSupport {

    private String jsonString;

    public String getJsonString() {
        return jsonString;
    }

    public void setJsonString(String jsonString)
            throws UnsupportedEncodingException {
```

```
            this.jsonString = new String(jsonString.getBytes("ISO8859-1"));
        }

        public String execute() {
            System.out.println(jsonString);
            JSONObject json = JSONObject.fromObject(jsonString);
            System.out.println("username=" + json.get("username"));
            System.out.println("username=" + json.get("password"));
            System.out.println("username=" + json.get("age"));
            System.out.println("username=" + json.get("BigText"));

            return null;

        }
    }
```

与 post 方式取值不同的是，get 提交时需要对客户端传递过来的数据进行编码，这样做的主要原因是针对中文的正确显示。

5.9.3 运行结果

程序的初始运行结果如图 5-15 所示。

图 5-15 程序的初始运行结果

单击"test"按钮后弹出转换后的 JSON 字符串，结果如图 5-16 所示。

图 5-16 转换后的 JSON 字符串

5.9.4 在控制台输出的数据

单击"确定"按钮，继续提交到服务器端，在控制台输出一些相关信息，结果如图 5-17 所示。

```
("username":"username","password":"password","age":10,"BigText":"我是高洪岩,我用提交的JSON转成Object对象了\r\n并且我有回车\r\n再来一个!")
username=username
username=password
username=10
username=我是高洪岩,我用GET提交的JSON转成Object对象了
并且我有回车
再来一个!
```

图 5-17　控制台中的信息

5.10　将数组转成 JSON 字符串提交到 Action 并解析(以 get 和 post 方式提交)

前面两个示例介绍的是将对象转成 JSON 字符串传递到服务器端，本节将演示如何将数组转换成 JSON 字符串传递到服务器端再正确地解析出来。

新建 Web 项目 struts5.9，新建 JSP 文件，代码如下。

```jsp
<%@ page language="java" import="java.util.*" pageEncoding="utf-8" %>
<!DOCTYPE HTML PUBLIC "-//W3C//DTD HTML 4.01 Transitional//EN">
<html>
    <head>
        <meta http-equiv="Content-Type" content="text/html; charset=utf-8">
        <script type="text/javascript" src="json2.js">
        </script>
        <script type="text/javascript">
            var myAjaxObject;
            
            function getMethod(){
                var valueArray = new Array();
                valueArray[0] = "高洪岩";
                valueArray[1] = 10;
                valueArray[2] = true;
                valueArray[3] = 100.99;
                
                var JSONString = JSON.stringify(valueArray);
                alert(JSONString);
                if (window.ActiveXObject) {
                    myAjaxObject = new ActiveXObject("Microsoft.XMLHTTP");
                }
                else {
                    myAjaxObject = new XMLHttpRequest();
                }
                
                var urlString = "jsonString=" + JSONString + "&timestamp=" + new Date().getTime();
                
                alert(urlString);
                myAjaxObject.open("GET", "getJSON.action?" + urlString, true);
                
                myAjaxObject.send(null);
                
            }
            
            function postMethod(){
                var valueArray = new Array();
                valueArray[0] = "高洪岩";
                valueArray[1] = 10;
                valueArray[2] = true;
```

```
                valueArray[3] = 100.99;

                var JSONString = JSON.stringify(valueArray);
                alert(JSONString);
                if (window.ActiveXObject) {
                    myAjaxObject = new ActiveXObject("Microsoft.XMLHTTP");
                }
                else {
                    myAjaxObject = new XMLHttpRequest();
                }

                var urlString = "jsonString=" + JSONString + "&timestamp=" + new Date().getTime();

                alert(urlString);
                myAjaxObject.open("POST", "postJSON.action", true);
                myAjaxObject.setRequestHeader("Content-Type", "application/x-www-form-urlencoded");

                myAjaxObject.send(urlString);
            }
        </script>
    </head>
    <body>
        <input type="button" value="GET" onclick="getMethod()"/>
        <br/>
        <input type="button" value="POST" onclick="postMethod()"/>
        <br/>
    </body>
</html>
```

在 JSP 文件中的 JavaScript 脚本代码中声明了两个方法，一个是以 get 提交的方法，另一个是以 post 提交的方法。

5.10.1 在服务器端用 get 方法解析 JSON 字符串

在服务器端创建两个 Action，代码分别如下。

```
package controller;

import java.io.UnsupportedEncodingException;

import net.sf.json.JSONArray;

import com.opensymphony.xwork2.ActionSupport;

public class GetJSON extends ActionSupport {

    private String jsonString;

    public String getJsonString() {
        return jsonString;
    }

    public void setJsonString(String jsonString)
            throws UnsupportedEncodingException {
```

```java
        this.jsonString = new String(jsonString.getBytes("ISO8859-1"));
    }

    public String execute() {
        System.out.println(jsonString);
        JSONArray json = JSONArray.fromObject(jsonString);
        System.out.println(json.get(0));
        System.out.println(json.get(1));
        System.out.println(json.get(2));
        System.out.println(json.get(3));

        return null;

    }
}
```

5.10.2 在服务器端用 post 方法解析 JSON 字符串

在服务器，可以用 post 方法解析 JSON 字符串，代码如下。

```java
package controller;

import java.io.UnsupportedEncodingException;

import net.sf.json.JSONArray;

import com.opensymphony.xwork2.ActionSupport;

public class PostJSON extends ActionSupport {

    private String jsonString;

    public String getJsonString() {
        return jsonString;
    }

    public void setJsonString(String jsonString)
            throws UnsupportedEncodingException {
        this.jsonString = jsonString;
    }

    public String execute() {
        System.out.println(jsonString);
        JSONArray json = JSONArray.fromObject(jsonString);
        System.out.println(json.get(0));
        System.out.println(json.get(1));
        System.out.println(json.get(2));
        System.out.println(json.get(3));

        return null;

    }
}
```

需要注意的是，在控制层 Action 的代码中使用了 JSONArray 类来解析 JSON 数组类型。

5.10.3 运行结果

程序运行的初始结果如图 5-18 所示。

图 5-18　程序运行的初始结果

5.10.4 在控制台输出的数据

依次先单击"GET"按钮，再单击"POST"按钮，在 MyEclipse 的控制台正确地输出解析后的值，结果如图 5-19 所示。

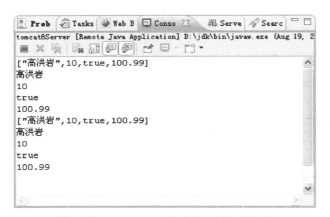

图 5-19　get 和 post 方式提交后解析的值

5.11 使用 Ajax 调用 Action 并生成 JSON 再传递到客户端（以 get 和 post 方式提交）

前面的示例都是在客户端向服务器端传递数据，本章节将介绍如何从服务器端向客户端传递数据。

本章的示例是在浏览器客户端发起一个 Ajax 请求，执行不同的 Action 来返回不同格式的 JSON 字符串，然后在页面中通过 JavaScript 脚本解析出来，本示例 JSP 代码虽然多一些，但主要有 4 个实例：

- 取得 List 中存 String 字符串的 JSON；

5.11 使用 Ajax 调用 Action 并生成 JSON 再传递到客户端(以 get 和 post 方式提交)

- 取得 List 中存 Bean 的 JSON；
- 取得 Map 中存 String 字符串的 JSON；
- 取得 Map 中存 Bean 的 JSON。

5.11.1 新建具有 Ajax 提交功能的 JSP

新建 Web 项目 struts5.10，新建 index.jsp 文件，代码如下。

```jsp
<%@ page language="java" import="java.util.*" pageEncoding="utf-8" %>
<!DOCTYPE HTML PUBLIC "-//W3C//DTD HTML 4.01 Transitional//EN">
<html>
    <head>
        <meta http-equiv="Content-Type" content="text/html; charset=utf-8">
        <script type="text/javascript" src="json2.js">
        </script>
        <script type="text/javascript">
            var myAjaxObject;
            //在 List 中存字符串
            function getListString(){
                if (window.ActiveXObject) {
                    myAjaxObject = new ActiveXObject("Microsoft.XMLHTTP");
                }
                else {
                    myAjaxObject = new XMLHttpRequest();
                }
                myAjaxObject.open("GET", "listString.action?date=" + new Date().getTime(), true);
                myAjaxObject.onreadystatechange = retrunListString;
                myAjaxObject.send();
            }

            function retrunListString(){
                if (myAjaxObject.readyState == 4) {
                    if (myAjaxObject.status == 200) {
                        var returnJSONString = myAjaxObject.responseText;
                        var returnJSON = JSON.parse(returnJSONString);
                        var showString = "";
                        for (var i = 0; i < returnJSON.length; i++) {
                            showString = showString + returnJSON[i] + " ";
                        }
                        alert(showString);
                    }
                }
            }

            //在 List 中存 Bean
            function getListBean(){
                if (window.ActiveXObject) {
                    myAjaxObject = new ActiveXObject("Microsoft.XMLHTTP");
                }
                else {
                    myAjaxObject = new XMLHttpRequest();
                }
```

```javascript
                myAjaxObject.open("GET", "listBean.action?date=" + new Date().getTime(), true);
                myAjaxObject.onreadystatechange = retrunListBean;
                myAjaxObject.send();
            }

            function retrunListBean(){
                if (myAjaxObject.readyState == 4) {
                    if (myAjaxObject.status == 200) {
                        var returnJSONString = myAjaxObject.responseText;
                        var returnJSON = JSON.parse(returnJSONString);
                        var showString = "";
                        for (var i = 0; i < returnJSON.length; i++) {
                            showString = showString + returnJSON[i].username + " " + returnJSON[i].password + " " + returnJSON[i].age + " " + returnJSON[i].createDate + "-----";
                        }
                        alert(showString);
                    }
                }
            }

            //在 Map 中存字符串
            function getMapString(){
                if (window.ActiveXObject) {
                    myAjaxObject = new ActiveXObject("Microsoft.XMLHTTP");
                }
                else {
                    myAjaxObject = new XMLHttpRequest();
                }

                myAjaxObject.open("GET", "mapString.action?date=" + new Date().getTime(), true);
                myAjaxObject.onreadystatechange = retrunMapString;
                myAjaxObject.send();
            }

            function retrunMapString(){
                if (myAjaxObject.readyState == 4) {
                    if (myAjaxObject.status == 200) {
                        var returnJSONString = myAjaxObject.responseText;
                        var returnJSON = JSON.parse(returnJSONString);
                        var showString = "";
                        for (var i in returnJSON[0]) {
                            showString = showString + "key=" + i + " value=" + returnJSON[0][i] + " --- ";
                        }
                        alert(showString);
                    }
                }
            }

            //在 Map 中存 Bean
            function getMapBean(){
                if (window.ActiveXObject) {
                    myAjaxObject = new ActiveXObject("Microsoft.XMLHTTP");
                }
                else {
```

```
                    myAjaxObject = new XMLHttpRequest();
                }
                myAjaxObject.open("GET", "mapBean.action?date=" + new Date().getTime(), true);
                myAjaxObject.onreadystatechange = retrunMapBean;
                myAjaxObject.send();

            }
            function retrunMapBean(){
                if (myAjaxObject.readyState == 4) {
                    if (myAjaxObject.status == 200) {
                        var returnJSONString = myAjaxObject.responseText;
                        var returnJSON = JSON.parse(returnJSONString);
                        var showString = "";
                        for (var i in returnJSON[0]) {
                            showString = showString + "key=" + i + " username=" + returnJSON[0][i].username + " password=" + returnJSON[0][i].password + " age=" + returnJSON[0][i].age + " createDate=" + returnJSON[0][i].createDate + "-----";
                        }
                        alert(showString);
                    }
                }
            }
        </script>
    </head>
    <body>
        <input type="button" value="返回 List 中存 String 类型的 JSON" onclick="getListString()"/>
        <br/>
        <input type="button" value="返回 List 中存 Bean 类型的 JSON" onclick="getListBean()"/>
        <br/>
        <input type="button" value="返回 Map 中存 String 类型的 JSON" onclick="getMapString()"/>
        <br/>
        <input type="button" value="返回 Map 中存 Bean 类型的 JSON" onclick="getMapBean()"/>
        <br/>
    </body>
</html>
```

5.11.2 在 Action 控制层创建 List 中存 String

新建 listString.java 文件，代码如下。

```java
package controller;

import java.io.IOException;
import java.io.PrintWriter;
import java.util.ArrayList;
import java.util.List;

import net.sf.json.JSONArray;

import org.apache.struts2.ServletActionContext;

import com.opensymphony.xwork2.ActionSupport;
```

```java
public class listString extends ActionSupport {

    public String execute() throws IOException {

        List listString = new ArrayList();
        listString.add("高洪岩1");
        listString.add("高洪岩2");
        listString.add("高洪岩3");
        listString.add("高洪岩4");

        JSONArray json = JSONArray.fromObject(listString);
        System.out.println(json.toString());

        ServletActionContext.getResponse().setContentType("text/html");
        ServletActionContext.getResponse().setCharacterEncoding("utf-8");
        ServletActionContext.getResponse().getWriter().print(json.toString());
        ServletActionContext.getResponse().getWriter().flush();
        ServletActionContext.getResponse().getWriter().close();

        return null;

    }
}
```

该类实现的功能是在 List 中存放字符串，然后再传给客户端。

5.11.3 在 Action 控制层创建 List 中存 Bean

新建 listBean.java 文件，代码如下。

```java
package controller;

import java.io.IOException;
import java.text.SimpleDateFormat;
import java.util.ArrayList;
import java.util.Date;
import java.util.List;

import net.sf.json.JSONArray;

import org.apache.struts2.ServletActionContext;

import com.opensymphony.xwork2.ActionSupport;

import entity.UserInfo;

public class listBean extends ActionSupport {

    public String execute() throws IOException {

        List listBean = new ArrayList();

        UserInfo UserInfo1 = new UserInfo();
        UserInfo1.setUsername("高洪岩1");
        UserInfo1.setPassword("密码1");
        UserInfo1.setAge(10);
        UserInfo1.setCreateDate(new SimpleDateFormat("yyyy-MM-dd hh-mm-ss")
```

```java
                .format(new Date()));

        UserInfo UserInfo2 = new UserInfo();
        UserInfo2.setUsername("高洪岩2");
        UserInfo2.setPassword("密码2");
        UserInfo2.setAge(10);
        UserInfo2.setCreateDate(new SimpleDateFormat("yyyy-MM-dd hh-mm-ss")
                .format(new Date()));

        listBean.add(UserInfo1);
        listBean.add(UserInfo2);

        JSONArray json = JSONArray.fromObject(listBean);
        System.out.println(json.toString());

        ServletActionContext.getResponse().setContentType("text/html");
        ServletActionContext.getResponse().setCharacterEncoding("utf-8");
        ServletActionContext.getResponse().getWriter().print(json.toString());
        ServletActionContext.getResponse().getWriter().flush();
        ServletActionContext.getResponse().getWriter().close();

        return null;

    }
}
```

该类实现的功能是在 List 中存 Bean，然后再传给客户端。

5.11.4 在 Action 控制层创建 Map 中存放的 String

新建 mapString.java 文件，代码如下。

```java
package controller;

import java.io.IOException;
import java.util.LinkedHashMap;

import net.sf.json.JSONArray;

import org.apache.struts2.ServletActionContext;

import com.opensymphony.xwork2.ActionSupport;

public class mapString extends ActionSupport {

    public String execute() throws IOException {

        LinkedHashMap mapString = new LinkedHashMap();
        mapString.put("1", "高洪岩1");
        mapString.put("2", "高洪岩2");
        mapString.put("3", "高洪岩3");
        mapString.put("4", "高洪岩4");
        mapString.put("5", "高洪岩5");

        JSONArray json = JSONArray.fromObject(mapString);
        System.out.println(json.toString());
```

```
ServletActionContext.getResponse().setContentType("text/html");
ServletActionContext.getResponse().setCharacterEncoding("utf-8");
ServletActionContext.getResponse().getWriter().print(json.toString());
ServletActionContext.getResponse().getWriter().flush();
ServletActionContext.getResponse().getWriter().close();

            return null;

    }
}
```

该类实现的功能是在 map 中存放 String，然后再传给客户端。

5.11.5 在 Action 控制层创建 Map 中存放的 Bean

新建 mapBean.java 文件，代码如下。

```
package controller;

import java.io.IOException;
import java.text.SimpleDateFormat;
import java.util.Date;
import java.util.LinkedHashMap;

import net.sf.json.JSONArray;

import org.apache.struts2.ServletActionContext;

import com.opensymphony.xwork2.ActionSupport;

import entity.UserInfo;

public class mapBean extends ActionSupport {

    public String execute() throws IOException {

        LinkedHashMap mapString = new LinkedHashMap();

        UserInfo UserInfo1 = new UserInfo();
        UserInfo1.setUsername("高洪岩 1");
        UserInfo1.setPassword("密码 1");
        UserInfo1.setAge(10);
        UserInfo1.setCreateDate(new SimpleDateFormat("yyyy-MM-dd hh-mm-ss")
                .format(new Date()));

        UserInfo UserInfo2 = new UserInfo();
        UserInfo2.setUsername("高洪岩 2");
        UserInfo2.setPassword("密码 2");
        UserInfo2.setAge(10);
        UserInfo2.setCreateDate(new SimpleDateFormat("yyyy-MM-dd hh-mm-ss")
                .format(new Date()));

        mapString.put("1", UserInfo1);
        mapString.put("2", UserInfo2);

        JSONArray json = JSONArray.fromObject(mapString);
        System.out.println(json.toString());
```

5.11 使用 Ajax 调用 Action 并生成 JSON 再传递到客户端(以 get 和 post 方式提交) 217

```
ServletActionContext.getResponse().setContentType("text/html");
ServletActionContext.getResponse().setCharacterEncoding("utf-8");
ServletActionContext.getResponse().getWriter().print(json.toString());
ServletActionContext.getResponse().getWriter().flush();
ServletActionContext.getResponse().getWriter().close();

return null;
    }
}
```

该类实现的功能是在 map 中存 Bean，然后再传给客户端。

5.11.6 单击不同的 button 按钮调用不同的 Action

初始运行结果如图 5-20 所示。

从上到下依次单击"button"按钮，alert()弹出的界面依次如图 5-21～图 5-24 所示。

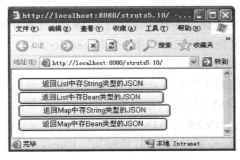
图 5-20 初始运行效果

图 5-21 返回 List 中存 String 类型的 JSON

对于图 5-21，控制台输出的 JSON 字符串如下。
["高洪岩1","高洪岩2","高洪岩3","高洪岩4"]

图 5-22 返回 List 中存 Bean 类型的 JSON

对于图 5-22，控制台输出的 JSON 字符串如下。
[{"age":10,"createDate":"2009-08-19 06-57-03","password":"密码1","username":"高洪岩1"},{"age":10,"createDate":"2009-08-19 06-57-03","password":"密码2","username":"高洪岩2"}]

图 5-23 返回 Map 中存 String 类型的 JSON

对于图 5-23，控制台输出的 JSON 字符串如下。

[{"1":"高洪岩1","2":"高洪岩2","3":"高洪岩3","4":"高洪岩4","5":"高洪岩5"}]

图 5-24　返回 Map 中存 Bean 类型的 JSON

对于图 5-24，控制台输出的 JSON 字符串如下。

[{"1":{"age":10,"createDate":"2009-08-19 06-57-38","password":"密码1","username":"高洪岩1"},"2":{"age":10,"createDate":"2009-08-19 06-57-38","password":"密码2","username":"高洪岩2"}}]

本示例已经涵盖了常用的 Ajax+JSON 使用的情况，JSON 的速度要比 XML 快很多。

5.12　jQuery、JSON 和 Struts 2

jQuery 是一个 JavaScript 语言的开源框架，它提供了非常全面的针对 DOM 对象方便操作的 API，并且在 jQuery 框架的基础上也提供了其他的第三方插件，比如 cookie 插件、灯箱插件、广告栏等实用的 Web 开发控件。它使用类似 SQL 语言的"select 语句"来选择想要操作的 DOM 对象，在 jQuery 中叫"选择器"，使用起来也非常方便，并且支持数组和列表对象上，提供了一些方便的 API 来供程序员调用，省略了以前大量写 for 语句的代码量。

需要说明的是，既然本章要讲解的是 jQuery、JSON 与 AJAX 进行联合开发，jQuery 基础的技能点不是本章的重点，所以读者需要掌握 jQuery 基础知识的，但还好，jQuery 有中文版的帮助文档。

5.12.1　jQuery 框架的 Ajax 功能介绍

在 jQuery 框架中，有针对 Ajax 功能调用的函数，jQuery 的全部 Ajax 功能函数如图 5-25 所示。

图 5-25　jQuery 框架的 Ajax 方法

在这些函数中比较常用的就是 jQuery.get (url, [data], [callback], [type])、jQuery.getJSON (url, [data], [callback])、jQuery.getScript (url, [callback])和 jQuery.post (url, [data], [callback], [type])。

通过方法就可以看到，在 jQuery 中仅仅通过调用一个方法就可以实现 Ajax 的功能，代码量非常少，而且使用起来也非常方便，后面的章节就为大家介绍这些内容。

5.12.2 用 jQuery 的 Ajax 功能调用远程 action（无返回结果）

本小节就实现一个使用 jQuery 框架的 Ajax 功能调用远程的 action，无返回结果的例子。

首先创建一个 Web 项目 struts7.3，项目结构如图 5-26 所示。

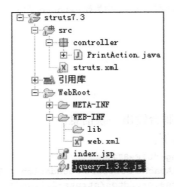

图 5-26 项目结构

在 WebRoot 目录中存放 jQuery 的框架 JavaScript 文件。

index.jsp 文件的代码如下。

```jsp
<%@ page language="java" import="java.util.*" pageEncoding="utf-8"%>
<!DOCTYPE HTML PUBLIC "-//W3C//DTD HTML 4.01 Transitional//EN">
<html>
    <head>
        <script type="text/javascript" src="jquery-1.3.2.js">
        </script>
        <script>
        function callAction()
        {
        $.get("printAction.action?t="+new Date().getTime());
        }

        </script>
    </head>
    <body>
        <input type="button" value="click me!" onclick="callAction()" />
    </body>
</html>
```

在 index.jsp 文件中可以看到 3 部分：
- 导入 jquery-1.3.2.js 文件；
- 自定义 JavaScript 函数 callAction()；
- 定义一个按钮去触发这个函数。

从代码中可以看到，callAction()方法中仅仅有一条语句。

```
$.get("printAction.action");
```

没有错！就是这一条语句，就可以实现远程调用 Struts 2 的 action 组件，这比前一章使用原生的 Ajax 对象调用要方便很多，并且不用判断浏览器的兼容性，所有这一切都在 jQuery 框架中进行内部处理。

远程的 Action 类 PrintAction.java 代码如下。

```java
package controller;

public class PrintAction {

    public String execute() {
        System.out.println("在 action 中执行 execute 了");

        return "";

    }

}
```

仅仅在 execute()方法中在控制台输出一条信息。

struts.xml 配置文件的代码如下。

```xml
<?xml version="1.0" encoding="UTF-8"?>
<!DOCTYPE struts PUBLIC "-//Apache Software Foundation//DTD Struts Configuration 2.1//EN"
        "http://struts.apache.org/dtds/struts-2.1.dtd">
<struts>
    <package name="default-package" extends="struts-default">
        <action name="printAction" class="controller.PrintAction">
        </action>
    </package>
</struts>
```

程序运行结果如图 5-27 所示。

图 5-27　正确调用远程 action

我们本章第一个示例，从简单开始，到简单结束。

下面再来分析一下 jQuery 对象的 Ajax 功能方法的结构。

5.12.3　jQuery 的 Ajax 方法结构

其实 jQuery 的 Ajax 方法的结构完全可以从帮助文档中得到。

打开帮助文档，在"jQuery.get(url, [data], [callback], [type])"结点中进行单击就可以看到 get 方法的参数，如图 5-28 所示。

5.12 jQuery、JSON 和 Struts 2

	jQuery.get(url, [data], [callback], [type])　返回值:XMLHttpRequest
概述	通过远程 HTTP GET 请求载入信息。 这是一个简单的 GET 请求功能以取代复杂 $.ajax。请求成功时可调用回调函数。如果需要在出错时执行函数，请使用 $.ajax。
参数	
url	String
	待载入页面的URL地址
data (可选)	Map
	待发送 Key/value 参数。
callback (可选)	Function
	载入成功时回调函数。
type (可选)	String
	返回内容格式，xml, html, script, json, text, default。

图 5-28　Ajax 函数的结构

函数 get 一共有 4 个参数，但第一个参数是必须要有值的，其他 3 个参数可以无值，是可选的状态。第 1 个参数是要 Ajax 调用远程资源的地址。第 2 个参数是将客户端的一些数据发送到服务器端传递的参数。第 3 个参数是回调，也就是将服务器端的数据再传给 IE 时调用的函数。第 4 个参数是返回的内容的类型，包括很常用的 JSON 类型。在上一小节中仅仅使用了第 1 个参数，即$.get("printAction.action?t="+new Date().getTime());。它实现的功能仅仅是调用远程的数据，但没有返回数据。下一小节就实现一个调用远程 action 并且返回数据到 IE，也就是加入了 Ajax 回调机制。

5.12.4　用 jQuery 的 Ajax 功能调用远程 action（有返回结果）

新建 Web 项目 struts7.5，项目结构如图 5-29 所示。

图 5-29　项目结构

JSP 文件 index.jsp 的代码如下。

```
<%@ page language="java" import="java.util.*" pageEncoding="utf-8" %>
 <!DOCTYPE HTML PUBLIC "-//W3C//DTD HTML 4.01 Transitional//EN">
<html>
    <head>
        <script type="text/javascript" src="jquery-1.3.2.js">
```

```
        </script>
        <script>
            function callAction(){
                $.get("printAction.action", function(data){
                    callBackFunction(data)
                });
            }

            function callBackFunction(dataParam){
                alert(dataParam);
            }
        </script>
    </head>
    <body>
        <input type="button" value="click me!" onclick="callAction()"/>
    </body>
</html>
```

代码 get 方法使用了两个参数:
- 第一个参数是 url 的访问地址;
- 第二个参数是一个回调函数,将从 action 回传过来的数据赋值参数 data,然后又将 data 参数传递给了方法 callBackFunction(),并且在 callBackFunction 方法中以警告方式显示这个数据。

PrintAction.java 的代码如下。

```java
package controller;

import java.io.IOException;

import org.apache.struts2.ServletActionContext;

public class PrintAction {

    public String execute() throws IOException {

        System.out.println("在 action 中执行 execute 了");

        ServletActionContext.getResponse().setContentType("text/html");
        ServletActionContext.getResponse().setCharacterEncoding("utf-8");
        ServletActionContext.getResponse().getWriter().print("我是高洪岩,有中文! ");
        ServletActionContext.getResponse().getWriter().flush();
        ServletActionContext.getResponse().getWriter().close();

        return null;

    }

}
```

在程序中通过使用 response 对象将中文的字符串进行输出,以便于前台 IE 的 Ajax 对象能够取到中文字符串,这个代码在上一章中有所体现,很容易理解。

运行结果如图 5-30 所示。

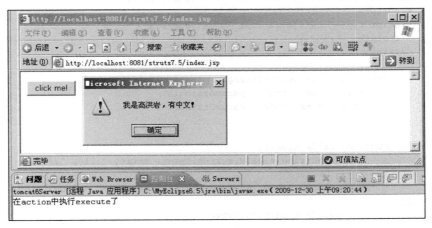

图 5-30 运行结果

从图 5-30 界面中可以看到，单击 IE 界面中的"click me!"按钮，去调用远程的 action，在控制台输出了一条信息，并且又在 IE 客户端弹出一个警告对话框，对话框中的文字正是 action 返回的正确中文字符串。

本示例实现了 jQuery 调用远程 action，并且返回数据的功能。

在本小节代码示例的基础上可以实现传递参数的功能，下一小节就来实现这个功能。

5.12.5 用 jQuery 的 Ajax 功能调用远程 action 并且传递 JSON 格式参数（有返回值）

新建 Web 项目 struts7.6，项目的文件结构如图 5-31 所示。

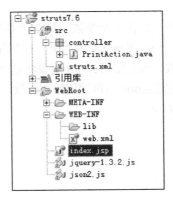

图 5-31 项目文件结构

导入了 2 个 JavaScript 文件：json2.jsp 和 jQuery 的框架文件。

index.jsp 代码如下。

```
<%@ page language="java" import="java.util.*" pageEncoding="utf-8" %>
<!DOCTYPE HTML PUBLIC "-//W3C//DTD HTML 4.01 Transitional//EN">
<html>
```

```
<head>
    <script type="text/javascript" src="jquery-1.3.2.js">
    </script>
    <script type="text/javascript" src="json2.js">
    </script>
    <script>
        function entity(paramString1, paramString2){
            this.paramString1 = paramString1;
            this.paramString2 = paramString2;
        }
        function callAction(){
            var JSONString = JSON.stringify(new entity("111", "222"));
            $.get("printAction.action", "transferString=" + JSONString, function(data){
                callBackFunction(data)
            });
        }
        function callBackFunction(dataParam){
            alert(dataParam);
        }
    </script>
</head>
<body>
    <input type="button" value="click me!" onclick="callAction()"/>
</body>
</html>
```

看一下核心函数，其代码如下。

```
$.get("printAction.action", "transferString=" + JSONString, function(data){
            callBackFunction(data)
        });
```

在上面的代码段中，get 方法一共有 3 个参数。第 2 个参数是传递给 action 的参数，参数的名称是 transferString，值是由一个 JavaScript 对象而来，这个对象的名称就是 entity() 函数对象。entity() 函数对象需要传递两个参数，然后使用 JSON 对象将这个对象转换成 JSON 字符串，然后再传递给远程的 action，最后执行完 action 后再调用回调函数，这就是本示例代码的执行流程。

PrintAction.java 文件的代码如下。

```
package controller;

import java.io.IOException;

import net.sf.json.JSONObject;

import org.apache.struts2.ServletActionContext;

public class PrintAction {

    private String transferString;

    public String getTransferString() {
        return transferString;
```

5.12 jQuery、JSON 和 Struts 2

```java
    }

    public void setTransferString(String transferString) {
        this.transferString = transferString;
    }

    public String execute() throws IOException {
        System.out.println("在 action 中执行 execute 了, transferString 变量值为: "
                + transferString);

        JSONObject jsonobject = JSONObject.fromObject(transferString);
        System.out.println("paramString1 值是: " + jsonobject.get("paramString1"));
        System.out.println("paramString2 值是: " + jsonobject.get("paramString2"));

        ServletActionContext.getResponse().setContentType("text/html");
        ServletActionContext.getResponse().setCharacterEncoding("utf-8");
        ServletActionContext.getResponse().getWriter().printf("我是高洪岩，有中文！");
        ServletActionContext.getResponse().getWriter().flush();
        ServletActionContext.getResponse().getWriter().close();

        return null;
    }
}
```

在 action 代码中首先定义了一个同名的私有属性：transferString，然后声明了属性的 get 和 set 方法，这样从 IE 端传递过来的同名参数就可以自动赋值。

在 execute() 方法中输出这个参数的值，然后使用 JSON 的 Java 类库将这个 JSON 字符串转换成 JSONObject 对象，然后从其属性中进行取值并且输出，代码如下。

```java
JSONObject jsonobject = JSONObject.fromObject(transferString);
System.out.println("paramString1 值是: " + jsonobject.get("paramString1"));
System.out.println("paramString2 值是: " + jsonobject.get("paramString2"));
```

运行结果如图 5-32 所示。

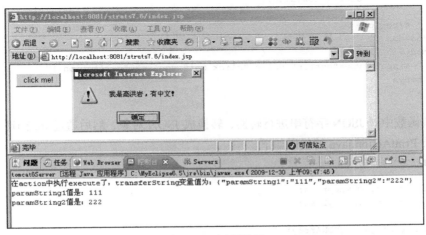

图 5-32 运行结果

正确在控制台中输出 transferString 变量的 JSON 字符串，正确输出 JSONObject 对象中的元素值，然后又在 IE 客户端弹出正确的中文字符。

到此，我们已经做了 3 个 jQuery 的 Ajax 示例，从这 3 个示例中可以发现，Ajax 的代码量非常少，jQuery 将尽可能多的精力放到业务层的设计中，而不是基础代码的建设上，这样就可以使程序员大大减少开发的时间，增加开发的效率。

5.12.6 用 jQuery 解析从 action 返回 List 中存 String 的 JSON 字符串

新建 Web 项目 struts7.7。

设计 index.jsp 的代码如下。

```jsp
<%@ page language="java" import="java.util.*" pageEncoding="utf-8" %>
<!DOCTYPE HTML PUBLIC "-//W3C//DTD HTML 4.01 Transitional//EN">
<html>
    <head>
        <script type="text/javascript" src="jquery-1.3.2.js">
        </script>
        <script type="text/javascript" src="json2.js">
        </script>
        <script>

            function callAction(){

                $.get("printAction.action", function(data){
                    callBackFunction(data)
                });
            }

            function callBackFunction(dataParam){
                var showValue = "";
                var jsonArrayObject = JSON.parse(dataParam);
                for (var i = 0; i < jsonArrayObject.length; i++) {
                    showValue = showValue + jsonArrayObject[i] + "\n";
                }
                alert(showValue);
            }
        </script>
    </head>
    <body>
        <input type="button" value="click me!" onclick="callAction()"/>
    </body>
</html>
```

在回调函数中对 JSON 字符串进行转换，转换成 JSON 对象，然后通过 for 循环语句输出。

控制层 PrintAction.java 的代码如下。

```java
package controller;

import java.io.IOException;
import java.util.ArrayList;

import net.sf.json.JSONArray;

import org.apache.struts2.ServletActionContext;
```

5.12 jQuery、JSON 和 Struts 2

```java
public class PrintAction {
    private ArrayList listString = new ArrayList();

    public ArrayList getListString() {
        return listString;
    }

    public void setListString(ArrayList listString) {
        this.listString = listString;
    }

    public String execute() throws IOException {

        listString.add("String1");
        listString.add("String2");
        listString.add("String3");
        listString.add("String4");
        listString.add("String5");
        listString.add("String6");

        JSONArray jsonobject = JSONArray.fromObject(listString);

        ServletActionContext.getResponse().setContentType("text/html");
        ServletActionContext.getResponse().setCharacterEncoding("utf-8");
        ServletActionContext.getResponse().getWriter().printf(
                jsonobject.toString());
        ServletActionContext.getResponse().getWriter().flush();
        ServletActionContext.getResponse().getWriter().close();

        return null;

    }

}
```

控制层负责生成 JSON 字符串，然后再通过 response 对象进行返回。

运行结果如图 5-33 所示。

图 5-33 运行结果

运行结果正确。

本示例其实也是前面原生 Ajax API 使用的简化版，具体用 JavaScript 操作 JSON 对象的方式几乎一模一样，比如以下操作：

- 用 jQuery 解析从 action 返回 List 中存 JavaBean 的 JSON 字符串；

- 用 jQuery 解析从 action 返回 Map 中存 String 的 JSON 字符串；
- 用 jQuery 解析从 action 返回 Map 中存 JavaBean 的 JSON 字符串；
- 用 jQuery 解析从 action 返回嵌套复杂类型的 JSON 字符串等场景。

都可以在前面章节中找到答案，JSON 对象在 Web 开发方面都是非常重要的技能知识点，掌握 JSON 也是掌握客户端与服务器端优良的通讯机制。另外 JSON 对象也可以在其他的语言中进行使用，相当于通用型技术。

5.13 在服务器端解析复杂结构的 JSON 对象

创建 Web 项目 lastProject，还要创建客户端 HTML 代码如下。

```html
<!DOCTYPE HTML PUBLIC "-//W3C//DTD HTML 4.01 Transitional//EN"
"http://www.w3.org/TR/html4/loose.dtd">
<html xmlns="http://www.w3.org/1999/xhtml">
    <head>
        <meta http-equiv="Content-Type" content="text/html; charset=utf-8" />
        <title>New Web Project</title>
        <script src="jquery-3.1.0.js">
        </script>
        <script src="json2.js">
        </script>
        <script>
            function sendAjax(){
                var userinfo = {
                    "myArray": [{
                        "username1": "usernameValue11"
                    }, {
                        "username2": "usernameValue22"
                    }, ["abc", 123, true, [123, 456]]],
                    "myObject": {
                        "username": "大中国"
                    },
                    "myObject1": {
                        "address": [{
                            "name": "name1"
                        }, {
                            "name": "name2"
                        }]
                    },
                };
                var jsonString = JSON.stringify(userinfo);
                $.post("test6.action?t=" + new Date().getTime(), {
                    "jsonString": jsonString
                });
            }
        </script>
    </head>
    <body>
        <input type="button" onclick="javascript:sendAjax()">
    </body>
</html>
```

服务器端的 Struts 2 的代码如下。

```java
package newcontroller;

import java.io.IOException;

import net.sf.json.JSONArray;
import net.sf.json.JSONObject;

public class Test6 {

    private String jsonString;

    public String getJsonString() {
        return jsonString;
    }

    public void setJsonString(String jsonString) {
        this.jsonString = jsonString;
    }

    public String execute() throws IOException {
        JSONObject jsonObject = JSONObject.fromObject(jsonString);
        JSONArray array1 = (JSONArray) jsonObject.get("myArray");
        JSONObject object1 = (JSONObject) array1.get(0);
        JSONObject object2 = (JSONObject) array1.get(1);
        JSONArray array2 = (JSONArray) array1.get(2);
        System.out.println(object1.get("username1"));
        System.out.println(object2.get("username2"));
        System.out.print(array2.get(0) + " " + array2.get(1) + " " + array2.get(2));
        JSONArray array3 = (JSONArray) array2.get(3);
        System.out.println(" " + array3.get(0) + " " + array3.get(1));
        JSONObject jsonObject2 = (JSONObject) jsonObject.get("myObject");
        System.out.println(jsonObject2.get("username"));
        JSONObject jsonObject3 = (JSONObject) jsonObject.get("myObject1");
        JSONArray array4 = (JSONArray) jsonObject3.get("address");
        JSONObject jsonObject4 = (JSONObject) array4.get(0);
        JSONObject jsonObject5 = (JSONObject) array4.get(1);
        System.out.println(jsonObject4.get("name") + " " + jsonObject5.get("name"));
        return null;
    }

}
```

程序运行后,单击"button"按钮,在控制台输出从 JSON 字符串解析出来的数据,结果如图 5-34 所示。

```
Tomcat v7.0 Server at localhost [Apache Tomcat
usernameValue11
usernameValue22
abc 123 true 123 456
大中国
name1 name2
```

图 5-34 成功解析出 JSON 字符串中的数据

第 6 章　用 Hibernate 5 操作数据库

如果说 MyBatis 是一种半自动化的 ORM 框架，那么 Hibernate 是一个接近于完全 ORM 的框架。使用 Hibernate 可以更加方便地操作数据表中的记录，更加具体地封装 CURD 操作，更加方便地管理事务，但它的入门却是非常简单。本章中读者应该着重掌握如下两点内容：

- 使用 MyEclipse 对数据表进行 Hibernate 逆向；
- 使用 Hibernate 操作数据库。

6.1 Hibernate 概述与优势

JBoss 公司的 Hibernate 开源项目为企业级开发中的数据持久层技术提供了非常好的解决方案，它不仅对跨数据库、事务封装、ORM 映射和延迟加载等这些细节技术提供了完善的解决方案，也对程序开发中一些容易忽略的问题进行了良好的封装和规划，包含以下内容。

- 通过使用 ORM(Object-Relational Mapping)映射，以实体类的形式操作数据库中的记录。
- 支持数据库连接池 Connection-Pool，使 CURD 的操作效率大大得到提升。
- Hibernate 支持 JPA（Java Persistence API-Java，持久化 API）规范，使程序员对 CURD 的 API 的掌握更加标准与统一。
- 减少那些大量重复的 JDBC 代码，使用 Hibernate 后，重复性的工作大大减少。
- 项目融入 Hibernate 后，由于省略了 SQL 语句，代码风格更加符合 OOP 特性。
- 更好的性能和移植性使 Hibernate 成为现阶段软件项目中最常使用的持久层技术框架。

ORM 框架的作用是将数据表中的数据记录与 JavaBean 互相转换并且带有很多附属功能的软件框架。

Hibernate 和 MyBatis 框架的功能与原理基本相似，都是对数据库进行操作并且可以转化成实体类 entity 的技术。只不过程序员在使用 MyBatis 时操作的是 SQL 语句，而 Hibernate 操作的是实体类，但 Hibernate 最终还是会将操作实体类的目的代码转换成 SQL 语句。

Struts 2 是 MVC 框架，Hibernate 是 ORM 框架，MyBatis 被官方称为 SQL Mapping 框架，也可称 MyBatis 是 ORM 框架。

Hibernate 框架的官方网站的界面如图 6-1 所示。

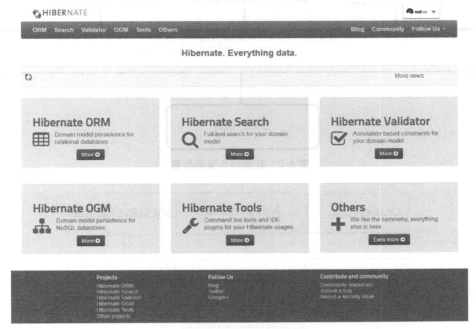

图 6-1　Hibernate 官方网站

在 Hibernate ORM 网页中，可以下载 Hibernate 框架的最新版。

6.2　持久层、持久化与 ORM

　　Hibernate 是一个 Java 数据持久化框架。
　　什么是数据持久化？数据持久化就是保存数据，可以将数据保存在可掉电的存储介质上。最常见的就是将数据保存在磁盘上，包括硬盘或闪存中。
　　什么是数据持久层？在软件体系结构中，Hibernate 也称为是一个数据持久层。在平常的处理数据的过程中，可以非常轻松地将数据通过 JDBC 代码存储进数据库，剩下的什么都不需要处理了。而这种心态恰恰是创建不可维护或后期不便于维护的软件项目的开始，简单的"保存"功能完全满足不了现在软件开发的模块性、可维护性、扩展性、分层性的原则，那么就需要一种技术框架，将业务层和数据库之间保存的操作做到可维护性、扩展性、分层性，"持久层"的概念便随着 Hibernate 的出现越来越得到人们的关注，数据持久层在软件体系中的位置如图 6-2 所示。
　　数据持久层的所在位置如图 6-2 所示，介于数据库与业务逻辑层之间。通过使用数据持久层，业务逻辑层只需要使用一套 API 即可在不同的数据库之间进行切换，大大增加软件的移植性。
　　如果不使用数据持久层框架，软件体系会是什么结构呢？结果如图 6-3 所示。

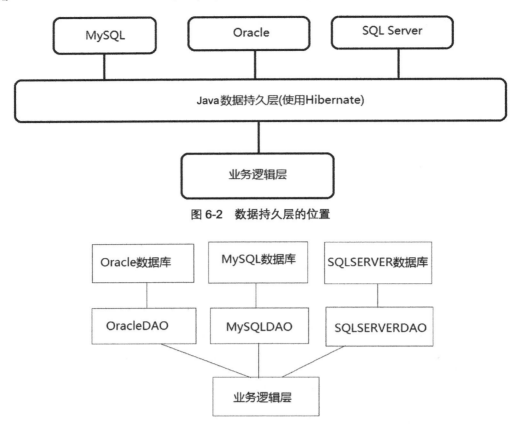

图 6-2 数据持久层的位置

图 6-3 每种数据库提供一个 DAO 类

若不使用数据持久层，那 DAO 类的数量会大幅增加，代码也要手动维护。这对软件项目的整体质量与开发进度有一定影响，软件移植性被大幅降低。

数据持久层的设计目标是为整个项目提供一个衔接高低层、统一、安全和并发的数据持久机制，完成对各种数据库进行持久化的编程工作，并为系统业务逻辑层提供服务。数据持久层提供了数据访问方法，能够使程序员避免手工编写程序访问数据持久层（Persistene Layer），使其专注于业务逻辑的开发，并且能够在不同项目中重用映射框架，大大简化了数据增、删、改、查等功能的开发过程，同时又不丧失多层结构的天然优势，继承延续 Java EE 特有的可伸缩性和可扩展性。

ORM 是通过映射的机制，把一条数据库中的记录转换成 Java 中的实体类，比如数据表具有如下 5 个字段。

```
id  name  address  school  email
```

那么通过 Hibernate 的 ORM 映射就可以将这个表中的字段映射成具有 5 个对应属性的一个 JavaBean。

```java
public class personalInfo implements java.io.Serializable {

    private String id;
    private String name;
    private String address;
```

```
    private String school;
    private String emial;

    public personalInfo(){}

    //省略属性的 get 和 set 方法
}
```

实体类要实现 implements java.io.Serializable 串行化接口，并且需要有一个参数为空的构造方法。

那么在以后的项目开发中，只需要操作这个类，效果就相当于用 JDBC 操作 DB 数据库一样。这样不仅使增、删、改、查的操作面向对象 OOP 编程，而且还使开发的重复性工作大大减少。Hibernate 已经将重复创建这些对象的操作进行非常好的封装，程序员将大部分的精力放在业务的具体实现上，注意力完全集中在软件的功能上，而不是那些 JDBC 中的对象。

6.3 用 MyEclipse 开发第一个 Hibernate 示例

在学习 Hibernate 起始阶段，为了实现一个完整的 Hibernate 存储数据的案例，我们首先要搭建 Hibernate 开发环境，需要做如下准备：

- 安装 Oracle 数据库；
- 准备 Oracle 的 JDBC 驱动程序；
- 准备 MyEclipse 集成开发工具；
- 在 MyEclipse 中测试连接数据库是否成功；
- 创建 Java/Java Web 项目；
- 开始逆向；
- 使用 Hibernate 进行数据持久化的操作。

6.3.1 在 MyEclipse 中创建 MyEclipse Database Explorer 数据库连接

将 MyEclipse 切换到 MyEclipse Database Explorer 透视图，结果如图 6-4 所示。

鼠标右键选择 New 菜单，结果如图 6-5 所示。

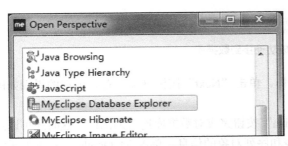

图 6-4　切换透视图

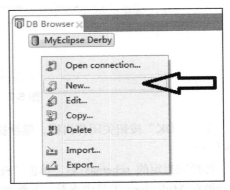

图 6-5　选择 New 菜单创建数据库连接

在弹出的窗口中设置连接数据库的基本信息，结果如图 6-6 所示。

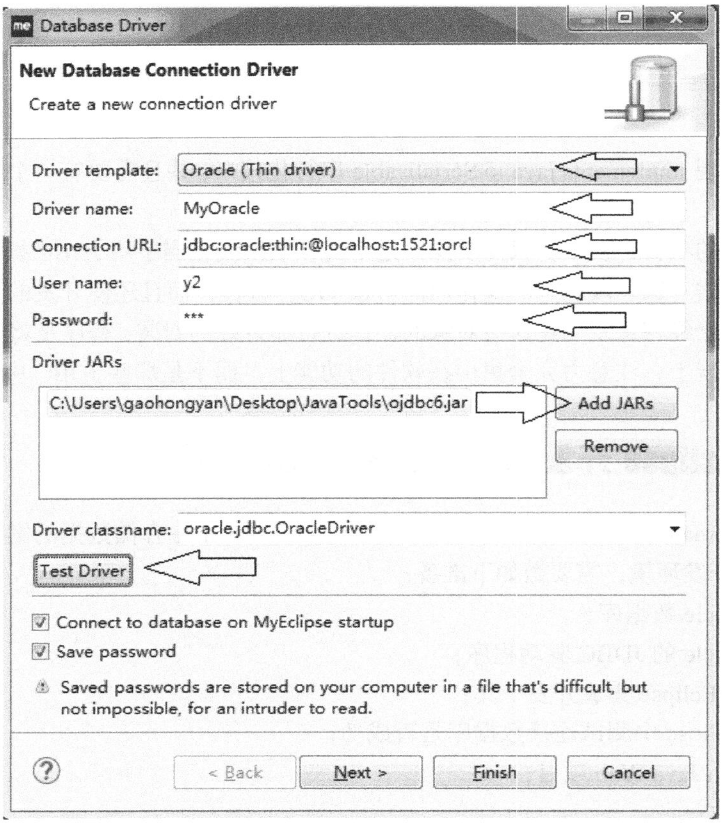

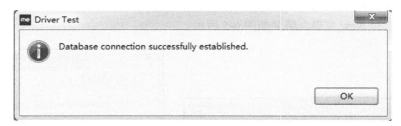

图 6-6　设置数据库基本信息

填写完毕后单击"Test Driver"按钮来测试一下是否能正常连接到 Oracle 数据库。如果成功连接到数据库，则出现的结果如图 6-7 所示。

图 6-7　成功连接到数据库

单击"OK"按钮关闭弹出框，继续操作，单击"Next"按钮继续配置，出现的界面如图 6-8 所示。

选择完指定的 schemas 后，单击"Finish"按钮完成对数据库连接的测试，这时还可以进一步地在 MyEclipse 工具中查看一下数据表和序列对象的信息，单击 MyOracle 节点下的子结点，结果如图 6-9 所示。

图 6-8　选择指定的 schemas

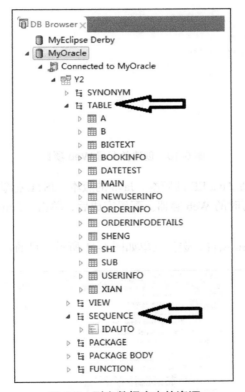

图 6-9　列出数据库中的资源

操作进行到这儿，说明数据库的环境是没有任何问题的。

6.3.2　创建 Web 项目并添加 Hibernate 框架

这一步就要创建 Web 项目，并对项目添加 Hibernate 框架支持的必要文件，包含 JAR 和 XML 文件。

创建 Web 项目 hibernate1，配置界面如图 6-10 所示。

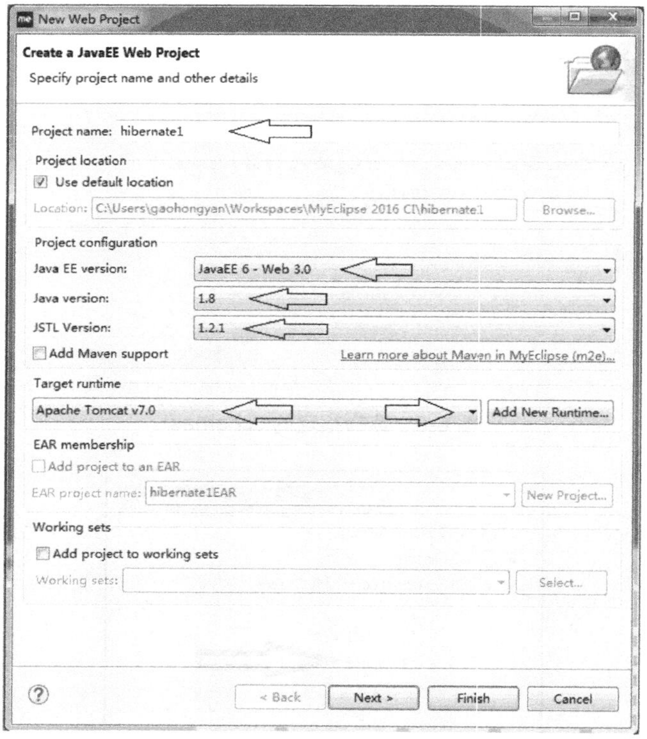

图 6-10 创建并配置 Web 项目

在此界面中，需要配置 Java EE 的版本、Java 的版本、JSTL 标签库的版本，以及要关联一个 Target runtime 目标运行时的 Web 容器。配置完成后，单击"Finish"按钮完成 Web 项目的创建。

使用鼠标右键单击 Web 项目，选择菜单如图 6-11 所示，目的是添加 Hibernate 框架。

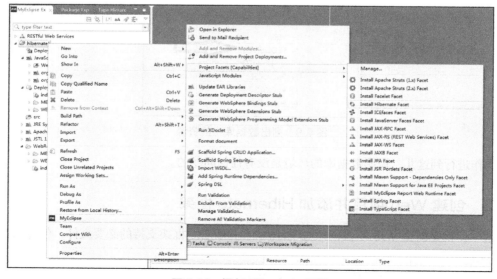

图 6-11 添加 Hibernate 框架

单击 Install Hibernate Facet 菜单后弹出的界面如图 6-12 所示。

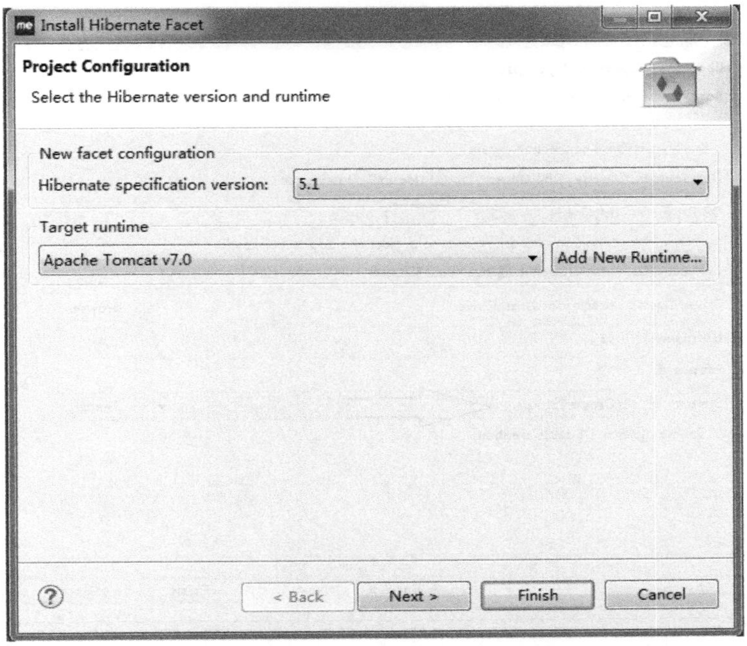

图 6-12 选择 Hibernate 框架版本及运行时软件

单击"Next"按钮继续配置，出现设置 Hibernate 框架文件所存放路径的界面。这里只需要设置 Java package 包即可，目的是将 Hibernate 逆向的文件存放在 orm 这个包中，结果如图 6-13 所示。

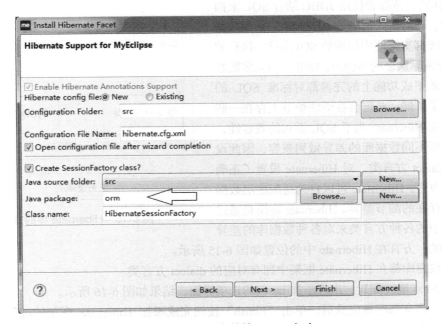

图 6-13 将资源文件放入 orm 包中

单击"Next"按钮开始设置 Hibernate 用什么参数去连接数据库，结果如图 6-14 所示。

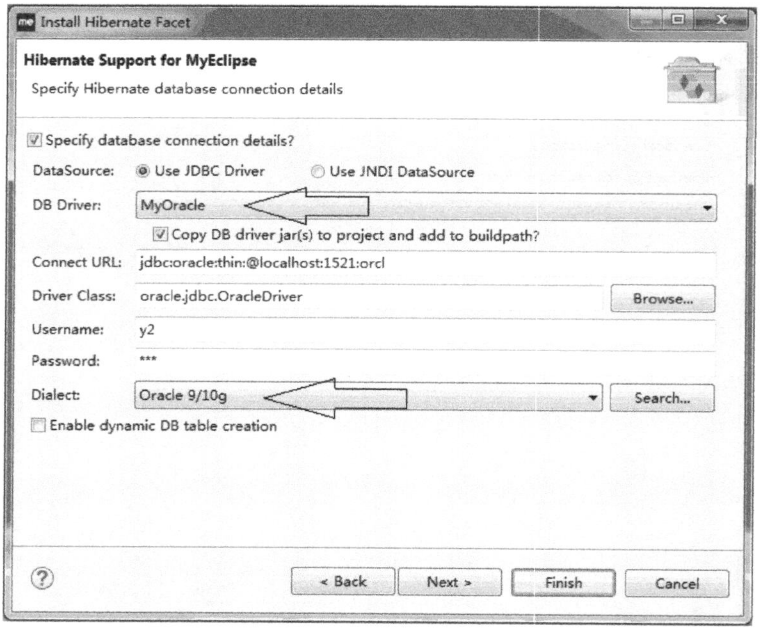

图 6-14　下拉中选择 MyOracle

在 DB Driver 下拉菜单中选择 MyOracle 即可，下方文本框的值会自动添充。MyOracle 是前面章节中创建的连接 Oracle 数据库的链接。

在此界面中，需要设置 Dialect 方言，什么是方言？Hibernate 虽然以 OOP 的方式在操作数据库，但其底层依然要使用 JDBC 结合 SQL 来操作数据库。现阶段大部分的 RDBMS 关系型数据库管理系统都支持使用标准的 SQL 语句，执行的 SQL 符合国际认证的 SQL92 标准。但大多数数据库为了效率或功能上的完善都对标准 SQL 的规范进行了扩展，必定会在语法细节上存在一些差异，因此 Hibernate 为了 SQL 语句的兼容性，也就要根据不同数据库的差异做到兼容，因此设计出了 Dialect 方言类。对 Hibernate 设置了正确的数据库方言，Hibernate 可以自动适配底层数据库访问所存在的细节差异。Hibernate 底层是通过 dialect 包中的各种方言类来对各种数据库的差异进行抽象的。方言在 Hibernate 中的位置如图 6-15 所示。

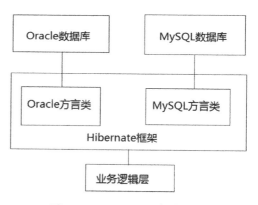

图 6-15　Hibernate 中的方言

主流数据库都在 Hibernate 框架中拥有对应的 dialect 方言类。

单击 Next 继续配置，出现添加 jar 包的选项界面，结果如图 6-16 所示。

以上步骤全部配置完成后，单击"Finish"按钮完成添加 Hibernate 框架的过程，添加完 Hibernate 框架的 Web 项目结构如图 6-17 所示。

6.3 用 MyEclipse 开发第一个 Hibernate 示例

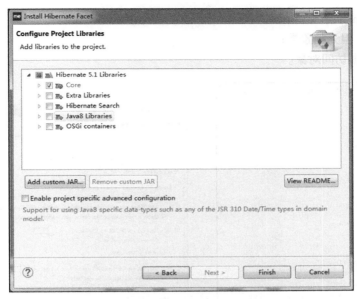

图 6-16　只选择 core 核心 jar 包即可

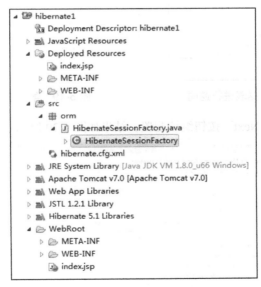

图 6-17　Web 项目结构

6.3.3　开始 Hibernate 逆向

这一步就要开始进行 Hibernate 逆向。逆向的作用就是将 Oracle 数据库中数据表的信息生成对应的 Java 类及 ORM 框架必需的 XML 映射文件。

回到 MyEclipse Database Explorer 透视图，在 userinfo 数据表上单击鼠标右键，选择菜单如图 6-18 所示。

弹出界面如图 6-19 所示。

在该界面中需要设置逆向 3 个主要元素：

- *.hbm.xml 映射文件；
- 实体类，但无抽象类；
- DAO 数据访问对象。

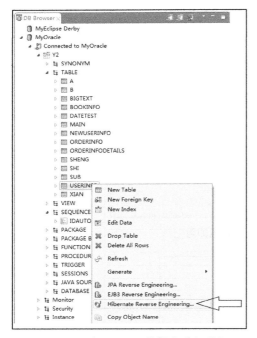

图 6-18 对 userinfo 数据表进行逆向

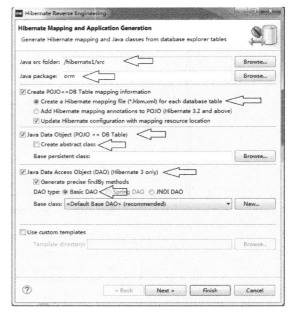

图 6-19 配置逆向的内容

设置完毕后，单击"Next"按钮继续配置，结果如图 6-20 所示。

图 6-20 设置主键值的生成策略为使用序列

在此界面中选择 sequence，因为 Oracle 中的主键并不是自增的，如果操作的数据库为 MySQL，则在下拉列表中可以选择 identity。

单击"Next"按钮继续配置，出现的界面如图 6-21 所示。

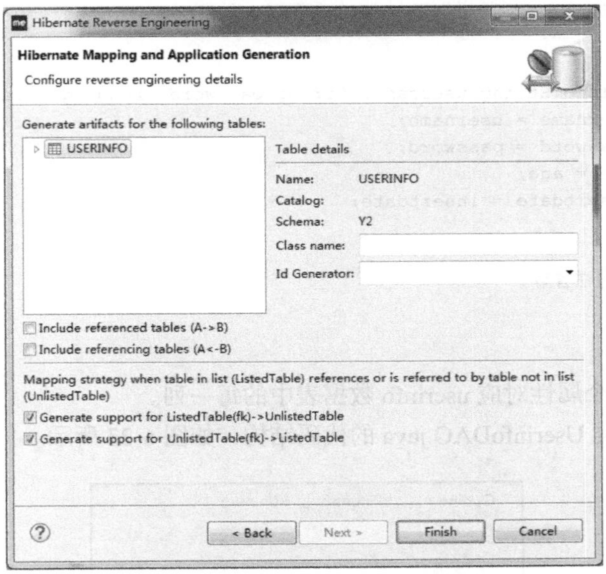

图 6-21　是否将主外关联的表进行逆向

在本实验中，只需要逆向 userinfo 这一个表，并没有其他具有主外关联关系的数据表。Hibernate 支持对具有主外关系的表进行操作，此知识点在下一个章节进行介绍。

单击"Finish"按钮完成逆向的操作。逆向成功后，src 节点下的文件如图 6-22 所示。

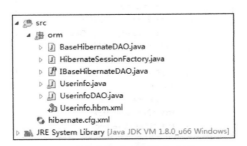

图 6-22　逆向后的文件

6.3.4　数据访问层 DAO 与实体类 entity 的代码分析

逆向生成的实体类 Userinfo.java 的代码如下。

```
package orm;

import java.util.Date;

public class Userinfo implements java.io.Serializable {

    private Long id;
```

```
        private String username;
        private String password;
        private Long age;
        private Date insertdate;

        public Userinfo() {
        }

        public Userinfo(String username, String password, Long age, Date insertdate) {
            this.username = username;
            this.password = password;
            this.age = age;
            this.insertdate = insertdate;
        }

        //set 和 get 方法省略
}
```

实体类中的每一个属性对应 userinfo 数据表中的每一列。

再来看看 DAO 类 UserinfoDAO.java 的代码结构，如图 6-23 所示。

图 6-23　DAO 类代码结构

从类的 API 结构来看，该类主要提供针对数据库的 CURD 操作方法。

6.3.5　使用 Hibernate 进行持久化

设计 Insert.java 类的代码如下。

```
package controller;

import java.util.Date;

import orm.Userinfo;
```

6.3 用 MyEclipse 开发第一个 Hibernate 示例

```
import orm.UserinfoDAO;

public class Insert {

    public static void main(String[] args) {

        Userinfo userinfo = new Userinfo();
        userinfo.setUsername("我是中国");
        userinfo.setPassword("我是大中国");
        userinfo.setAge(50L);
        userinfo.setInsertdate(new Date());

        UserinfoDAO userinfoDAO = new UserinfoDAO();
        userinfoDAO.getSession().beginTransaction();
        userinfoDAO.attachDirty(userinfo);
        userinfoDAO.getSession().getTransaction().commit();
        userinfoDAO.getSession().close();

    }
}
```

当运行 Insert.java 类后，控制台出现了异常，如图 6-24 所示。

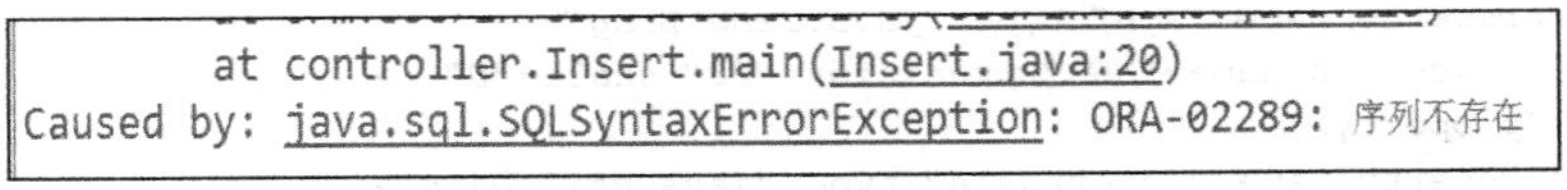

```
            at controller.Insert.main(Insert.java:20)
Caused by: java.sql.SQLSyntaxErrorException: ORA-02289: 序列不存在
```

图 6-24　提示序列不存在

出现异常的原因是如果数据库中有很多序列，那么应该使用哪个序列来生成主键值呢？在实验中并没有配置。下面来更改一下 Userinfo.hbm.xml 文件中的配置代码，更改后的代码如下。

```
<generator class="sequence">
    <param name="sequence_name">idauto</param>
</generator>
```

在配置文件中，指明要使用名称为 idauto 这个序列对象生成主键值。

再次运行程序，并没有出现异常。数据表 userinfo 中成功添加了新的记录，结果如图 6-25 所示。

| 2042493 | 我是中国 | 我是大中国 | | 50 | 2015-4-21 1:29:08 |

图 6-25　数据表 userinfo 中新添加的一条记录

成功在数据表 userinfo 中插入了一条记录。从程序代码中可以看到，没有任何的 JDBC 类对象，只使用了 Hibernate 的 DAO 和实体类就可以实现无 SQL 语句、无 JDBC 类对象的数据保存功能。可见 Hibernate 在 ORM 对象的封装上做得非常好。让程序员开发软件时将大部分的注意力放在如何设计与实现业务逻辑上，而放弃重复乏味的 JDBC 操作代码。

6.3.6　映射文件 Userinfo.hbm.xml 的代码分析

前面章节保存一个 Userinfo 类的对象即可转变成数据表中的一条记录，其原理主要就是源于 Userinfo.hbm.xml 映射文件。

映射文件从大的方向上提供了类与表的映射，从小的方向上提供了类中的属性与表中列的映射。

项目中的 Userinfo.hbm.xml 映射文件的代码如下。

```xml
<?xml version="1.0" encoding="utf-8"?>
<!DOCTYPE hibernate-mapping PUBLIC "-//Hibernate/Hibernate Mapping DTD 3.0//EN"
"http://www.hibernate.org/dtd/hibernate-mapping-3.0.dtd">
<!-- Mapping file autogenerated by MyEclipse Persistence Tools -->
<hibernate-mapping>
    <class name="orm.Userinfo" table="USERINFO" schema="Y2">
        <id name="id" type="java.lang.Long">
            <column name="ID" precision="18" scale="0" />
            <generator class="sequence">
                <param name="sequence_name">idauto</param>
            </generator>
        </id>
        <property name="username" type="java.lang.String">
            <column name="USERNAME" length="50" />
        </property>
        <property name="password" type="java.lang.String">
            <column name="PASSWORD" length="50" />
        </property>
        <property name="age" type="java.lang.Long">
            <column name="AGE" precision="18" scale="0" />
        </property>
        <property name="insertdate" type="java.util.Date">
            <column name="INSERTDATE" length="7" />
        </property>
    </class>
</hibernate-mapping>
```

文件 Userinfo.hbm.xml 是 Hibernate 中的映射文件，映射文件的功能就是将 O 对象与 R 关系进行对照。通过这个配置文件，可以将数据表中的字段与 Java 中的属性进行有机结合，还可以知道 O 对象中的哪个属性对应 R 数据表中的哪个字段。Userinfo.hbm.xml 中的 hbm 代表 hibernate mapping。

映射文件 Userinfo.hbm.xml 的主文件名为 Userinfo，它代表的是对哪个实体类进行 ORM 关系映射。本项目中的映射文件名是 Userinfo.hbm.xml，所以是对 Userinfo.java 类进行 ORM 映射。当然映射文件可以改成任意的名称，但要在 hibernate.cfg.xml 配置文件中使用<mapping resource="orm/Userinfo.hbm.xml" />标签进行注册。

<class>标记中的 name 属性代表要映射哪个实体类，当前的值为 orm 包中的 Userinfo 实体类。class 标记中的 table 属性值代表要对数据库中的哪个表进行映射，当前值为 USERINFO 数据表。class 中的 schema 属性代表在 Oracle 中使用哪个用户下的 USERINFO 数据表，当前值为 Y2，也就是使用 Y2 用户下的 USERINFO 数据表与 orm 包中 Userinfo.java 类进行关系映射。

<id>标记代表主键的映射关系，其中 name 值代表使用实体类中的哪个属性来作为主键，当前的值为 orm.Userinfo 类中的 id 属性。<id>标记中的 type 属性值代表主键属性的数据类型，这里是 Long——长整数类型。<id>的子标记<column>的 name 属性代表数据表中的主键字段的名称，当前值为 ID，属性 precision="18" scale="0"确定主键值的精度。由于数据表中的 id 列存储的是整数，没有小数点，所以 scale 值为 0。使用<generator>来配置主键信息，由 Oracle 数据

库的序列提供主键的值,所以本项目主键生成策略是 sequence。但是这里有一个重要的问题,序列在 Oracle 数据库中可以有多个,主键值到底是由哪个序列来生成值呢?所以还需要添加代码来确定由哪个序列对象来提供主键 ID 的值,添加的完整代码如下。

```xml
<generator class="sequence">
    <param name="sequence_name">idauto</param>
</generator>
```

只需要加一个 param 标记即可确认是由 IDAUTO 这个序列来提供主键 ID 的值。

如果使用 MySQL 数据库,则配置代码更改如下。

```xml
<generator class="identity">
</generator>
```

<property>标记中的 name 代表实体类 orm.Userinfo 中的某个属性的映射。当前值为 username 属性,数据类型为 String 字符串类型,<property>标记的子标记 column 的 name 属性值代表为实体类中的属性对应到数据表中的某一个字段,当前值为 USERNAME,容量大小为 50。

知识点介绍到这里,可以看到,在 Hibernate 中的 ORM 关系映射主要的文件是*.hbm.xml 映射文件,所有的 ORM 映射的关联全在该配置文件中,掌握*.hbm.xml 文件的知识在以后的开发中将起到很重要的作用。映射文件尽量不要手写,使用工具进行逆向生成。Hibernate 借助于*.hbm.xml 映射文件中的关系映射,结合 DOM 与反射技术,最终生成正确可执行的 SQL 语句,也就达到了完全使用实体类 entity 去操作数据库的目的了。

6.3.7 查询—修改—删除的操作代码

其他功能的示例代码如下。

查询全部功能的示例代码如下。

```java
package test1;

import java.util.List;

import orm.Userinfo;
import orm.UserinfoDAO;

public class SelectAll {

    public static void main(String[] args) {
        UserinfoDAO userinfoDao = new UserinfoDAO();
        userinfoDao.getSession().beginTransaction();
        List<Userinfo> listUserinfo = userinfoDao.findAll();
        for (int i = 0; i < listUserinfo.size(); i++) {
            Userinfo userinfo = listUserinfo.get(i);
            System.out.println(userinfo.getId() + " " + userinfo.getUsername() + " " + userinfo.getPassword() + " "
                    + userinfo.getAge());
        }
        userinfoDao.getSession().getTransaction().commit();
        userinfoDao.getSession().close();
    }
```

}

根据 ID 查询功能的示例代码如下。

```java
package test1;

import orm.Userinfo;
import orm.UserinfoDAO;

public class SelectById {

    public static void main(String[] args) {
        UserinfoDAO userinfoDao = new UserinfoDAO();
        userinfoDao.getSession().beginTransaction();
        Userinfo userinfo = userinfoDao.findById(2032495L);
        System.out.println(userinfo.getId() + " " + userinfo.getUsername() + " " + userinfo.getPassword() + " "
                + userinfo.getAge());
        userinfoDao.getSession().getTransaction().commit();
        userinfoDao.getSession().close();
    }

}
```

修改功能的示例代码如下。

```java
package test1;

import orm.Userinfo;
import orm.UserinfoDAO;

public class UpdateById {
    public static void main(String[] args) {
        UserinfoDAO userinfoDao = new UserinfoDAO();
        userinfoDao.getSession().beginTransaction();
        Userinfo userinfo = userinfoDao.findById(2032495L);
        userinfo.setUsername("new zzzzzzzzz");
        userinfoDao.attachDirty(userinfo);
        userinfoDao.getSession().getTransaction().commit();
        userinfoDao.getSession().close();
    }
}
```

删除功能的示例代码如下。

```java
package test1;

import orm.Userinfo;
import orm.UserinfoDAO;

public class DeleteById {
    public static void main(String[] args) {
        UserinfoDAO userinfoDao = new UserinfoDAO();
        userinfoDao.getSession().beginTransaction();
        Userinfo userinfo = userinfoDao.findById(2032495L);
        userinfoDao.delete(userinfo);
        userinfoDao.getSession().getTransaction().commit();
        userinfoDao.getSession().close();
    }
```

}

6.3.8 其他类解释

在项目中还有 3 个 Java 类没有介绍，这 3 个 Java 类的关系如图 6-26 所示。

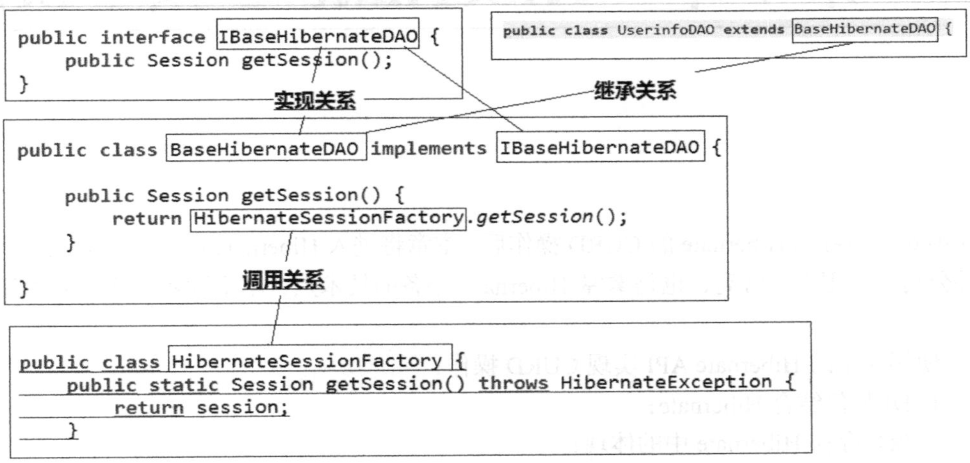

```
public interface IBaseHibernateDAO {
    public Session getSession();
}

public class UserinfoDAO extends BaseHibernateDAO {

public class BaseHibernateDAO implements IBaseHibernateDAO {
    public Session getSession() {
        return HibernateSessionFactory.getSession();
    }
}

public class HibernateSessionFactory {
    public static Session getSession() throws HibernateException {
        return session;
    }
}
```

图 6-26 3 个类的关系

各个 Java 类的功能与作用解释如下。

- IBaseHibernateDAO.java：该文件是一个接口文件，主要提供获取 Session 的 API 规范，方法如下

```
public Session getSession();
```

- BaseHibernateDAO.java：该文件是 IBaseHibernateDAO.java 接口的实现类，实现 getSession()方法，获得 Session 对象，示例代码如下。

```
public Session getSession() {
    return HibernateSessionFactory.getSession();
}
```

- HibernateSessionFactory.java：真正获取 Session 对象的是 HibernateSessionFactory.java，在类中提供访问 Hibernate 的 API 代码来获得 Session 对象，最终 BaseHibernateDAO.java 类获得 Session 对象。
- UserinfoDAO.java：数据访问对象需要使用 Session 对象，所以该类继承自 BaseHibernateDAO.java。

本章从一个简单的目的——使用 Hibernate 操作数据库出发，从数据库的环境配置、开发工具的使用及代码的实现去演示使用 Hibernate 持久化一个实体类以变成数据表中记录的功能。通过这个示例，读者应该掌握使用 Hibernate 开发数据库应用程序由哪几部分结构组成，比如 *.hbm.xml 映射配置文件，cfg.xml 连接配置文件，DAO 和实体类及创建 Session 的工厂类等。但在细节上读者可能对某些类的使用和类之间的关系不太理解，下面的章节就详细分析这些内容。

第 7 章 Hibernate 5 核心技能

在第 6 章学习完 Hibernate 的 CURD 操作后，本章将进入 Hibernate 的核心功能。这些功能在软件公司中使用率非常高，也是掌握 Hibernate 必备的技术点，本章读者应该着重掌握如下内容：

- 使用原生的 Hibernate API 实现 CURD 操作；
- JNDI 如何结合 Hibernate；
- 一级缓存在 Hibernate 中的体现；
- 双向一对多的 CURD 操作；
- 用 Hibernate 操作 CLOB 类型的字段。

7.1 工厂类 HibernateSessionFactory.java 中的静态代码块

在 MyEclipse 中自动生成了一个工厂文件 HibernateSessionFactory.java，该文件的主要作用就是创建 Session 接口，静态代码块中的代码如下。

```
static {
    try {
        configuration.configure();
        serviceRegistry = new StandardServiceRegistryBuilder().configure().build();
        try {
            sessionFactory = new MetadataSources(serviceRegistry).buildMetadata().buildSessionFactory();
        } catch (Exception e) {
            StandardServiceRegistryBuilder.destroy(serviceRegistry);
            e.printStackTrace();
        }
    } catch (Exception e) {
        System.err.println("%%%% Error Creating SessionFactory %%%%");
        e.printStackTrace();
    }
}
```

上面的静态代码块的作用是通过联合使用 StandardServiceRegistryBuilder 和 MetadataSources 类创建 SessionFactory 工厂对象，这也是 Hibernate5 新版本的写法。此种写法不要死记硬背，因为 Hibernate 的较新版本会减化 API 的调用结构。

将 static{}中的代码：

```
configuration.configure();
```

注释掉程序也能正常执行，读取 hibernate.cfg.xml 配置文件的任务由 StandardServiceRegistryBuilder.java 类负责。

Hibernate 默认加载的配置文件名为 hibernate.cfg.xml，如果更改默认配置文件的名称在运行时会出现异常。如果想自定义配置文件的名称，可以使用如下 static{}块中的代码。

```
static {
    try {
        serviceRegistry = new StandardServiceRegistryBuilder().configure
("hibernateABCABC.cfg.xml").build();
        try {
            sessionFactory = new MetadataSources(serviceRegistry).buildMetadata().
buildSessionFactory();
        } catch (Exception e) {
            StandardServiceRegistryBuilder.destroy(serviceRegistry);
            e.printStackTrace();
        }
    } catch (Exception e) {
        System.err.println("%%%% Error Creating SessionFactory %%%%");
        e.printStackTrace();
    }
}
```

在最新版本的 Hibernate 框架中获取 SessionFactory 对象还可以使用如下代码。

```
Configuration config = new Configuration();
config.configure("/hibernate.cfg.xml");
SessionFactory factory = config.buildSessionFactory();
```

代码不同，但都可以获得 SessionFactory 对象。

7.2 SessionFactory 介绍

前面通过 DAO 保存数据的功能就是使用 Session 接口实现的，而 Session 实例是由 SessionFactory 接口创建而来的。SessionFactory 不止具有创建 Session 对象的功能，还担任着持有所有映射关系及二级缓存的维护等工作。SessionFactory 最大的特点就是"重量级"，这里的重量级指的就是在使用 SessionFactory 时消耗的内存很大，创建的过程非常复杂，使用的时间也很长。所以基于这个特性，在创建完 SessionFactory 后，SessionFactory 就是一个独立的内存对象，不与其他对象产生任何的关系。

SessionFactory 采用了线程安全的设计，所以如果有多个线程，可以非常安全地创建 Session 对象。

7.3 Session 介绍

Hibernate 中的 Session 对象相当于 MyBatis 中的 SqlSession 对象，使用 Session 接口可以进行数据的 CURD（create update read delete）操作。Session 接口也是 Hibernate 使用最频繁的工

具类。掌握 Hibernate 的基础和 Session 的使用是很重要的技能点。

Session 由于在系统中被频繁地调用，所以在框架的设计上使用了"非线程安全"。将非线程安全的 Session 对象放入 HibernateSessionFactory.java 类中的代码如下所示。

```
private static final ThreadLocal<Session> threadLocal = new ThreadLocal<Session>();
```

将 Session 对象放入 ThreadLocal 对象中，在每一个客户端的线程访问数据库时都有自己独有的 Session 对象，使用 ThreadLocal 类非常完美地解决了 Session 非线程安全的问题，那么我们现在使用 HibernateSessionFactory.java 类中的 Session 对象完全是"线程安全"的。

Session 对象也是"轻量级"的，所以 Session 对象可以频繁地被创建或销毁，性能上的影响非常小。

7.4 使用 Session 实现 CURD 功能

在下面的任务中，我们一起来实现一个使用 Hibernate 原生 API 中的 Session 接口进行 CURD 的操作示例。本示例也是一个非常简单，并且非常实用的案例，通过学习此示例可以知道 UserinfoDAO.java 是如何进行数据操作封装的，对理解 Hibernate 有非常大的帮助。

7.4.1 数据表 userinfo 结构与映射文件

继续使用 userinfo 数据表来实现 Session 接口的使用案例。userinfo 数据表一共有 5 个字段，这 5 个字段类型是开发时常用的类型，非常具有代表性。Id 为主键整型，username 和 password 类型为 varchar2 类型，大小为 50，age 为 Number 整型，而 insertDate 注册时间是 Date 类型。

新建项目 hibernate12.3，添加 Hibernate 框架支持。

对 userinfo 表进行逆向工程，创建测试用的 Servlet 对象。

实体类和 DAO 还有 hibernate.cfg.xml 等文件都和前面章节的代码大体相同，在这里只需要关注最重要的映射文件：Userinfo.hbm.xml，代码如下。

```xml
<?xml version="1.0" encoding="utf-8"?>
<!DOCTYPE hibernate-mapping PUBLIC "-//Hibernate/Hibernate Mapping DTD 3.0//EN"
"http://www.hibernate.org/dtd/hibernate-mapping-3.0.dtd">
<!-- Mapping file autogenerated by MyEclipse Persistence Tools -->
<hibernate-mapping>
    <class name="orm.Userinfo" table="USERINFO" schema="Y2">
        <id name="id" type="java.lang.Long">
            <column name="ID" precision="18" scale="0" />
            <generator class="sequence">
                <param name="sequence_name">idauto</param>
            </generator>
        </id>
        <property name="username" type="java.lang.String">
            <column name="USERNAME" length="50" />
        </property>
        <property name="password" type="java.lang.String">
            <column name="PASSWORD" length="50" />
        </property>
```

```xml
        <property name="age" type="java.lang.Long">
            <column name="AGE" precision="18" scale="0" />
        </property>
        <property name="insertdate" type="java.util.Date">
            <column name="INSERTDATE" length="7" />
        </property>
    </class>
</hibernate-mapping>
```

7.4.2 创建 SessionFactory 工厂类

手动创建 SessionFactory 工厂类，代码如下。

```java
package hibernateapi.test;

import org.hibernate.SessionFactory;
import org.hibernate.boot.MetadataSources;
import org.hibernate.boot.registry.StandardServiceRegistryBuilder;
import org.hibernate.service.ServiceRegistry;

public class GetSessionFactory {
    public static SessionFactory getSessionFactory() {
        ServiceRegistry serviceRegistry = new StandardServiceRegistryBuilder().configure().build();
        SessionFactory sessionFactory = new MetadataSources(serviceRegistry).buildMetadata().buildSessionFactory();
        return sessionFactory;
    }
}
```

也可以使用如下类获得 SessionFactory 对象。

```java
package hibernateapi.test;

import org.hibernate.SessionFactory;
import org.hibernate.cfg.Configuration;

public class GetSessionFactory2 {
    public static SessionFactory getSessionFactory() {
        Configuration config = new Configuration();
        config.configure("/hibernate.cfg.xml");
        SessionFactory sessionFactory = config.buildSessionFactory();
        return sessionFactory;
    }
}
```

7.4.3 添加记录

添加记录的代码如下。

```java
package hibernateapi.test;

import java.util.Date;

import org.hibernate.Session;
import org.hibernate.SessionFactory;
```

```java
import orm.Userinfo;

public class Insert {
    public static void main(String[] args) {
        Userinfo userinfo = new Userinfo();
        userinfo.setUsername("中国");
        userinfo.setPassword("中国人");
        userinfo.setAge(200L);
        userinfo.setInsertdate(new Date());

        SessionFactory sessionFactory = GetSessionFactory.getSessionFactory();
        Session session = sessionFactory.openSession();
        session.beginTransaction();
        session.save(userinfo);
        session.getTransaction().commit();
        session.close();
    }
}
```

7.4.4 查询单条记录

查询单条记录的代码如下。

```java
package hibernateapi.test;

import org.hibernate.Session;
import org.hibernate.SessionFactory;

import orm.Userinfo;

public class SelectById {
    public static void main(String[] args) {
        SessionFactory sessionFactory = GetSessionFactory.getSessionFactory();
        Session session = sessionFactory.openSession();
        session.beginTransaction();
        Userinfo userinfo = (Userinfo) session.load(Userinfo.class, 2052495L);
        System.out.println(userinfo.getId() + " " + userinfo.getUsername() + " " + userinfo.getPassword() + " "
                + userinfo.getAge() + " " + userinfo.getInsertdate());
        session.getTransaction().commit();
        session.close();
    }
}
```

在 Hibernate 操作数据库的过程中，还可以打印 SQL 语句，做为调试或查看内部运行过程的一个参考。输出 SQL 语句的操作步骤是双击 hibernate.cfg.xml 文件，然后会弹出一个编辑窗口，在 Properties 面板中单击 "Add" 按钮进行 show_sql 属性的添加，然后再重启 tomcat 即可，操作的面板界面如图 7-1 所示。

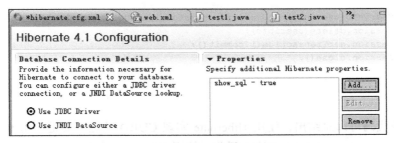

图 7-1 show_sql 的所在面板

7.4.5 更改记录

更改记录的代码如下。

```java
package hibernateapi.test;

import org.hibernate.Session;
import org.hibernate.SessionFactory;

import orm.Userinfo;

public class UpdateById {

    public static void main(String[] args) {
        SessionFactory sessionFactory = GetSessionFactory.getSessionFactory();
        Session session = sessionFactory.openSession();
        session.beginTransaction();
        Userinfo userinfo = (Userinfo) session.load(Userinfo.class, 2052495L);
        userinfo.setUsername("最新版的 username");
        session.update(userinfo);
        session.getTransaction().commit();
        session.close();
    }

}
```

7.4.6 删除记录

删除记录的代码如下。

```java
package hibernateapi.test;

import org.hibernate.Session;
import org.hibernate.SessionFactory;

import orm.Userinfo;

public class DeleteById {

    public static void main(String[] args) {
        SessionFactory sessionFactory = GetSessionFactory.getSessionFactory();
        Session session = sessionFactory.openSession();
        session.beginTransaction();
```

```
        Userinfo userinfo = (Userinfo) session.load(Userinfo.class, 2052495L);
        session.delete(userinfo);
        session.getTransaction().commit();
        session.close();
    }
}
```

从上面的代码可以总结出，使用 Hibernate 实现 CURD 功能主要是使用 Session 接口的 save()、update()、get()、delete()方法。到此，使用 Hibernate 来实现一个真实的 CURD 增、删、改、查的操作就介绍完了，Hibernate 框架给我们带来了很大的便利性，无论是事务的管理还是持久化的管理。CURD 是很简单的操作，可以使软件的设计完全转向 OOP 的设计，大大增加了软件的健壮性。

7.5 Hibernate 使用 JNDI 技术

JNDI 基于连接池的数据源技术广泛应用于中大型的项目中，它不仅是 Java EE 的核心技术，而且还被认证的 Web 容器所支持。但在 Hibernate 中如何应用这种强大的 JNDI 技术呢？其实它在 Hibernate 下的配置和 JDBC 中的配置大体一致，下面会介绍具体步骤。

7.5.1 备份 Tomcat/conf 路径下的配置文件

先把 Tomcat\conf 路径下的 context.xml 和 web.xml 文件备份，此步非常重要！

7.5.2 更改配置文件 context.xml

然后更改 context.xml 文件的内容，具体如下。

```xml
<Resource name="jdbc/ghy_shop" type="javax.sql.DataSource"
    driverClassName="oracle.jdbc.OracleDriver"
url="jdbc:oracle:thin:@localhost:1521:orcl"
    username="y2" password="123" maxActive="200" maxIdle="10" maxWait="-1" />
```

从配置中可以看出，JNDI 的数据源的名称为 jdbc/ghy_shop，使用的是 Oracle 数据库驱动，数据库的连接 url 也被指向到 ghy 数据库，还有用户名和密码等选项。

7.5.3 更改配置文件 web.xml

然后更改 web.xml 文件的内容。文件 web.xml 内容很多，只需要把下面的代码放到 web.xml 文件的最后面即可。

```xml
<resource-ref>
    <res-ref-name>jdbc/ghy_shop</res-ref-name>
    <res-type>javax.sql.DataSource</res-type>
</resource-ref>
```

在 web.xml 文件中配置所关联的数据源名称，以让应用程序能找到指定的数据源。

7.5.4 添加 Hibernate 框架配置的关键步骤

新建一个 Web 项目，名称是 hibernate12.4，在添加 hibernate 框架支持的关键步骤中，设置如图 7-2 所示。

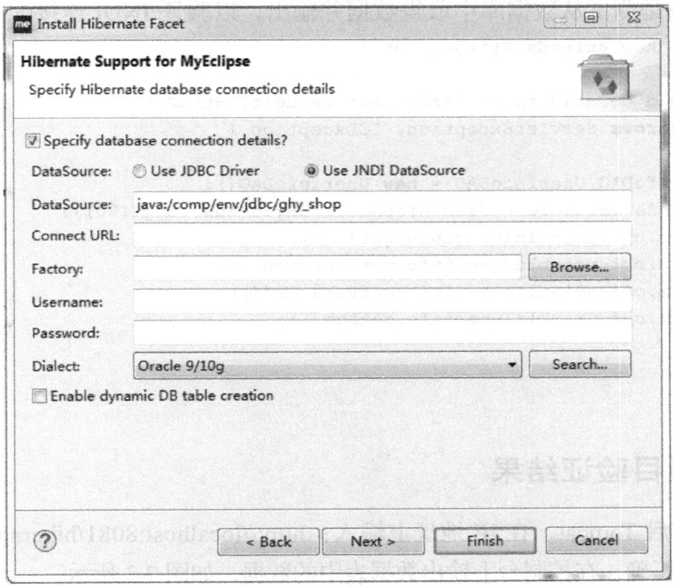

图 7-2 使用 JNDI 连接数据库

在添加 hibernate 框架的过程中，指定数据连接方式时需要选择 Use JNDI DataSource 方式，并且只需要设置 DatSource 和 Dialect 选项即可，单击 Next 继续按正常步骤配置。

7.5.5 逆向工程

针对数据库中的 userinfo 表进行逆向工程。

7.5.6 支持 JNDI 的 hibernate.cfg.xml 配置文件内容

文件 hibernate.cfg.xml 配置文件的关键内容如下。

```xml
<?xml version='1.0' encoding='UTF-8'?>
<!DOCTYPE hibernate-configuration PUBLIC
        "-//Hibernate/Hibernate Configuration DTD 3.0//EN"
        "http://www.hibernate.org/dtd/hibernate-configuration-3.0.dtd">
<!-- Generated by MyEclipse Hibernate Tools. -->
<hibernate-configuration>
    <session-factory>
        <property name="dialect">
            org.hibernate.dialect.Oracle9Dialect
        </property>
        <property name="connection.datasource">
            java:/comp/env/jdbc/ghy_shop
        </property>
        <mapping resource="orm/Userinfo.hbm.xml" />
    </session-factory>
```

```
</hibernate-configuration>
```

7.5.7 创建查询数据的 Servlet

Servlet 的代码是测试从数据库中取得数据并输出，以验证 JNDI 连接配置的正确性。

```java
public class show extends HttpServlet {

    public void doGet(HttpServletRequest request, HttpServletResponse response)
            throws ServletException, IOException {

        UserinfoDAO UserinfoDAO = new UserinfoDAO();
        Userinfo userinfo = UserinfoDAO.findById(new Long(100));
        System.out.println(userinfo.getId());
        System.out.println(userinfo.getUsername());
        System.out.println(userinfo.getPassword());
        System.out.println(userinfo.getAge());
    }

}
```

7.5.8 部属项目验证结果

部属项目，重启 Tomcat，在 IE 地址上输入：http://localhost:8081/hibernate12.4/show。如果环境配置正确，在控制台上输出数据表中的数据，如图 7-3 所示。

经过上面的 8 个步骤，可以看到在 Hibernate 中使用 JNDI 技术是非常简单的，配置的代码量也较少。

7.6 缓存与实体状态

图 7-3 JNDI 配置打印的结果

在 Hibernate 框架中有一些自己独有的术语，熟悉这些术语的含义对理解 Hibernate 的原理有重要作用。

7.6.1 Hibernate 的 OID 与缓存

在 Hibernate 中标识一个对象的方法是使用 OID（对象标识符），即实体类中的 ID 属性。Hibernate 认为 ID 值一样的实体类在 Hibernate 框架中是同一个对象。

Hibernate 中的缓存是由 SessionFactory 进行管理的，Hibernate 中的缓存存放的是最近查询到的数据，以方便以后查询相同的数据时，直接从内存中获取，而不用直接操作数据库。

下面通过一个示例，来验证 Hibernate 中的 OID 与缓存在实际开发中的知识点，更改 useinfo 数据表的字段值为如图 7-4 所示。

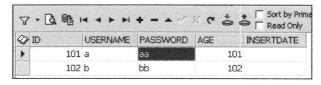

图 7-4 userinfo 数据表的字段值

7.6 缓存与实体状态

表中一共有两条记录。

创建项目 idTest，控制层 Servlet 的代码如下。

```java
public class test extends HttpServlet {

    public void doGet(HttpServletRequest request, HttpServletResponse response)
            throws ServletException, IOException {
        UserinfoDAO userinfoDAO = new UserinfoDAO();
        userinfoDAO.getSession().beginTransaction();
        // 重复从数据库中查询数据，目的是测试 hibernate 的缓存的功用
        Userinfo u1 = userinfoDAO.findById(new Long(101));
        Userinfo u2 = userinfoDAO.findById(new Long(102));
        Userinfo u3 = userinfoDAO.findById(new Long(101));
        userinfoDAO.getSession().getTransaction().commit();
        userinfoDAO.getSession().close();
    }

}
```

部属项目，启动 Tomcat，调用 Servlet 程序，运行结果如图 7-5 所示。

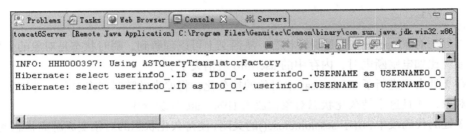

图 7-5　项目运行结果

从结果中可以看到，Hibernate 一共执行了两次 select 查询，然而程序中却执行了 3 次 findById 查询，为什么？这就是 Hibernate 高效的一级缓存机制。Hibernate 能通过实体类的 OID 来识别出缓存中是否有相同的对象，如果有，则从缓存中取出，如果在一级缓存中找不到，则到二级缓存中继续寻找，如果还找不到，则到数据库中进行数据记录的查询。

程序中第 1 个 findById 查询执行时，Hibernate 先在缓存中找是否有这个对象。因为是第 1 次执行，缓存中没有，则用一条 select 语句从数据库中进行查询，将查询到的 Userinfo 对象放入缓存中。在执行第 2 个 findById 时，Hibernate 从缓存中还没有找到相同 OID 值的 Userinfo 对象，然后就再执行一条 select 语句。而执行第 3 个 findById 时缓存中有相同 OID 的 Userinfo 对象，则使用缓存中的 Userinfo 对象，不发起执行一条新的 select 查询语句，所以会出现执行两条 SQL 语句的情况。

Hibernate 的一级缓存存放的实体类，这些实体类是占用内存的，但在执行了 session.close() 方法后这些实体类随时就被 GC 垃圾回收，以释放内存资源。但如果在执行 session.close() 方法之前操作了海量的实体类，完全有可能会出现内存溢出的情况，所以这时可以使用 session.clear() 方法来清除持久化上下文中的实体类，也就是一级缓存，以释放内存资源，模拟代码如下。

```java
List<Userinfo> list = dao.findAll();
for (int i = 0; i < list.size(); i++) {
    Userinfo userinfo = list.get(i);
    // 操作 userinfo 的代码
```

```
        if (i % 1000 == 0) {
            dao.getSession().clear();
        }
    }
```

7.6.2　Hibernate 中的对象状态：瞬时状态、持久化状态和游离状态

实体类在 Hibernate 框架中的不同操作阶段具有不同的状态，用文字来解释概念是非常抽象的，用代码来解释会好一些，看下面的代码段。

```
Userinfo userinfo = new Userinfo ();
userinfo.setUsername("高红岩");
userinfo.setAge(99);//1

Session session = HibernateSessionFactory.getSession();
Transaction tra = session.beginTransaction();
session.save(userinfo); //2
tra.commit();
session.close();

//3
```

在 1 处，对象 userinfo 就是瞬时状态，从语义上可以理解到，瞬时状态是对象的生命随时可以销毁。比如电脑断电时，内存中的 userinfo 就消失了。

而在 2 处由于进行了 session.save() 对象的持久化，将原来的瞬时状态对象立即转变成持久化状态对象，并且这个持久化状态对象被放入 Hibernate 的缓存中。

在 3 处的代码段中，由于 session.close() 的关闭，使持久化状态的对象变成游离状态的对象。

Hibernate 中的对象状态也就是生命周期的一个过程：创建、操作、关闭。

实体类还有另外一个状态，就是"删除状态"，删除状态的产生是执行了 session.delete(userinfo) 方法。该方法使实体类变成删除状态了，所以 Hibernate 中的实体类有 4 种状态：

- 瞬时状态/新建状态：new Userinfo();
- 持久化状态：session.save(userinfo);
- 游离状态：session.close();
- 删除状态：session.delete(userinfo)

实体类的状态在 Hibernate 框架中体现的不是特别明显。在 EJB 跨容器开发分布式应用软件中，实体类的状态是必须要知道的知识点。因为实体类在不同的容器中进行传输时，对状态是有严格要求的，如果状态出错，则有可能在 update() 或 save() 时出现异常，而这种要求在 Hibernate 框架中不需要考虑。

7.7　双向一对多在 MyEclipse 中的实现

什么是双向一对多？一对多是指一个对象关联多个对象，而双向的意思是不管从主表还是从从表都能取得到对方所关联的数据。在 Hibernate 框架中就是不管是主表还是从表生成的实体类，都有方法取到对方所关联的数据。

在项目开发中双向一对多的情况举不胜数，但比较经典而且容易理解的就是省市二级关联

了，下面的双向一对多的 Hibernate 实践就用这个最经典不过的示例来介绍。

创建项目 hibernate12.6。

在图 7-6 所示的界面中选择主表 MAIN 准备进行逆向。

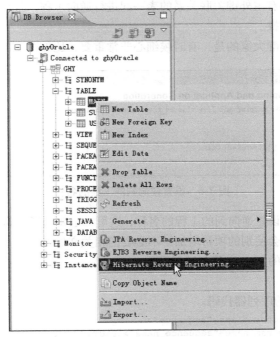

图 7-6 从主表逆向工程

在进行到设置主从表的关联关系界面时，需要将两个选项打上勾，目的是使用 MyEclipse 工具来帮助程序员以自动化的方式分析出哪些表有物理主外关系，如图 7-7 所示。

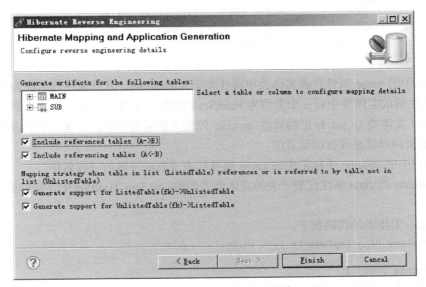

图 7-7 设置主从表的互关联

这两个选项如下所示。
- Include referenced tables (A->B);
- Include referencing tables (A<-B)。

勾选之后就可以将有主外键关联关系的表一起进行逆向工程，也就是我们想要创建的双向一对多关联关系的逆向操。

不过在这里需要提醒大家的是，有时候细心非常重要，比如图 7-8 所示的设置。

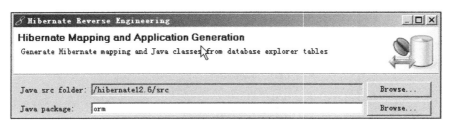

图 7-8　设置逆向到的工程的名称

逆向的第 1 步就是设置逆向到的工程的名称，这里一定要选择正确，不然逆向到别的项目中不仅是错误的，而且会使别的项目增加了不必要的相关文件。

下面开始分析逆向的 Java 文件与映射配置。

（1）main 主方。

手动增加 ID 生成的序列器代码。

```xml
<generator class="sequence">
    <param name="sequence_name">IDAUTO</param>
</generator>
```

逆向工程后在项目中创建了 Main.hbm.xml 配置文件，其中增加了 set 标记。

```xml
<set name="subs" inverse="true">
    <key>
        <column name="MAIN_ID" precision="18" scale="0" />
    </key>
    <one-to-many class="orm.Sub" />
</set>
```

set 标记中的 name 属性代表实体类中属性类型为 HashSet，并且属性的名称为 subs。也就是，在主表生成的实体类中有一个类型为 HashSet 的属性，名称为 subs，里面存放的是主外建关联的记录（实体类）。set 标记的属性 inverse 代表关联关系的维护是通过子表进行维护的，inverse 的中文翻译就是反方向反方的。

key 标记中的 column 子标记中的 name 属性代表子表的外键的字段名称。

one-to-many 的 class 属性代表子表的实体类的名称 orm.Sub，HashSet 中存储的就是 Sub.java 对象。

Main.java 实体类的代码如下。

```java
public class Main implements java.io.Serializable {
    private Long id;
    private String title;
    private String content;
    private Date createtime;
```

```
            private Set subs = new HashSet(0);
```

从代码片段中可以看到，只需要通过：

```
            private Set subs = new HashSet(0);
```

Set 集合就可以获得关联的子表中的实体类数据。

（2）sub 子方。

手动增加 ID 生成的序列器代码。

```
                <generator class="sequence">
                    <param name="sequence_name">IDAUTO</param>
                </generator>
```

逆向工程后在项目中创建了 Sub.hbm.xml 配置文件，其中增加了 many-to-one 标记。

```
                <many-to-one name="main" class="orm.Main" fetch="select">
                    <column name="MAIN_ID" precision="18" scale="0" />
                </many-to-one>
```

many-to-one 标记中的 main 属性值代表在子表实体类 Sub 中有一个 class 类型为 orm.Main 的变量名称 main，也就是 Sub 子表所关联的主表记录的信息。

Column 标记中的 name 属性代表真实数据表 sub 中外键的名称。

Sub.java 文件的代码片段如下。

```
public class Sub implements java.io.Serializable {

    private Long id;
    private Main main;
    private String subContent;
    private Date createtime;
```

从程序中可以看到，子表可以通过属性 Main main 来得到与之对应的主表的数据。

在 Hibernate 的双向一对多的关系中可以看到：
- 在一方通过使用 HashSet 来取得多方；
- 在多方通过使用一方的实体类属性来取得对应的一方。

7.7.1 添加主表记录

添加主表记录的代码如下。

```
public class InsertMain {

    public static void main(String[] args) {
        Main main = new Main();
        main.setMainname("中国");

        MainDAO dao = new MainDAO();
        dao.getSession().beginTransaction();
        dao.save(main);
        dao.getSession().getTransaction().commit();
        dao.getSession().close();

    }
```

}

7.7.2 添加子表记录

添加子表记录的代码如下。

```java
public class InsertSub {

    public static void main(String[] args) {

        MainDAO maindao = new MainDAO();
        SubDAO subdao = new SubDAO();

        maindao.getSession().beginTransaction();
        Main main = maindao.findById(2062495L);

        Sub sub = new Sub();
        sub.setSubname("中国的子表数据");
        sub.setMain(main);

        subdao.save(sub);

        maindao.getSession().getTransaction().commit();
        maindao.getSession().close();

    }

}
```

7.7.3 更改主表数据

更改主表数据的代码如下。

```java
public class UpdateMain {

    public static void main(String[] args) {
        MainDAO dao = new MainDAO();
        dao.getSession().beginTransaction();
        Main main = dao.findById(1445482L);
        main.setMainname("最新版的username");
        dao.attachDirty(main);
        dao.getSession().getTransaction().commit();
        dao.getSession().close();

    }

}
```

7.7.4 更改子表数据

更改子表数据的代码如下。

```java
public class UpdateSub {

    public static void main(String[] args) {

        SubDAO subdao = new SubDAO();
```

```
        MainDAO maindao = new MainDAO();
        maindao.getSession().beginTransaction();
        Main main = maindao.findById(1445481L);

        Sub sub = subdao.findById(2062496L);
        sub.setSubname("81的子表数据");
        sub.setMain(main);

        subdao.attachDirty(sub);

        maindao.getSession().getTransaction().commit();
        maindao.getSession().close();
    }
}
```

7.7.5 删除子表数据

删除子表数据的代码如下。

```
public class DeleteSub {

    public static void main(String[] args) {

        SubDAO subdao = new SubDAO();

        subdao.getSession().beginTransaction();
        Sub sub = subdao.findById(1445480L);

        subdao.delete(sub);

        subdao.getSession().getTransaction().commit();
        subdao.getSession().close();
    }
}
```

7.7.6 删除主表 main 数据

删除主表数据的代码如下。

```
package test;

import orm.Main;
import orm.MainDAO;

public class DeleteMain {

    public static void main(String[] args) {
        MainDAO dao = new MainDAO();
        dao.getSession().beginTransaction();
        Main main = dao.findById(1445481L);
        dao.delete(main);
        dao.getSession().getTransaction().commit();
```

```
            dao.getSession().close();
        }
    }
```

程序运行后却在控制台输出了出错的信息，如图 7-9 所示。

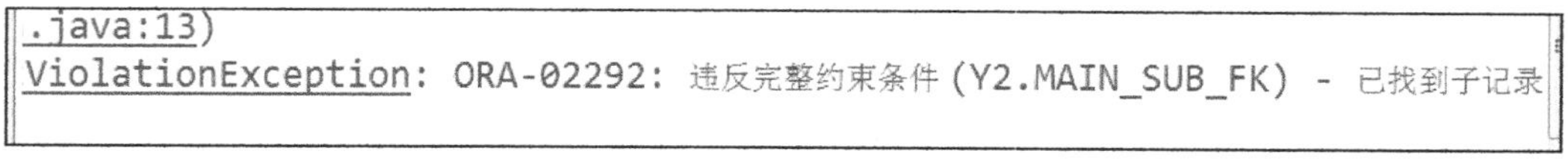

```
.java:13)
ViolationException: ORA-02292: 违反完整约束条件 (Y2.MAIN_SUB_FK) - 已找到子记录
```

图 7-9 报异常

出错的大体原因是删除主表的数据时，Oracle 检测到有子表的数据与其进行主外关联，所以出错，如何解决这个问题呢？

在主表的 set 标记中加入属性 cascade="all"即可，cascade 的中文翻译为级联、一起，all 代表主表有 save 或 update 操作时将子表的数据一同更新，当然也包括 delete 的操作。在本示例中使用的是 delete 操作，所以结合 cascade="all"就是级联删除。

```
<set name="subs" inverse="true" cascade="all">
    <key>
        <column name="MAINID" precision="18" scale="0" />
    </key>
    <one-to-many class="orm.Sub" />
</set>
```

再次运行程序后成功删除了主表及子表的数据。如果使用 show_sql=true 属性则在控制台输出执行的 SQL 语句，从 SQL 语句的执行过程中可以发现是先删除的子表然后再删除的主表，并且数据表 main、sub 中的记录都已经在数据表中被删除。

7.7.7 通过主表获取子表数据

通过主表获取子表数据的示例代码如下。

```java
public class PrintMoreSub {

    public static void main(String[] args) {
        MainDAO dao = new MainDAO();
        dao.getSession().beginTransaction();
        Main main = dao.findById(1445483L);
        System.out.println(main.getId() + " " + main.getMainname());
        Set<Sub> subSet = main.getSubs();
        Iterator<Sub> iterator = subSet.iterator();
        while (iterator.hasNext()) {
            Sub sub = iterator.next();
            System.out.println(" " + sub.getId() + " " + sub.getSubname());
        }
        dao.getSession().getTransaction().commit();
        dao.getSession().close();
    }

}
```

到这一步，在 Hibernate 中的双向一对多的操作步骤就介绍到这里。其实 Hibernate 的映射关系还有单向一对多、一对一、多对多等关系，但在软件项目中应用非常少，如果读者有兴趣可以参阅相关的资料。

虽然双向一对多的关联映射相比单向一对多、一对一在软件项目中的应用率高，但大部分的项目在使用 Hibernate 框架时都避开使用双向一对多的关联。虽然双向一对多的关联能够在主从表双方的操作上减少代码量，但却增加了程序员维护的工作量，并且新加入的程序员要把注意力放在 hbm.xml 的映射关系上，所以大部分的软件项目都使用无关联的 Hibernate 映射。即只使用 Hibernate 的 ORM 部分，把 set 和 many-to-one 这样的功能弃之不用，这样软件在数据库 ORM 部分变得更加灵活。关系的维护需要单独写一个 Java 类，这样的实践已经在很多软件公司的大中型项目中被应用。

7.8 Hibernate 备忘知识点

- 永远不要手动修改 OID(Object ID)，OID 就是实体类中的 id 属性，随意更改 OID 会造成缓存中的数据不准确。
- Session 接口中的 update 方法是将一个游离对象转变成持久化对象的方案，也是更新的作用，示例代码如下。

```java
public class Test1 {
    public static void main(String[] args) {
        Session session = HibernateSessionFactory.getSession();
        session.beginTransaction();
        Main main = session.load(Main.class, 2062495L);
        main.setMainname("xxxxxxxxxxx");
        session.update(main);
        session.getTransaction().commit();
        session.close();
    }
}
```

- Session 接口中的 saveOrUpdate()方法，如果参数传进去的是瞬时对象就执行 save，如果传入的是游离/托管对象，就执行 update。

对 saveOrUpdate()方法传入新建状态的实体类就是 insert 操作，示例代码如下。

```java
public class Test2 {

    public static void main(String[] args) {
        Session session = HibernateSessionFactory.getSession();
        session.beginTransaction();
        Main main = new Main();//新建状态就是 insert 操作
        main.setMainname("xxxxxxxxxxx");
        session.saveOrUpdate(main);
        session.getTransaction().commit();
        session.close();
    }
}
```

修改托管状态的示例代码如下。

```java
public class Test3 {

    public static void main(String[] args) {
```

```
        Session session = HibernateSessionFactory.getSession();
        session.beginTransaction();
        Main main = session.load(Main.class, 2062495L);//托管状态就是update操作
        main.setMainname("zzzzzzzzzzzzzzz");
        session.saveOrUpdate(main);
        session.getTransaction().commit();
        session.close();
    }

}
```

修改游离状态的示例代码如下。

```
public class Test4 {

    public static void main(String[] args) {
        Session session = HibernateSessionFactory.getSession();
        session.beginTransaction();
        Bookinfo bookinfo = session.load(Bookinfo.class, 5L);
        System.out.println("旧值: " + bookinfo.getBookname());
        session.getTransaction().commit();
        session.close();
//close()执行，session 关闭，所有实体类都变成游离状态
//所谓的游离状态就是实体类不受 Hibernate 管理，长期不再使用会被 GC 垃圾回收

        bookinfo.setBookname("新的书名");// 游离状态

        session = HibernateSessionFactory.getSession();
        session.beginTransaction();
        session.update(bookinfo);
        session.getTransaction().commit();
        session.close();

    }

}
```

- 如果正在使用延迟加载，并且在 session.close()之后要取得子表中的数据时会提示出错：Exception in thread "main" org.hibernate.LazyInitializationException: failed to lazily initialize a collection of role: orm.Main.subs, could not initialize proxy - no Session。堆栈异常信息提示 session.close()之后不能再访问数据库了，因为 Connection 已经关闭，解决办法是在 session.close()之前激活子表中的数据，将子表中的数据放入内存中，随时进行获取即可。
激活机制只存在于物理主从表的关系中，激活的只是子表中的数据，没有主表。

核心代码如下。
```
main.getSubs().iterator().next();
```

这时在 session.close()之后的代码也能正确获取子表中的数据了，完整示例代码如下。
```
public class PrintMoreSub3 {

    public static void main(String[] args) {
        MainDAO dao = new MainDAO();
        dao.getSession().beginTransaction();
        Main main = dao.findById(1445483L);
```

```
System.out.println(main.getId() + " " + main.getMainname());
// 使用next()方法激活子表数据,将子表的数据放入内存中
main.getSubs().iterator().next();//////////////////////////////////////
dao.getSession().getTransaction().commit();
dao.getSession().close();

Set<Sub> subSet = main.getSubs();
Iterator<Sub> iterator = subSet.iterator();
while (iterator.hasNext()) {
    Sub sub = iterator.next();
    System.out.println(" " + sub.getId() + " " + sub.getSubname());
}
}
}
```

- 使用 Hibernate 的 ORM 技术需要根据实际情况,超大数据量的软件项目不要为了 Hibernate 而 Hibernate,该框架只是适用于数据量比较小的情况,大数据量可以参考使用 MyBatis 和原生的 JDBC API。

7.9 对主从表结构中的 HashSet 进行排序

在主从表的环境中想输出 HashSet 集合时按照排序的情况该如何解决呢?其实 Hibernate 自动提供这样的功能,只需要在 set 端的 hbm.xml 文件中加入 order-by 属性即可,代码如下。

```
<set name="subs" inverse="true" order-by="id desc">
    <key>
        <column name="MAIN_ID" precision="18" scale="0" />
    </key>
    <one-to-many class="orm.Sub" />
</set>
```

经过这样的简单配置,查询结果放入到 HashSet 对象中,并且 HashSet 中的实体是按着 ID 倒序排列的。

7.10 延迟加载与 load() 和 get() 的区别

什么是延迟加载?延迟加载是一种数据加载的机制,它可以在还没有使用数据时不将数据放入内存中,使用时再加载到内存中,以减少内存占用率。延迟加载的优势主要体现在,在具有物理主外关系的表结构中,比如有一个"城市表",还有一个"短信表",如果在查询城市时,若将当前城市下的所有短信信息放入内存中就比较占用内存了,这时就可以使用延迟加载,当使用短信数据时再加载到内存中。

前面的示例中一直在介绍 Hibernate 中的延迟加载是一种主表与代理类之间的交互,那么本节将用一个示例来说明延迟加载具体是什么形式的。

7.10.1 主从表表结构的设计

使用 sheng 与 shi 数据表进行延迟加载的实验。

创建测试用的 Web 项目 lazyTest。

7.10.2　对省表和市表内容的添充

添加 sheng 表数据内容。
sheng 表数据内容如图 7-10 所示
shi 表数据内容如图 7-11 所示。

　　图 7-10　sheng 表数据内容　　　　图 7-11　shi 表数据内容

7.10.3　更改映射文件

更改 Sheng.hbm.xml 映射文件，代码如下。

```xml
<set name="shis" lazy="true" inverse="true">
    <key>
        <column name="SHENGID" precision="18" scale="0" />
    </key>
    <one-to-many class="orm.Shi" />
</set>
```

添加了 lazy="true" 的配置代码，含义是设置 Hibernate 加载行为是"延迟加载"。

7.10.4　新建测试用的 Servlet 对象

Servlet 程序的代码如下。

```java
public void doGet(HttpServletRequest request, HttpServletResponse response)
    throws ServletException, IOException {
    Session session = HibernateSessionFactory.getSession();
    session.beginTransaction().begin();
    Sheng sheng = (Sheng) session.load(Sheng.class, 1L);
    System.out.println(sheng.getShengname());
    session.getTransaction().commit();
    session.close();
}
```

程序运行后，控制台的输出结果如图 7-12 所示。

图 7-12　仅仅输出 1 条 SQL 语句

7.10.5 更改映射文件 Sheng.hbm.xml

更改映射文件 Sheng.hbm.xml，代码如下。

```xml
<set name="shis" lazy="false" inverse="true">
    <key>
        <column name="SHENGID" precision="18" scale="0" />
    </key>
    <one-to-many class="orm.Shi" />
</set>
```

lazy 属性值为 false 的含义是立即加载，重启 Tomcat 后，程序运行后的结果图 7-13 所示。

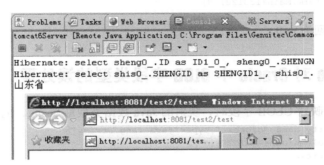

图 7-13　打印了 2 条语句

通过使用 lazy 属性的 true 或 false 值即可控制 load 函数的加载方式。

方法 load()如果结合 lazy="true"，与在映射文件中不添加此属性时效果是一样的，执行的 SQL 语句数量为 1 个，只是发起查询主表数据的 SQL 语句具有延迟加载的行为。

方法 load()如果结合 lazy="false"，执行的 SQL 语句数量为 2 个，一个是查询主表的，另一个是查询从表的。

上面的文字总结的是具有物理主外关系的延迟加载特性。在查询数据时，Hibernate 还提供了另外一个方法 get()，它也能实现与 load()方法同样的功能。它们之间的区别是在查询数据时，load()方法默认使用延迟加载，而 get()方法默认使用立即加载。

方法 load()的示例代码如下。

```java
public class Test6 {
    public static void main(String[] args) {
        Session session = HibernateSessionFactory.getSession();
        session.beginTransaction();
        Main main = (Main) session.load(Main.class, 1445483L);
        session.getTransaction().commit();
        session.close();
    }
}
```

程序运行后在控制台并没有输出 SQL 语句，因为并没有取得 main 对象中的数据。

方法 get()的示例代码如下。

```java
public class Test7 {
    public static void main(String[] args) {
        Session session = HibernateSessionFactory.getSession();
        session.beginTransaction();
        Main main = (Main) session.get(Main.class, 1445483L);
```

```
            session.getTransaction().commit();
            session.close();
        }
    }
```

程序运行后，在控制台显示出 SQL 语句，虽然没有取得 main 对象中的数据，但数据还是从数据库中取出并放入内存，具有立即加载数据的特性。

另外它们的其他不同点是 load（）方法根据 id 如果找不到实体类时返回一个代理实体类，输出这个代理实体类时报如下异常信息。

```
Exception in thread "main" org.hibernate.ObjectNotFoundException: No row with the given
identifier exists: [orm.Main#123123123]
```

而方法 get()返回为 null。

7.11　Hibernate 对 Oracle 中 CLOB 字段类型的读处理

在 Hibernate 中映射 CLOB 字段时，映射的 Java 数据类型是 String，配置代码如下。

```
<property name="bigtext" type="java.lang.String">
    <column name="BIGTEXT" />
</property>
```

7.12　Hibernate 中的 inverse 与 cascade 的测试

```
inverse="true" cascade="all"
```

上面的属性是大多数 Hibernate 学习者感到比较苦恼的部分，其实这两个属性的功能是风牛马不相及的关系：

- inverse 的功能是将外键值维护让给对方，由对方进行外键值的维护。这样在双向一对多的情况下，比如主贴和回贴，让回贴维护主从关系的效率就相对较高；
- cascade 的功能是级联，常用的取值是 all 和 save-update，意思是当某方进行数据操作时，比如进行添加和删除操作，采用什么样的级联方式对对方的数据进行操作，这个级联的方式就是 save、update 或 delete 操作。前面的实验中已经实现了级联删除，后面的实验将实现级联添加。

在数据库中创建 sheng 和 shi 表，设置 sheng 和 shi 表具有物理主外键关系，创建名为 sheng_shi_test 的 Web 项目，添加 Hibernate 框架。

逆向 sheng 表的默认映射文件代码如下。

```
<hibernate-mapping>
    <class name="orm.Sheng" table="sheng" schema="dbo" catalog="ghydb">
        <id name="id" type="java.lang.Integer">
            <column name="id" />
            <generator class="identity" />
        </id>
        <property name="shengname" type="java.lang.String">
            <column name="shengname" length="50" />
        </property>
        <set name="shis" inverse="true">
            <key>
```

7.12 Hibernate 中的 inverse 与 cascade 的测试

```
            <column name="shengid" />
        </key>
        <one-to-many class="orm.Shi" />
    </set>
    </class>
</hibernate-mapping>
```

从代码中可以发现，set 标签中自动添加了 inverse="true"的代码。inverse 属性表示控制权交给谁，值为 true 时，把物理主外键的关系控制权交给对方，也就是 Shi 对象来处理；值为 false 交给自己，也就是 Sheng 对象来进行处理。

创建名为 test1 的 Servlet，核心代码如下。

```
public class test1 extends HttpServlet {

    public void doGet(HttpServletRequest request, HttpServletResponse response)
        throws ServletException, IOException {

        Sheng sheng = new Sheng();
        sheng.setShengname("山东省");

        Shi shi1 = new Shi();
        shi1.setShiname("山东省市1");
        shi1.setSheng(sheng);

        Session session = HibernateSessionFactory.getSession();
        session.beginTransaction();
        session.save(sheng);
        session.save(shi1);
        session.getTransaction().commit();
        session.close();
    }

}
```

运行成功后在控制台输出的 SQL 语句如图 7-14 所示。

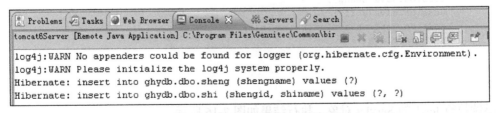

图 7-14　初次打印的 sql 语句

数据表中的内容如图 7-15 所示。

从此结果中可以看到，使用<set name="shis" inverse="true">配置代码，插入 shi 数据时，自动在 shengid 字段中插入了值为 3 的 int 类型。

那如果把代码改成<set name="shis" inverse="false">呢？配置代码如下。

```
            <set name="shis" inverse="false">
                <key>
                    <column name="shengid" />
                </key>
                <one-to-many class="orm.Shi" />
```

```
</set>
```

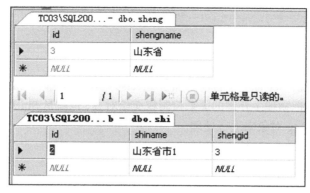

图 7-15 成功插入省和市

继续测试，清空数据表 sheng 和 shi 中的内容，并更改 test1 的 Servlet 核心代码如下。

```java
public class test1 extends HttpServlet {

    public void doGet(HttpServletRequest request, HttpServletResponse response)
            throws ServletException, IOException {

        Sheng sheng = new Sheng();
        sheng.setShengname("山东省");

        Shi shi1 = new Shi();
        shi1.setShiname("山东省市1");
        shi1.setSheng(sheng);

        sheng.getShis().add(shi1);
//由于 Sheng 要维护关系，所以在 Sheng 中添加 Shi

        Session session = HibernateSessionFactory.getSession();
        session.beginTransaction();
        session.save(sheng);
        session.save(shi1);
        session.getTransaction().commit();
        session.close();
    }

}
```

继续执行 test1 的 Servlet 对象，执行结果如图 7-16 所示。

图 7-16 由 sheng 管理关联关系

7.12 Hibernate 中的 inverse 与 cascade 的测试

通过上面的两个实验可以看出，当使用配置<set name="*shis*" inverse="*false*">时，由 sheng 表维护关系，而多增加执行一条 update 语句；当使用配置<set name="*shis*" inverse="*true*">时，是由 shi 表维护这种关联关系，不需要多执行一条 update 语句，提高了运行效率。所以，大多数的情况下都使用<set name="shis" inverse="true">此种配置写法。

而属性 cascade 的作用就是设置主外联级关系的方式。

创建名为 test2 的 Servlet，清空 sheng 和 shi 数据表中的内容。

更改 Sheng.java 类对应的映射文件，使用如下配置代码。

```
<set name="shis" inverse="true">
```

Servlet 的核心代码如下。

```java
public class test2 extends HttpServlet {
    public void doGet(HttpServletRequest request, HttpServletResponse response)
            throws ServletException, IOException {

        Sheng sheng = new Sheng();
        sheng.setShengname("山东省");

        Shi shi1 = new Shi();
        shi1.setShiname("山东省市1");
        shi1.setSheng(sheng);

        sheng.getShis().add(shi1);

        Session session = HibernateSessionFactory.getSession();
        session.beginTransaction();
        session.save(sheng);
        session.getTransaction().commit();
        session.close();
    }
}
```

路径为 test2 的 Servlet 代码只使用一个 save()方法保存 sheng 对象，这和前面的示例有一些区别。

程序运行后的结果如图 7-17 所示。

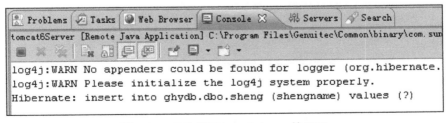

图 7-17　只执行了 insert sheng 的 SQL

数据表中的内容如图 7-18 所示。

这样的结果是在 Servlet 中只执行了如下代码。

```
session.save(sheng);
```

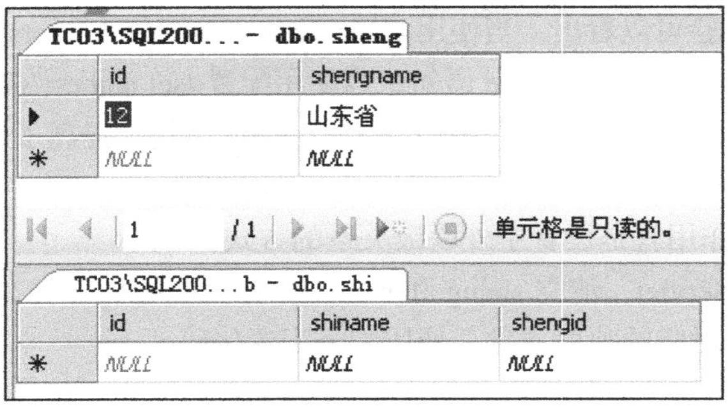

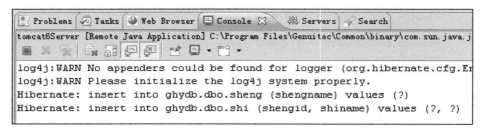

图 7-18 只插入了 Sheng 表数据

虽然在 sheng 中添加了 shi 对象，但是在 shi 中又关联了 sheng 对象，代码如下。

```
Shi shi1 = new Shi();
shi1.setShiname("山东省市1");
shi1.setSheng(sheng);
sheng.getShis().add(shi1);
```

现在的情况是 sheng 和 shi 对象已经在 Java 实体类中进行了关联设置，想实现 save(sheng) 时自动把 shi 也插入到数据表，解决的办法就是更改配置如下。

```
<set name="shis" inverse="true" cascade="all">
```

清空 sheng 和 shi 数据表内容，运行 Servlet 后的结果如图 7-19 所示。

图 7-19 save(sheng)时也把 shi 顺便插入了

第 8 章 Hibernate 5 使用 HQL 语言进行检索

虽然 Hibernate 用 ORM 技术来操作数据表中的数据，但在查询方面上单纯使用 DAO 类中的方法并不能实现复杂的查询。为了弥补这个缺点 Hibernate 设计了 HQL 查询语言。HQL 查询语言非常类似于 SQL 语言，所以如果学习过 SQL 语言将会非常容易掌握 HQL 语言，本章中读者应该着重掌握如下内容：

- 参数绑定在 HQL 语言中的应用；
- where 语句在 HQL 语言中的应用；
- 分组在 HQL 语言中的应用；
- list()和 iterator()方法的区别。

8.1 Hibernate 的检索方式

Hibernate 检索方式大体有两种：**HQL（Hibernate Query Language）** 语言和基于面向对象方式的 **QBC（Query By Curiteria）** 检索。但由于 QBC 方式的检索在灵活性上没有 HQL 高，所以 HQL 语言在 Hibernate 的使用中占据了很大的份额。本章也就主要使用 HQL 语言来进行数据的查询检索。

在数据库中创建表 main13 和 sub13，数据库 main13 的结构如图 8-1 所示。

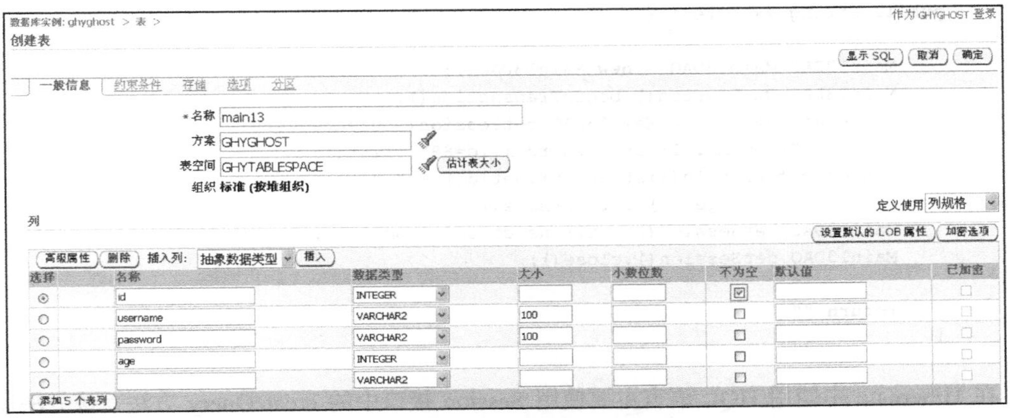

图 8-1　main13 数据表的结构

创建完 main13 表再继续创建一个主键对象。

sub13 数据表结构如图 8-2 所示。

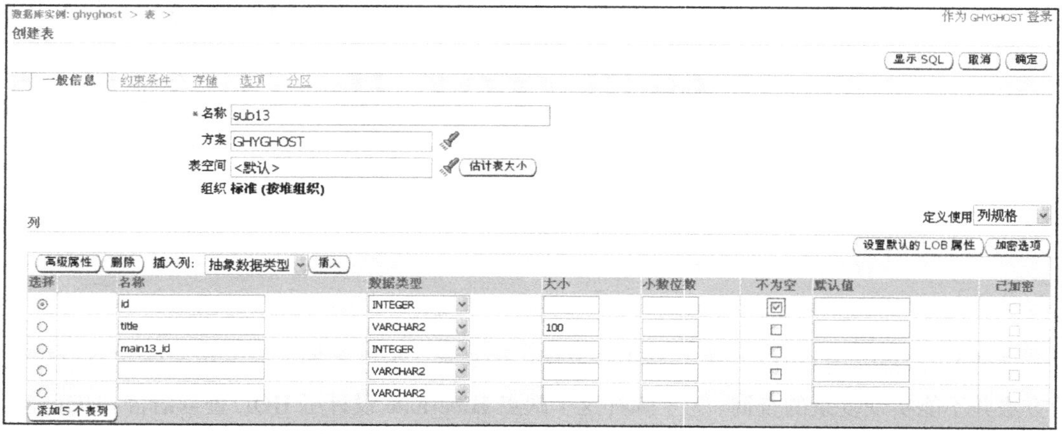

图 8-2 sub13 数据表结构

创建完 sub13 表再继续创建一个主键对象，也就是创建 main13 表和 sub13 表的主外关联关系。

在 sub13 的表结构中可以看出，这是典型的主从关系表。在下面的章节中将使用 HQL 语言对这主从关系表进行比较详实的检索，以让读者对 HQL 语言的掌握更加透析，并且可以熟练地应用到实际的软件项目开发中。

8.1.1 HQL 表别名

Hibernate 中的 HQL 语言和 SQL 非常类似，只不过 SQL 是直接将数据从数据表中取出，而 HQL 并不是这样。它不止从数据表中取出数据，还要进行 ORM 映射，把数据表中的 1 条记录转化成 1 个实体类或者是 1 个 Object[]对象数组。

在 SQL 语言中，可以在语句中使用 as 关键字来为表或字段名进行简化，比如 user_info 表可以使用 as 简写成：user_info as ui，那么 ui 就是数据表 user_info 的别名，使用 ui 就代表 user_info 数据表。在 HQL 语言中也支持 as 关键字的使用，代码如下。

```java
public String execute() {

    Main13DAO Main13DAO = new Main13DAO();
    Main13DAO.getSession().beginTransaction();
    List<Main13> list = Main13DAO.getSession().createQuery(
            "from Main13 as m where m.id=89").list();
    System.out.println(list.get(0).getId() + " "
            + list.get(0).getUsername());
    Main13DAO.getSession().getTransaction().commit();
    Main13DAO.getSession().close();

    return "";
}
```

在 Hibernate 中使用 HQL 语言主要使用 Session 接口中的 createQuery 方法，该方法通过 String 参数传入 HQL 语句，createQuery 方法返回的数据类型是 Query 接口。Query 接口中的 list()

方法返回 List 集合的实体类列表。

使用 as 关键字来为使用 HQL 语言的程序设计简化，比如例程中就将表名 Main13 使用 as 关键字简化成表名为 m 了，这样在涉及代码行数多的 HQL 语句中很常用。由于 HQL 的语句使用的是 from Main13，那么就相当于把 Main13 表中的所有字段取出来，也就是相当于取出了 Main13 实体类。所以可以将 List 设计成泛型的 List，这样在取出数据时就不需要数据类型强转。

程序运行结果如图 8-3 所示。

图 8-3 as 示例运行结果

如果 HQL 语句以 from 开头，则 List 中存储的就是实体类。

如果以 select 开头，并且只查询一列，则 List 中存储的是该列的数据类型，示例代码如下。

```
public class Test7_ {
    public static void main(String[] args) {
        Session session = HibernateSessionFactory.getSession();
        session.beginTransaction();
        Query query1 = session.createQuery("select u.id from Userinfo u");
        List<Long> listMain = query1.list();
        for (int i = 0; i < listMain.size(); i++) {
            Long eachId = listMain.get(i);
            System.out.println(eachId);
        }
        session.getTransaction().commit();
        session.close();
    }
}
```

如果 select 语句后面要查询两列以上，则 List 中存储的就是 Object[]对象数组，两个示例代码如下。

```
public class Test8_ {
    public static void main(String[] args) {
        Session session = HibernateSessionFactory.getSession();
        session.beginTransaction();
        Query query1 = session.createQuery("select u.id,u.username from Userinfo u");
        List<Object[]> listMain = query1.list();
        for (int i = 0; i < listMain.size(); i++) {
            Object[] objectArray = (Object[]) listMain.get(i);
            System.out.println(objectArray[0] + " " + objectArray[1]);
        }
        session.getTransaction().commit();
        session.close();
    }
}
```

查询 3 列的示例代码如下。

```
public class Test9_ {
    public static void main(String[] args) {
```

```java
            Session session = HibernateSessionFactory.getSession();
            session.beginTransaction();
            Query query1 = session.createQuery("select u.id,u.username,u.age from Userinfo u");
            List<Object[]> listMain = query1.list();
            for (int i = 0; i < listMain.size(); i++) {
                Object[] objectArray = (Object[]) listMain.get(i);
                System.out.println(objectArray[0] + " " + objectArray[1] + " " + objectArray[2]);
            }
            session.getTransaction().commit();
            session.close();
        }
    }
```

8.1.2 HQL 对结果进行排序与 list()和 iterator()方法的区别

可以在 HQL 语言中使用 order by 子句对结果集进行正排序 asc 和倒排序 desc，程序代码如下。

```java
            Main13DAO Main13DAO = new Main13DAO();
            Main13DAO.getSession().beginTransaction();
            List<Main13> list_orderby = Main13DAO.getSession().createQuery(
                    "from Main13 as m order by m.id desc").list();
            for (int i = 0; i < list_orderby.size(); i++) {
                System.out.println(list_orderby.get(i).getId() + " "
                        + list_orderby.get(i).getUsername());
            }

            Main13DAO.getSession().getTransaction().commit();
            Main13DAO.getSession().close();
```

程序输结果如下所示。

90 ghy_username2
89 ghy_username1

正排序为 order by m.id asc，与 desc 相反。

前面的示例中，查询都要使用 list()方法。其实 Hibernate 中还有一个方法也可以实现查询的功能，它就是 iteartor()，它们之间到底有什么区别呢？

创建名称为的 Web 项目，Servlet 核心代码如下。

```java
public class test extends HttpServlet {

    public void doGet(HttpServletRequest request, HttpServletResponse response)
            throws ServletException, IOException {
        Session session = HibernateSessionFactory.getSession();
        session.beginTransaction();

        List<Userinfo> listUserinfo = session.createQuery("from Userinfo")
                .list();
        for (int i = 0; i < listUserinfo.size(); i++) {
            Userinfo userinfo = listUserinfo.get(i);
            System.out.println(userinfo);
        }

        session.getTransaction().commit();
        session.close();
```

        }
    }

程序运行后在控制台输出如下结果。

```
Hibernate: select userinfo0_.id as id0_, userinfo0_.username as username0_, userinfo0_.password as password0_, userinfo0_.age as age0_ from ghydb.dbo.userinfo userinfo0_
    orm.Userinfo@b65bb8
    orm.Userinfo@1a704f4
    orm.Userinfo@365d27
```

上面的运行结果使用一条 select 查询语句查询出所有的 userinfo 信息。

Servlet 核心代码的更改如下。

```java
public class test extends HttpServlet {

    public void doGet(HttpServletRequest request, HttpServletResponse response)
            throws ServletException, IOException {
        Session session = HibernateSessionFactory.getSession();
        session.beginTransaction();

        Iterator<Userinfo> iteratorUserinfo = session.createQuery(
                "from Userinfo").iterate();
        while (iteratorUserinfo.hasNext()) {
            Userinfo userinfo = iteratorUserinfo.next();
            System.out.println(userinfo);
        }

        session.getTransaction().commit();
        session.close();

    }
}
```

程序运行后在控制台输出如下结果。

```
Hibernate: select userinfo0_.id as col_0_0_ from ghydb.dbo.userinfo userinfo0_
Hibernate: select userinfo0_.id as id0_0_, userinfo0_.username as username0_0_, userinfo0_.password as password0_0_, userinfo0_.age as age0_0_ from ghydb.dbo.userinfo userinfo0_ where userinfo0_.id=?
    orm.Userinfo@1052a99
Hibernate: select userinfo0_.id as id0_0_, userinfo0_.username as username0_0_, userinfo0_.password as password0_0_, userinfo0_.age as age0_0_ from ghydb.dbo.userinfo userinfo0_ where userinfo0_.id=?
    orm.Userinfo@18f5630
Hibernate: select userinfo0_.id as id0_0_, userinfo0_.username as username0_0_, userinfo0_.password as password0_0_, userinfo0_.age as age0_0_ from ghydb.dbo.userinfo userinfo0_ where userinfo0_.id=?
    orm.Userinfo@15a792d
```

从程序来看，使用 iterator()方法先是查询出 userinfo 表中的所有 id，再根据此 id 值重新发起新的 SQL 语句来进行查询 userinfo 表中的数据。

继续更改 Servlet 核心代码，如下所示。

```java
public class test extends HttpServlet {

    public void doGet(HttpServletRequest request, HttpServletResponse response)
```

```
        throws ServletException, IOException {
    Session session = HibernateSessionFactory.getSession();
    session.beginTransaction();

    List<Userinfo> listUserinfo = session.createQuery("from Userinfo")
            .list();
    for (int i = 0; i < listUserinfo.size(); i++) {
        Userinfo userinfo = listUserinfo.get(i);
        System.out.println(userinfo);
    }

    Iterator<Userinfo> iteratorUserinfo = session.createQuery(
            "from Userinfo").iterate();
    while (iteratorUserinfo.hasNext()) {
        Userinfo userinfo = iteratorUserinfo.next();
        System.out.println(userinfo);
    }

    session.getTransaction().commit();
    session.close();
        }
    }
```

程序运行后的结果如下。

```
Hibernate: select userinfo0_.id as id0_, userinfo0_.username as username0_,
userinfo0_.password as password0_, userinfo0_.age as age0_ from ghydb.dbo.userinfo userinfo0_
orm.Userinfo@d8d958
orm.Userinfo@f7657b
orm.Userinfo@27a30c
Hibernate: select userinfo0_.id as col_0_0_ from ghydb.dbo.userinfo userinfo0_
orm.Userinfo@d8d958
orm.Userinfo@f7657b
orm.Userinfo@27a30c
```

从上面的结果来看，iterator()方法并没有根据 id 重新发起新的查询请求，而是使用 Session 对象中缓存的 userinfo 对象。

继续更改 Servlet 代码，如下所示。

```
public class test extends HttpServlet {

    public void doGet(HttpServletRequest request, HttpServletResponse response)
            throws ServletException, IOException {
        Session session = HibernateSessionFactory.getSession();
        session.beginTransaction();

        Iterator<Userinfo> iteratorUserinfo = session.createQuery(
                "from Userinfo").iterate();
        while (iteratorUserinfo.hasNext()) {
            Userinfo userinfo = iteratorUserinfo.next();
            System.out.println(userinfo);
        }

        List<Userinfo> listUserinfo = session.createQuery("from Userinfo")
                .list();
        for (int i = 0; i < listUserinfo.size(); i++) {
            Userinfo userinfo = listUserinfo.get(i);
```

```
            System.out.println(userinfo);
        }

        session.getTransaction().commit();
        session.close();

    }
}
```

程序运行后输出如下结果。

```
    Hibernate: select userinfo0_.id as col_0_0_ from ghydb.dbo.userinfo userinfo0_
    Hibernate: select userinfo0_.id as id0_0_, userinfo0_.username as username0_0_,
userinfo0_.password as password0_0_, userinfo0_.age as age0_0_ from ghydb.dbo.userinfo
userinfo0_ where userinfo0_.id=?
    orm.Userinfo@eff179
    Hibernate: select userinfo0_.id as id0_0_, userinfo0_.username as username0_0_,
userinfo0_.password as password0_0_, userinfo0_.age as age0_0_ from ghydb.dbo.userinfo
userinfo0_ where userinfo0_.id=?
    orm.Userinfo@7d5b1a
    Hibernate: select userinfo0_.id as id0_0_, userinfo0_.username as username0_0_,
userinfo0_.password as password0_0_, userinfo0_.age as age0_0_ from ghydb.dbo.userinfo
userinfo0_ where userinfo0_.id=?
    orm.Userinfo@1f10c35
    Hibernate: select userinfo0_.id as id0_, userinfo0_.username as username0_,
userinfo0_.password as password0_, userinfo0_.age as age0_ from ghydb.dbo.userinfo userinfo0_
    orm.Userinfo@eff179
    orm.Userinfo@7d5b1a
    orm.Userinfo@1f10c35
```

从上面的运行结果来看，list()方法发起了 SQL 查询，将查询出来的结果和缓存中的结果进行对比，发现是同一条记录，则复用缓存中的 userinfo 对象。

方法 list()执行一条 SQL，方法 iterator()执行多条 SQL，这两个方法都可以复用缓存中的数据。

8.1.3 HQL 索引参数绑定

在 JDBC 中的接口 PreparedStatement 可以在执行 SQL 语句时以参数的形式进行设计，在 HQL 语言中也支持这种以 "?" 问号为参数形式的语句，示例如下。

```java
            Main13DAO Main13DAO = new Main13DAO();
            Main13DAO.getSession().beginTransaction();
            Query query_wenhaoParam = Main13DAO.getSession().createQuery(
                    "from Main13 as m where m.id =?");
            query_wenhaoParam.setInteger(0, 89);
            List<Main13> list_wenhaoParam = query_wenhaoParam.list();
            for (int i = 0; i < list_wenhaoParam.size(); i++) {
                System.out.println(list_wenhaoParam.get(i).getId() + " "
                        + list_wenhaoParam.get(i).getUsername());
            }
            Main13DAO.getSession().getTransaction().commit();
            Main13DAO.getSession().close();
```

程序运行的顺序首先使用 createQuery 方法来准备 HQL 语句，HQL 语句查询 Main13 表，但条件是 m.id=?，这个 "?" 代表传的是参数，具有"占位"的功能，createQuery 方法的返回值是一个 Query 对象，那么这个具体的参数值是如何传进去的？是通过 Query 接口中的

setInteger 方法来达到的，setInteger 方法的第 1 个参数是“?”的索引，是从 0 开始的，0 代表第一个占位符参数（问号），通过对“?”参数传值，然后通过 Query 接口中的 list()方法返回一个 List 集合对象，再通过 for 语句输出，程序的执行结果如下所示。

```
89 ghy_username1
```

8.1.4 HQL 命名参数绑定与安全性

前面章节介绍的是使用“?”占位的方式进行传递参数，那么这种方式很简单地解决了使用 HQL 参数传递的问题。但是如果在复杂的 HQL 语句，并且参数数目较多，而且参数的位置经常变动的情况下，单纯的使用“?”占位传递参数的方式就有些不适合正规的软件项目，HQL 的命名参数正是解决这样的问题。它可以将参数由“?”转变成一个参数名，然后根据参数名进行相关的传值。这样不仅可以解决参数位置混乱的问题，也有助于软件的后期维护，示例如下。

```
Main13DAO Main13DAO = new Main13DAO();
Main13DAO.getSession().beginTransaction();
Query query_wenhaoParam = Main13DAO.getSession().createQuery(
        "from Main13 as m where m.id =:pk_id");
query_wenhaoParam.setInteger("pk_id", 90);
List<Main13> list_wenhaoParam = query_wenhaoParam.list();
for (int i = 0; i < list_wenhaoParam.size(); i++) {
    System.out.println(list_wenhaoParam.get(i).getId() + " "
            + list_wenhaoParam.get(i).getUsername());
}

Main13DAO.getSession().getTransaction().commit();
Main13DAO.getSession().close();
```

程序运行结果如下。

```
90 ghy_username2
```

命名参数的格式如下。

```
:参数名称
```

参数名称前面要加上一个冒号以区分这是一个命名式的参数。

为了防止 SQL 注入，所以在设计程序代码时作者建议使用传参的形式来增加后台数据库内容的安全性。

命名参数还有一种使用方法，就是参数可以用一个实体类进行封装，但参数的名称必须和实体类的属性名一致，以便于反射调用。

创建数据表 userinfo，表内容如图 8-4 所示。

TC03\SQL200... dbo.userinfo | TC03\SQL200...b - dbo.shi

	id	username	password	age
	1	高洪岩1	高洪岩11	1
▶	2	高洪岩2	高洪岩22	2
	3	abc	xyz	3
*	NULL	NULL	NULL	NULL

图 8-4 userinfo 数据表内容

创建实体类 UserinfoParam.java 代码结构，如图 8-5 所示。

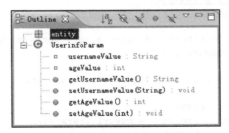

图 8-5　代码结构

创建 Servlet，核心代码如下。

```java
public class test extends HttpServlet {
    public void doGet(HttpServletRequest request, HttpServletResponse response)
            throws ServletException, IOException {

        UserinfoParam upRef = new UserinfoParam();
        upRef.setUsernameValue("%高洪岩%");
        upRef.setAgeValue(2);

        Session session = HibernateSessionFactory.getSession();
        session.beginTransaction();
        Query query = session
                .createQuery("from Userinfo u where u.username like :usernameValue and u.age=:ageValue");
        query.setProperties(upRef);
        List<Userinfo> listUserinfo = query.list();
        for (int i = 0; i < listUserinfo.size(); i++) {
            Userinfo userinfo = listUserinfo.get(i);
            System.out
                    .println(userinfo.getUsername() + " " + userinfo.getAge());
        }
        session.getTransaction().commit();
        session.close();
    }

}
```

运行结果如图 8-6 所示。

图 8-6　查询出 1 条记录

查询条件也可以封装进 Map 中，示例代码如下。

```java
public class Test19 {

    public static void main(String[] args) {
```

```java
            Map map = new HashMap();
            map.put("idKey", 2032496L);
            map.put("usernameKey", "a");

            Session session = HibernateSessionFactory.getSession();
            session.beginTransaction();
            Query query1 = session.createQuery("from Userinfo u where u.id=:idKey and u.username=:usernameKey");
            query1.setProperties(map);
            Iterator<Userinfo> iterator = query1.iterate();
            while (iterator.hasNext()) {
                Userinfo userinfo = iterator.next();
                System.out.println(userinfo.getId() + " " + userinfo.getUsername() + " " + userinfo);
            }
            session.getTransaction().commit();
            session.close();
        }
    }
```

8.1.5 HQL 方法链的使用

前面章节使用命名参数的形式进行 HQL 的查询，传递参数时每次执行 query.setType()这样的方法，这时可以使用方法链的功能来进行简化开发，程序代码如下。

```java
            Main13DAO Main13DAO = new Main13DAO();
            Main13DAO.getSession().beginTransaction();
            List<Main13> list_wenhaoParam = Main13DAO.getSession().createQuery(
                    "from Main13 as m where m.id =:pk_id").setInteger("pk_id", 89)
                    .list();
            for (int i = 0; i < list_wenhaoParam.size(); i++) {
                System.out.println(list_wenhaoParam.get(i).getId() + " "
                        + list_wenhaoParam.get(i).getUsername());
            }
            Main13DAO.getSession().getTransaction().commit();
            Main13DAO.getSession().close();
```

在程序中使用如下代码。

```java
Main13DAO.getSession().createQuery(
    "from Main13 as m where m.id =:pk_id")
.setInteger("pk_id", 89).list();
```

连贯的方法调用可以节省大量的开发工作，但同时也带来软件代码不易读的反作用，这也是技术有好处也有坏处的体现。善用每一个技术点增加软件开发的方便性。

8.1.6 HQL 中的 uniqueResult 方法的使用

正常的情况下，如果按照以前学习过的 HQL 的写法，查询的结果只有一条记录时，也放入 List 中，然后再通过 List.get(0)的形式来返回实体类。其实，在 Hibernate 中，如果确认结果集中只有一条记录，那么完全可以使用 uniqueResult 方法来直接取出一个实体类，示例如下。

```java
            Main13DAO Main13DAO = new Main13DAO();
```

```
Main13DAO.getSession().beginTransaction();
Main13 main13_bean = (Main13) Main13DAO.getSession().createQuery(
        "from Main13 as m where m.id =:pk_id").setInteger("pk_id", 89)
        .uniqueResult();
System.out.println(main13_bean.getId() + " "
        + main13_bean.getUsername());

Main13DAO.getSession().getTransaction().commit();
Main13DAO.getSession().close();
```

从程序中可以看到，根据主键 id 进行查询结果集肯定只有一条记录，那么就可以使用 uniqueResult()方法来直接取出一个实体类然后再输出。

8.1.7　HQL 中的 where 子句与查询条件

HQL 语言也支持 where 条件子句，可以支持的运算包括：=、<>、>、>=、<、<=、is null、is not null、in、not in、between、not between、like、and、or 和 not。

下面将用一个综合性的程序示例，来演示一下这些查询条件的用法。

程序示例如下。

```
// >98 开始
Main13DAO Main13DAO = new Main13DAO();
Main13DAO.getSession().beginTransaction();
List<Main13> list_dayu = Main13DAO.getSession().createQuery(
        "from Main13 as m where m.age >89").list();
for (int i = 0; i < list_dayu.size(); i++) {
    System.out.println(">98 的示例：" + list_dayu.get(i).getId() + " "
            + list_dayu.get(i).getUsername() + " "
            + list_dayu.get(i).getAge());
}
// >98 结束

// <>99 开始
List<Main13> list_budengyu = Main13DAO.getSession().createQuery(
        "from Main13 as m where m.age <>90").list();
for (int i = 0; i < list_budengyu.size(); i++) {
    System.out.println("<>99 的示例：" + list_budengyu.get(i).getId() + " "
            + list_budengyu.get(i).getUsername() + " "
            + list_budengyu.get(i).getAge());
}
// <>99 结束

// is not null 开始
List<Main13> list_feikong = Main13DAO.getSession().createQuery(
        "from Main13 as m where m.password is not null").list();
if (list_feikong.size() != 0) {
    for (int i = 0; i < list_feikong.size(); i++) {
        System.out.println("is not null 的示例：" 
                + list_feikong.get(i).getId() + " "
                + list_feikong.get(i).getUsername() + " "
                + list_feikong.get(i).getAge() + " "
                + list_feikong.get(i).getPassword());
    }
} else {
    System.out.println("is not null 的示例　没有找到匹配的记录");
```

```java
        }
        // is not null 结束

        // in 开始
        List<Main13> list_in = Main13DAO.getSession().createQuery(
                "from Main13 as m where m.age in (89,90)").list();
        for (int i = 0; i < list_in.size(); i++) {
            System.out.println("in 的示例: " + list_in.get(i).getId() + " "
                    + list_in.get(i).getUsername() + " "
                    + list_in.get(i).getAge() + " "
                    + list_in.get(i).getPassword());
        }
        // in 结束

        // notin 开始
        List<Main13> list_notin = Main13DAO.getSession().createQuery(
                "from Main13 as m where m.age not in (89,90)").list();
        for (int i = 0; i < list_notin.size(); i++) {
            System.out.println("notin 的示例: " + list_notin.get(i).getId() + " "
                    + list_notin.get(i).getUsername() + " "
                    + list_notin.get(i).getAge() + " "
                    + list_notin.get(i).getPassword());
        }
        // notin 结束

        // between 开始
        List<Main13> list_between = Main13DAO.getSession().createQuery(
                "from Main13 as m where m.age between 89 and 90").list();
        for (int i = 0; i < list_between.size(); i++) {
            System.out.println("between 的示例: " + list_between.get(i).getId()
                    + " " + list_between.get(i).getUsername() + " "
                    + list_between.get(i).getAge() + " "
                    + list_between.get(i).getPassword());
        }
        // between 结束

        // not between 开始
        List<Main13> list_notbetween = Main13DAO.getSession().createQuery(
                "from Main13 as m where m.age not between 90 and 91").list();
        for (int i = 0; i < list_notbetween.size(); i++) {
            System.out.println("not between 的示例: "
                    + list_notbetween.get(i).getId() + " "
                    + list_notbetween.get(i).getUsername() + " "
                    + list_notbetween.get(i).getAge() + " "
                    + list_notbetween.get(i).getPassword());
        }
        // not between 结束

        // not between 开始
        List<Main13> list_like = Main13DAO.getSession().createQuery(
                "from Main13 as m where username like '%ghy%'").list();
        for (int i = 0; i < list_like.size(); i++) {
            System.out.println("like 的示例: " + list_like.get(i).getId() + " "
                    + list_like.get(i).getUsername() + " "
                    + list_like.get(i).getAge() + " "
                    + list_like.get(i).getPassword());
        }
        // not between 结束
```

```java
// _开始
List<Main13> list__ = Main13DAO.getSession().createQuery(
        "from Main13 as m where age like '9_'").list();
for (int i = 0; i < list__.size(); i++) {
    System.out.println("_的示例: " + list__.get(i).getId() + " "
            + list__.get(i).getUsername() + " "
            + list__.get(i).getAge() + " "
            + list__.get(i).getPassword());
}
// _结束

// and 开始
List<Main13> list_and = Main13DAO
        .getSession()
        .createQuery(
                "from Main13 as m where age like '9_' and username like '%name1%'")
        .list();
for (int i = 0; i < list_and.size(); i++) {
    System.out.println("and的示例: " + list_and.get(i).getId() + " "
            + list_and.get(i).getUsername() + " "
            + list_and.get(i).getAge() + " "
            + list_and.get(i).getPassword());
}
// and 结束

Main13DAO.getSession().getTransaction().commit();
Main13DAO.getSession().close();
```

程序主要使用了常用的运算符,程序运行结果如下。

```
>98 的示例: 89 ghy_username1 100
>98 的示例: 90 ghy_username2 200
<>99 的示例: 89 ghy_username1 100
<>99 的示例: 90 ghy_username2 200
is not null 的示例: 89 ghy_username1 100 ghy_password1
is not null 的示例: 90 ghy_username2 200 ghy_password2
notin 的示例: 89 ghy_username1 100 ghy_password1
notin 的示例: 90 ghy_username2 200 ghy_password2
not between 的示例: 89 ghy_username1 100 ghy_password1
not between 的示例: 90 ghy_username2 200 ghy_password2
like 的示例: 89 ghy_username1 100 ghy_password1
like 的示例: 90 ghy_username2 200 ghy_password2
```

读者可以根据自己的要求来将运算符进行组合查询以达到功能更灵活的目的。HQL 语句虽然没有 SQL 语言那样功能强大,但如果在一般的项目中,比如查询的操作时,使用 HQL 还是非常方便快捷的。

8.1.8 查询日期——字符串格式

Hibernate 也支持日期区间查询。本示例的日期来自于字符串,示例代码如下。

```java
package test2;

import java.sql.Timestamp;
import java.text.ParseException;
import java.text.SimpleDateFormat;
```

```
import java.util.Date;
import java.util.List;

import org.hibernate.Query;
import org.hibernate.Session;

import orm.HibernateSessionFactory;

public class Test27 {

    public static void main(String[] args) throws ParseException {
        String beginDateString = "2000-1-1";
        String endDateString = "2000-1-1";

        beginDateString = beginDateString + " 00:00:00";
        endDateString = endDateString + " 09:10:11";

        SimpleDateFormat format = new SimpleDateFormat("yyyy-MM-dd hh:mm:ss");
        Date beginDateObject = format.parse(beginDateString);
        Date endDateObject = format.parse(endDateString);

        Session session = HibernateSessionFactory.getSession();
        session.beginTransaction();
        Query query = session
                .createQuery("from Userinfo u where u.insertdate between :beginDate and :endDate order by id asc");
        query.setTimestamp("beginDate", new Timestamp(beginDateObject.getTime()));
        query.setTimestamp("endDate", new Timestamp(endDateObject.getTime()));
        List list = query.list();
        System.out.println(list.size());
        session.getTransaction().commit();
        session.close();
    }
}
```

8.1.9 查询日期——数字格式

Hibernate 也支持日期区间查询，本示例的日期来自于数字，示例代码如下。

```
package test2;

import java.sql.Timestamp;
import java.text.ParseException;
import java.util.Calendar;
import java.util.Date;
import java.util.List;

import org.hibernate.Query;
import org.hibernate.Session;

import oracle.sql.TIMESTAMP;
import orm.HibernateSessionFactory;

public class Test29 {

    public static void main(String[] args) throws ParseException {

        Calendar calendarBegin = Calendar.getInstance();
```

```java
            calendarBegin.set(Calendar.YEAR, 2000);
            calendarBegin.set(Calendar.MONTH, 0);
            calendarBegin.set(Calendar.DAY_OF_MONTH, 1);
            calendarBegin.set(Calendar.HOUR_OF_DAY, 0);
            calendarBegin.set(Calendar.MINUTE, 0);
            calendarBegin.set(Calendar.SECOND, 0);
            calendarBegin.set(Calendar.MILLISECOND, 0);

            Calendar calendarEnd = Calendar.getInstance();
            calendarEnd.set(Calendar.YEAR, 2000);
            calendarEnd.set(Calendar.MONTH, 0);
            calendarEnd.set(Calendar.DAY_OF_MONTH, 1);
            calendarEnd.set(Calendar.HOUR_OF_DAY, 23);
            calendarEnd.set(Calendar.MINUTE, 59);
            calendarEnd.set(Calendar.SECOND, 59);
            calendarEnd.set(Calendar.MILLISECOND, 0);

            Date beginDate = calendarBegin.getTime();
            Date endDate = calendarEnd.getTime();

            System.out.println(beginDate.toLocaleString());
            System.out.println(endDate.toLocaleString());

            Session session = HibernateSessionFactory.getSession();
            session.beginTransaction();
            Query query = session
                    .createQuery("from Userinfo u where u.insertdate between :beginDate and :endDate order by id asc");

            query.setTimestamp("beginDate", new Timestamp(beginDate.getTime()));
            query.setTimestamp("endDate", new Timestamp(endDate.getTime()));

            List list = query.list();
            System.out.println(list.size());
            session.getTransaction().commit();
            session.close();
        }
    }
```

8.1.10 分页的处理

在 Hibernate 中也支持对查询结果进行分页，示例代码如下。

```java
package test2;

import java.text.ParseException;
import java.util.List;

import org.hibernate.Query;
import org.hibernate.Session;

import orm.HibernateSessionFactory;
import orm.Userinfo;

public class Test30 {

    public static void main(String[] args) throws ParseException {
        String gotoPage = "3";
```

```java
            int gotoPageInt = 1;
            try {
                gotoPageInt = Integer.parseInt(gotoPage);
                if (gotoPageInt <= 0) {
                    gotoPageInt = 1;
                }
            } catch (NumberFormatException e) {
                gotoPageInt = 1;
            }

            int pageSize = 3;
            int beginPosition = (gotoPageInt - 1) * pageSize;

            Session session = HibernateSessionFactory.getSessionFactory().openSession();
            session.beginTransaction();
            Query query = session.createQuery("from Userinfo u order by id asc");
            query.setFirstResult(beginPosition);
            query.setMaxResults(pageSize);
            List<Userinfo> listUserinfo = query.list();
            for (int i = 0; i < listUserinfo.size(); i++) {
                Userinfo userinfo = listUserinfo.get(i);
                System.out.println(userinfo.getId() + " " + userinfo.getUsername());
            }
            session.getTransaction().commit();
            session.close();
        }
    }
```

8.1.11 HQL 中的聚集函数：distinct-count-min-max-sum-avg

设计代码如下。

```java
package controller;

import java.util.List;

import org.hibernate.Transaction;

import orm.Main13;
import orm.Main13DAO;
import orm.Sub13DAO;

public class Test {

    public String execute() {

        Sub13DAO Sub13DAO = new Sub13DAO();
        Sub13DAO.getSession().beginTransaction();
        List list_distinct = Sub13DAO.getSession().createQuery(
                "select distinct main13 from Sub13").list();
        for (int i = 0; i < list_distinct.size(); i++) {
            System.out.println("distinct 示例: "
                    + ((Main13) list_distinct.get(i)).getId());
        }
        Sub13DAO.getSession().getTransaction().commit();
        Sub13DAO.getSession().close();

        // 在 distinct 示例代码中，通过使用 HQL 语句：select distinct main13 from Sub13
        // 来取得在 Sub13 子表中关联主表字段的列表（被过滤），在 list 中每一个元素的数据类型是 Main13
```

```java
// 所以从 list 中取出数据就相当的容易，当然 list 里面存储具体是什么类型是通过调试查出来的
// distinct 也可以过滤掉简单类型的字段，比如年龄这些

Main13DAO Main13DAO = new Main13DAO();
Main13DAO.getSession().beginTransaction();
List list_count = Main13DAO.getSession().createQuery(
        "select count(*) from Main13").list();
for (int i = 0; i < list_count.size(); i++) {
    System.out.println("count 示例: " + list_count.get(i));
}
Main13DAO.getSession().getTransaction().commit();
Main13DAO.getSession().close();
// HQL 语句: select count(*) from Main13
// 查询数据表中具体有多少条记录
// 在 HQL 语句中关于打印的方式到底是直接打印 list 中的元素，还是打印 list 中的 Object[] 数组对象
// 是根据 HQL 语句的写法得来的，比如语句: select count(*) from Main13 它投影出一个字段
// 那么就可以直接打印 list 中的值了，因为 list 中的每一个元素就是要取对象类型的值

Main13DAO Main13DAO_all = new Main13DAO();
Main13DAO_all.getSession().beginTransaction();
List list_all = Main13DAO_all.getSession().createQuery(
        "select min(age),max(age),sum(age),avg(age) from Main13")
        .list();
for (int i = 0; i < list_all.size(); i++) {
    System.out.println("min 示例: " + ((Object[]) list_all.get(i))[0]);
    System.out.println("max 示例: " + ((Object[]) list_all.get(i))[1]);
    System.out.println("sum 示例: " + ((Object[]) list_all.get(i))[2]);
    System.out.println("avg 示例: " + ((Object[]) list_all.get(i))[3]);
}
Main13DAO_all.getSession().getTransaction().commit();
Main13DAO_all.getSession().close();
// HQL 语句: select min(age),max(age),sum(age),avg(age) from Main13
// 和 HQL 语句: select count(*) from Main13 在结构上的不同即是前者查询的是多个值
// 而后者是一个值，这样的情况根据上一个示例的代码解释中可以得到答案，
// 取出其中的值根据 select 子句的写法而定
// 比如 select min(age),max(age),sum(age),avg(age) from Main13 这样的写法
// list 中存储的肯定是 Object[] 数组对象类型了
// 而 select count(*) from Main13 中 list 中元素存储的就是对象
// 所以读者在以后的 HQL 代码设计中，取值的方式决定于 select 的写法

return "";

    }

}
```

部属项目，启动 Tomcat，输入指定的网址后，控制台输出如图 8-7 所示的结果。

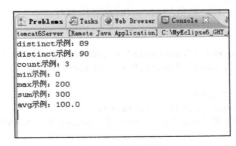

图 8-7　打印结果

8.1.12　HQL 中的分组查询

设计代码如下。

```java
package controller;

import java.util.List;

import org.hibernate.Transaction;

import orm.Main13DAO;

public class Test {

    public String execute() {

        Main13DAO Main13DAO = new Main13DAO();
        Transaction tra_group_count = Main13DAO.getSession().beginTransaction();
        List list_group_count = Main13DAO
                .getSession()
                .createQuery(
                        "select m.username,count(id) from Main13 as m group by m.username")
                .list();
        for (int i = 0; i < list_group_count.size(); i++) {

            System.out.println("username:"
                    + ((Object[]) list_group_count.get(i))[0]);
            System.out.println("count:"
                    + ((Object[]) list_group_count.get(i))[1]);
        }
        tra_group_count.commit();
        Main13DAO.getSession().close();
        // 上面代码中通过对 Main13 表中的 username 用户名属性进行分组来统计出有多少位是重名的
        // 由于 HQL 语句中的 select 子句：select m.username,count(id) 中的字段数目大于 1
        // 所以 list 中存储的是 Object[] 对象数组类型，强转后取到值并且输出

        // 下面的代码是查询出相同姓名的数量，并且对分组进行了条件限制
        Main13DAO Main13DAO_having = new Main13DAO();
        Transaction tra_group_having = Main13DAO_having.getSession()
                .beginTransaction();
        List list_having = Main13DAO
                .getSession()
                .createQuery(
                        "select m.username,count(id) from Main13 as m group by m.username having (count(id)>1)")
                .list();
        for (int i = 0; i < list_having.size(); i++) {

            System.out
                    .println("username:" + ((Object[]) list_having.get(i))[0]);
            System.out.println("count:" + ((Object[]) list_having.get(i))[1]);
        }
        tra_group_having.commit();
        Main13DAO_having.getSession().close();
        // 由于 having by 的显示条件是 count>1 才显示记录，所以在这里控制台不输出任何信息

        return "";
```

		}
	}

控制台打印如图 8-8 所示。

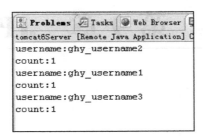

图 8-8　输出结果

第 9 章　JPA 核心技能

本章将介绍 JPA 接口的使用。JPA 可以规范化 ORM 框架的 API，有助于项目写法的统一性。JPA 是官方制订的 Java 数据持久化标准，主流的第三方 ORM 框架大部分都实现了 JPA 接口。学习 JPA 就是在学 EntityManager 接口的使用，该接口提供了对数据库的基本操作，与 Hibernate 中的 Session 接口以及 MyBatis 中的 SqlSession 接口作用一样。

9.1　什么是 JPA 以及为什么要使用 JPA

JPA（Java Persistence API-Java，持久化 API）是由 Sun 发布的数据持久化接口标准。在没有 JPA 的情况下，程序员需要掌握多种 ORM 框架的 API，如图 9-1 所示。

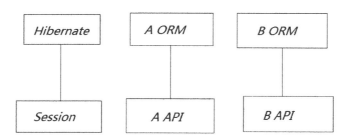

图 9-1　程序员需要掌握 3 套 API

而使用 JPA 规范后，程序员只需要掌握 1 套 API 即可。这样即规范了程序员的写法，也规范了各大开源厂商的 API 标准，两全之事，如图 9-2 所示。

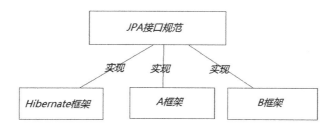

图 9-2　JPA 规范与实现

通过使用 JPA 规范，程序员只需要掌握一套 API 即可，但实现操作数据库的 ORM 框架却是不同的。这些 ORM 框架都实现了 JPA 规范，这就使程序员的学习成本大大降低，提高了软件项目中代码写法的统一性。

以前学习过 JDBC 接口规范，那么 JDBC 和 JPA 有什么区别呢？区别如图 9-3 所示。

在图中可以发现两者所在的位置是不同的，作用当然也不一样：

- JDBC 的作用就是规范使用 Java 访问数据库的 API 标准；
- JPA 是规范 ORM 框架的 API 标准。

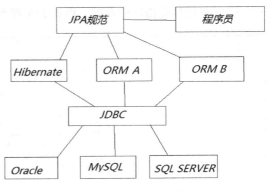

图 9-3　JDBC 和 JPA 的区别

9.2　搭建 JPA 开发环境与逆向

创建 Web 项目 jpaTest，配置界面如图 9-4 所示。

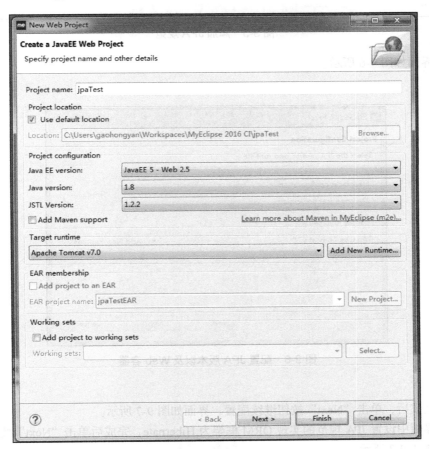

图 9-4　创建 Web 项目

配置完成后，单击"Finish"按钮，完成创建 Web 项目。

对 Web 项目点击鼠标右键，添加 JPA 规范，菜单如图 9-5 所示。

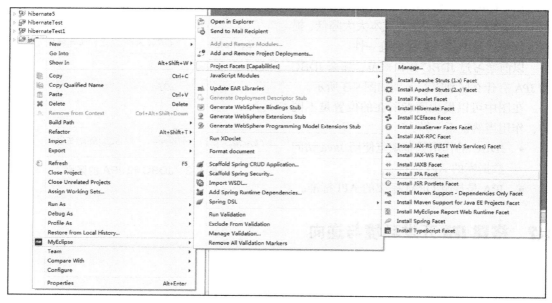

图 9-5　添加 JPA 规范

弹出界面如图 9-6 所示。

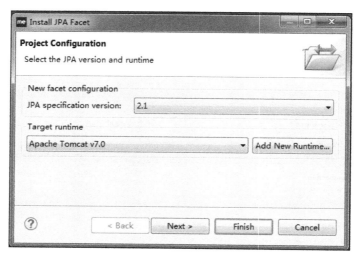

图 9-6　配置 JPA 版本以及 Web 容器

配置完成后，单击"Next"按钮继续配置，界面如图 9-7 所示。

在该界面中设置 JPA 规范的实现 ORM 框架为 Hibernate，完成后单击"Next"按钮继续配置，界面如图 9-8 所示。

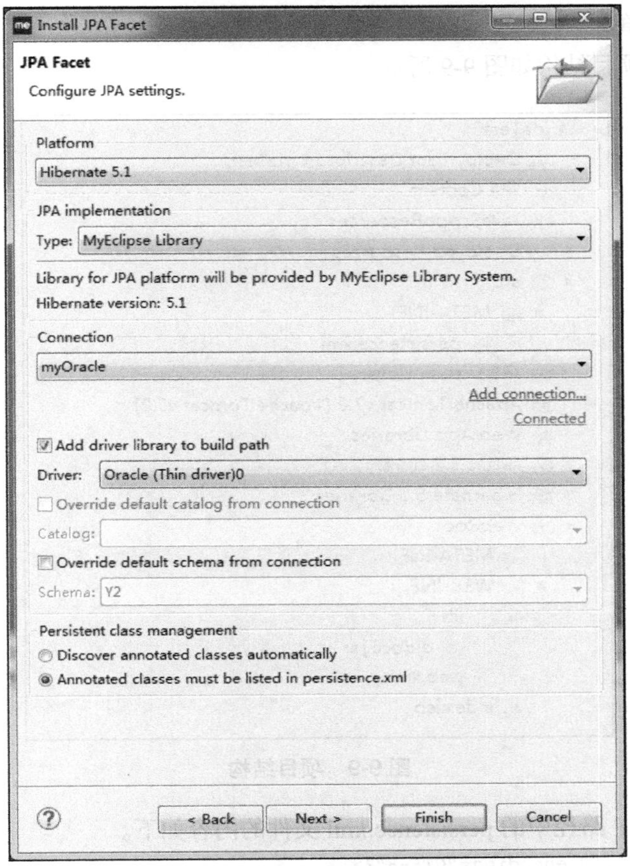

9.2 搭建 JPA 开发环境与逆向

图 9-7 配置 ORM 相关的参数

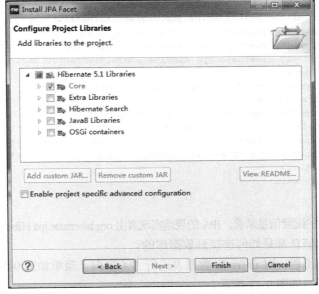

图 9-8 添加核心 JAR 包

配置结束后，单击"Finish"按钮，完成 JPA 的添加操作。
添加完 JPA 的项目结构如图 9-9 所示。

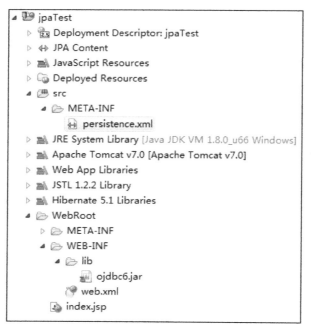

图 9-9　项目结构

在 src\META-INF 路径中的 persistence.xml 文件的内容如下。

```xml
<?xml version="1.0" encoding="UTF-8"?>
<persistence version="2.1"
    xmlns="http://xmlns.jcp.org/xml/ns/persistence"
    xmlns:xsi="http://www.w3.org/2001/XMLSchema-instance"
    xsi:schemaLocation="http://xmlns.jcp.org/xml/ns/persistence
http://xmlns.jcp.org/xml/ns/persistence/persistence_2_1.xsd">
    <persistence-unit name="jpaTest" transaction-type="RESOURCE_LOCAL">
        <provider>org.hibernate.jpa.HibernatePersistenceProvider</provider>
        <properties>
            <property name="hibernate.connection.driver_class" value="oracle.jdbc.OracleDriver" />
            <property name="hibernate.connection.url" value="jdbc:oracle:thin:@localhost:1521:orcl" />
            <property name="hibernate.connection.username" value="y2" />
            <property name="hibernate.connection.password" value="123" />
        </properties>
    </persistence-unit>
</persistence>
```

从<provider>标签的配置信息来看，JPA 的规范实现者由 org.hibernate.jpa.HibernatePersistenceProvider 类进行提供，其他的信息都是如何连接到数据库的。

下一步就是逆向工程，对 userinfo 数据表进行 JPA 逆向，菜单如图 9-10 所示。
弹出界面并配置，如图 9-11 所示。

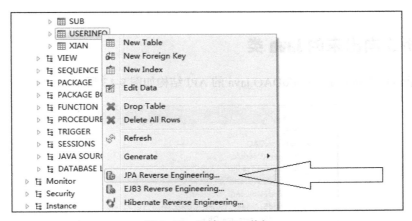

图 9-10 开始 JPA 逆向

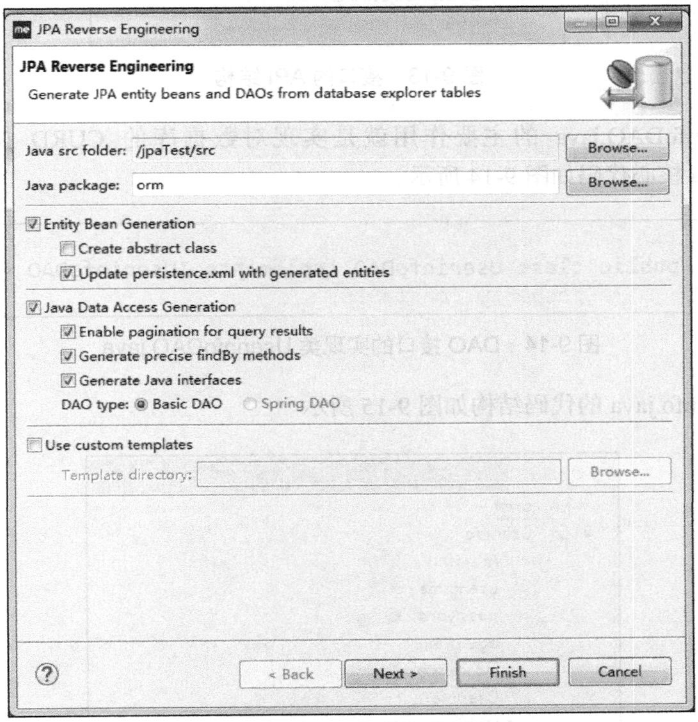

图 9-11 配置逆向选项

单击"Finish"按钮，完成 JPA 逆向工程。

逆向成功后在项目中的 src 结点下出现若干 Java 类，如图 9-12 所示。

逆向成功之后，在 persistence.xml 文件中将实体类 Userinfo.java 进行注册，代码如下。

```
<class>orm.Userinfo</class>
```

接下来将分析这些 Java 类的作用与关系。

图 9-12 逆向出来的 Java 类

9.3 分析逆向出来的 Java 类

数据访问 DAO 接口 IUserinfoDAO.java 的 API 结构如图 9-13 所示。

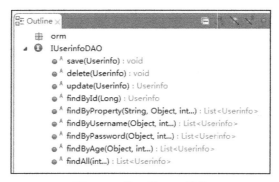

图 9-13 接口的 API 结构

接口 IUserinfoDAO.java 的主要作用就是实现对数据库的 CURD 操作，其实现类 UserinfoDAO.java 核心代码如图 9-14 所示。

```
18   */
19  public class UserinfoDAO implements IUserinfoDAO {
20      // property constants
```

图 9-14 DAO 接口的实现类 UserinfoDAO.java

实体类 Userinfo.java 的代码结构如图 9-15 所示。

图 9-15 实体类 Userinfo.java 的代码结构

9.4 使用 IUserinfoDAO.java 接口中的方法

实体类 Userinfo.java 中注解配置解释如下。

```
// @Entity：代表本类是一个实体类，与数据表进行映射
// @Table：定义该实体类与哪个数据表进行映射
// schema = "Y2"：在 Oracle 中是用户
// name = "USERINFO"：Userinfo.java 映射 USERINFO 数据表
@Entity
@Table(name = "USERINFO", schema = "Y2")
public class Userinfo implements java.io.Serializable {
```

在实体类的每个 get()方法的上面也有相关的注解，解释如下。

```
// @Column：当前属性映射数据表中的列
// name = "USERNAME"：映射 USENRAME 列
// length = 50：该列只能存储最大 50 个字符
@Column(name = "USERNAME", length = 50)
public String getUsername() {
    return this.username;
}
```

在实体类中使用@Id 注解来对主键进行映射，示例代码如下。

```
@Id
@Column(name = "ID", unique = true, nullable = false, precision = 18, scale = 0)
public Long getId() {
    return this.id;
}
```

代表当前 Userinfo.java 类中的 id 属性映射 userinfo 表中的 id 列。

下面来看一下最重要的 EntityManagerHelper.java 类，该类的 API 结构如图 9-16 所示。

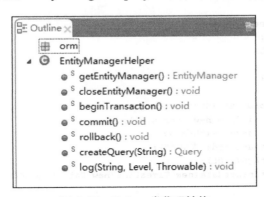

图 9-16 Helper 类代码结构

类 EntityManagerHelper.java 提供了获取 EntityManager 对象、提交事务、回滚事务、关闭连接等方法。

下面开始介绍接口中方法的基本使用。

9.4 使用 IUserinfoDAO.java 接口中的方法

DAO 接口中的方法提供了操作数据库的基本方式，对初步掌握 JPA 技术有着重要作用，方法列表如图 9-17 所示。

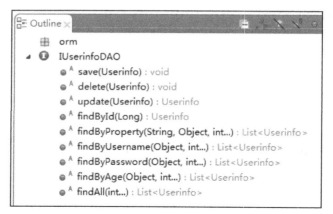

图 9-17　DAO 接口的 API 结构

建议在使用这些方法的同时，研究一下它们内部的源代码是如何实现的，这对学习 JPA 的 API 有相当大的好处。

9.4.1　方法 public void save(Userinfo entity)的使用

示例代码如下。

```
package test;

import java.sql.Timestamp;
import java.util.Date;

import orm.EntityManagerHelper;
import orm.Userinfo;
import orm.UserinfoDAO;

public class Test1 {

    public static void main(String[] args) {
        Userinfo userinfo = new Userinfo();
        userinfo.setUsername("中国");
        userinfo.setPassword("大中国");
        userinfo.setAge(100L);
        userinfo.setInsertdate(new Timestamp(new Date().getTime()));

        UserinfoDAO dao = new UserinfoDAO();
        EntityManagerHelper.beginTransaction();

        dao.save(userinfo);

        EntityManagerHelper.commit();
        EntityManagerHelper.closeEntityManager();

    }

}
```

程序运行后在控制台出现的异常信息如下。

javax.persistence.PersistenceException: org.hibernate.id.IdentifierGenerationException:

9.4　使用 IUserinfoDAO.java 接口中的方法

```
ids for this class must be manually assigned before calling save(): orm.Userinfo
```

从异常信息中可以得知，是因为 ID 值并没有受到管理。因为操作的数据库是 Oracle，必须得通过序列进行主键值的生成，解决这个异常得在 Userinfo.java 实体类中的 getId()方法上面添加注解即可，完整的示例代码如下。

```java
@Id
@Column(name = "ID", unique = true, nullable = false, precision = 18, scale = 0)
// @SequenceGenerator 作用：在 JPA 环境中要使用 Oracle 中的序列
// 所以要使用这个注解
// sequenceName = "idauto"：使用 Oracle 名称为 idauto 的注解
// name = "idautoRef"：在 JPA 环境中要使用 Oracle 中的 idauto 注解
// idautoRef 是 idauto 序列的别名
@SequenceGenerator(name = "idautoRef", sequenceName = "idauto")
// @GeneratedValue：生成主键 id 值，如何生成呢？有多种策略！
// 使用 strategy = GenerationType.SEQUENCE 的含义就是使用序列这种策略来生成 id 值，
// 数据库中的序列有很多，使用哪个序列呢？由 generator = "idautoRef"别名来确定！
@GeneratedValue(strategy = GenerationType.SEQUENCE, generator = "idautoRef")
public Long getId() {
    return this.id;
}
```

重新运行程序，成功添加了一条记录，控制台输出信息如图 9-18 所示。

```
信息: saving Userinfo instance
                        orm.EntityManagerHelper log
信息: save successful
```

图 9-18　成功添加一条记录

9.4.2　方法 public Userinfo findById(Long id)的使用

示例代码如下。

```java
public class Test2 {
    public static void main(String[] args) {
        UserinfoDAO dao = new UserinfoDAO();
        EntityManagerHelper.beginTransaction();
        Userinfo userinfo = dao.findById(2082444L);
        System.out.println(userinfo.getId() + " " + userinfo.getUsername() + " " + userinfo.getPassword() + " "
                + userinfo.getInsertdate());
        EntityManagerHelper.commit();
        EntityManagerHelper.closeEntityManager();
    }
}
```

程序运行后，控制台输出信息如图 9-19 所示。

```
2082444  中国 大中国              21:18:40.0
```

图 9-19　控制台输出 Userinfo 信息

9.4.3 方法 public List<Userinfo> findByProperty(String propertyName, final Object value, final int... rowStartIdxAndCount)的使用

示例代码如下。

```java
public class Test3 {
    public static void main(String[] args) {
        UserinfoDAO dao = new UserinfoDAO();
        EntityManagerHelper.beginTransaction();
        List<Userinfo> listUserinfo = dao.findByProperty("username", "中国", 0, 100);
        for (int i = 0; i < listUserinfo.size(); i++) {
            Userinfo userinfo = listUserinfo.get(i);
            System.out.println(userinfo.getId() + " " + userinfo.getUsername() + " " + userinfo.getPassword() + " "
                    + userinfo.getInsertdate());
        }
        EntityManagerHelper.commit();
        EntityManagerHelper.closeEntityManager();
    }
}
```

9.4.4 方法 public List<Userinfo> findByUsername(Object username, int... rowStartIdxAndCount)的使用

示例代码如下。

```java
public class Test5 {
    public static void main(String[] args) {
        UserinfoDAO dao = new UserinfoDAO();
        EntityManagerHelper.beginTransaction();
        List<Userinfo> listUserinfo = dao.findByUsername("中国", 0, 100);
        for (int i = 0; i < listUserinfo.size(); i++) {
            Userinfo userinfo = listUserinfo.get(i);
            System.out.println(userinfo.getId() + " " + userinfo.getUsername() + " " + userinfo.getPassword() + " "
                    + userinfo.getInsertdate());
        }
        EntityManagerHelper.commit();
        EntityManagerHelper.closeEntityManager();
    }
}
```

9.4.5 方法 public List<Userinfo> findByPassword(Object password, int... rowStartIdxAndCount)的使用

示例代码如下。

```java
public class Test6 {
    public static void main(String[] args) {
        UserinfoDAO dao = new UserinfoDAO();
        EntityManagerHelper.beginTransaction();
        List<Userinfo> listUserinfo = dao.findByPassword("中国人", 0, 100);
        for (int i = 0; i < listUserinfo.size(); i++) {
            Userinfo userinfo = listUserinfo.get(i);
            System.out.println(userinfo.getId() + " " + userinfo.getUsername() + " " +
```

```
userinfo.getPassword() + " "
                            + userinfo.getInsertdate());
        }
        EntityManagerHelper.commit();
        EntityManagerHelper.closeEntityManager();
    }
}
```

9.4.6 方法 public List<Userinfo> findByAge(Object age, int... rowStartIdxAndCount)的使用

示例代码如下。

```
public class Test6 {
    public static void main(String[] args) {
        UserinfoDAO dao = new UserinfoDAO();
        EntityManagerHelper.beginTransaction();
        List<Userinfo> listUserinfo = dao.findByAge(100L, 0, 100);
        for (int i = 0; i < listUserinfo.size(); i++) {
            Userinfo userinfo = listUserinfo.get(i);
            System.out.println(userinfo.getId() + " " + userinfo.getUsername() + " " +
userinfo.getPassword() + " "
                            + userinfo.getInsertdate());
        }
        EntityManagerHelper.commit();
        EntityManagerHelper.closeEntityManager();
    }
}
```

9.4.7 方法 public List<Userinfo> findAll(final int... rowStartIdxAndCount)的使用

示例代码如下。

```
public class Test7 {
    public static void main(String[] args) {
        UserinfoDAO dao = new UserinfoDAO();
        EntityManagerHelper.beginTransaction();
        List<Userinfo> listUserinfo = dao.findAll(0, 100);
        for (int i = 0; i < listUserinfo.size(); i++) {
            Userinfo userinfo = listUserinfo.get(i);
            System.out.println(userinfo.getId() + " " + userinfo.getUsername() + " " +
userinfo.getPassword() + " "
                            + userinfo.getInsertdate());
        }
        EntityManagerHelper.commit();
        EntityManagerHelper.closeEntityManager();
    }
}
```

9.4.8 方法 public Userinfo update(Userinfo entity)的使用

示例代码如下。

```
public class Test8 {
    public static void main(String[] args) {
```

```java
        UserinfoDAO dao = new UserinfoDAO();
        EntityManagerHelper.beginTransaction();
        Userinfo userinfo = dao.findById(2082444L);
        userinfo.setUsername("xxxxxxxxxx");
        dao.update(userinfo);
        System.out.println(userinfo.getId() + " " + userinfo.getUsername() + " " +
userinfo.getPassword() + " "
                + userinfo.getInsertdate());
        EntityManagerHelper.commit();
        EntityManagerHelper.closeEntityManager();
    }
}
```

9.4.9 方法 public void delete(Userinfo entity)的使用

示例代码如下。

```java
public class Test9 {
    public static void main(String[] args) {
        UserinfoDAO dao = new UserinfoDAO();
        EntityManagerHelper.beginTransaction();
        Userinfo userinfo = dao.findById(2082444L);
        dao.delete(userinfo);
        EntityManagerHelper.commit();
        EntityManagerHelper.closeEntityManager();
    }
}
```

前面的示例都是针对数据表 userinfo 进行增删改查的操作。这些 main()方法中的代码并不是重点，重点的知识是在 UserinfoDAO.java 类源代码中，要知道 UserinfoDAO.java 类是如何实现 CURD 操作的。

9.5 JPA 核心接口介绍

在使用 JPA 规范时，常用的有 3 个核心类或接口：Persistence，EntityManagerFactory 和 EntityManager。

9.5.1 类 Persistence

类 Persistence 的主要作用就是根据 persistence.xml 文件中的信息，创建出 EntityManagerFactory 对象，类 Persistence 的 API 结构如图 9-20 所示。

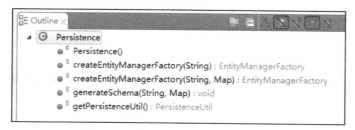

图 9-20 类 Persistence 的 API 结构

使用 Persistence 类创建 EntityManagerFactory 的代码如下。
```
EntityManagerFactory emfObject = Persistence.createEntityManagerFactory("jpaTest");
```

类 Persistence 的 createEntityManagerFactory()方法需要传递一个 persistence-unit 持久化单元名称，这个持久化单元名称就在 persistence.xml 文件中进行定义，该配置文件完整代码如下。

```xml
<?xml version="1.0" encoding="UTF-8"?>
<persistence version="2.1"
    xmlns="http://xmlns.jcp.org/xml/ns/persistence"
xmlns:xsi="http://www.w3.org/2001/XMLSchema-instance"
    xsi:schemaLocation="http://xmlns.jcp.org/xml/ns/persistence
http://xmlns.jcp.org/xml/ns/persistence/persistence_2_1.xsd">
    <persistence-unit name="jpaTest" transaction-type="RESOURCE_LOCAL">
        <provider>org.hibernate.jpa.HibernatePersistenceProvider</provider>
        <class>orm.Userinfo</class><properties>
            <property name="hibernate.connection.driver_class" value="oracle.jdbc.OracleDriver" />
            <property name="hibernate.connection.url" value="jdbc:oracle:thin:@localhost:1521:orcl" />
            <property name="hibernate.connection.username" value="y2" />
            <property name="hibernate.connection.password" value="123" />
        </properties>
    </persistence-unit>
</persistence>
```

标签<persistence-unit>的 name 属性值是什么，方法 createEntityManagerFactory()就要传递什么值。persistence-unit 代表了连接数据库的基本信息，还有一些通用性的相关配置。

9.5.2 JPA 中的事务类型

标签<persistence-unit>的 transaction-type 属性值为 RESOURCE_LOCAL，代表由程序员处理数据库的事务。如果值为 JPA，示例代码如下。
```
transaction-type="JTA"
```

则代表数据库的事务由 Web 容器进行处理，不需要程序员进行干预。

总结一下这个知识点，JPA 支持两种事务管理类型。

- RESOURCE_LOCAL 本地资源事务，使用本地资源事务指的是使用 JDBC 中的连接来管理当前活动的事务对象。程序员必须以手动开启和手动提交或手动回滚的方式来对当前活动的事务进行处理，这样的方式主要用在 Java SE 类型的项目中。
- JTA 事务，也叫容器型事务，在使用容器型事务中，Java EE 容器自动对事务进行开启、关闭或回滚，程序员不需要用代码的方式进行显式的调用，完全自动化，而且支持多个数据源。Java EE 容器自动管理事务生命周期，支持分布式的 JTA 进行参与。使用分布式事务的优点在于，在同一个 Java EE 容器中，同时访问多个不同的数据库，这些数据库分散在不同的物理位置，如果有一个数据库操作失败，则所有对其他数据库的操作都要回滚。这样的功能由事务管理器进行处理，如图 9-21 所示。

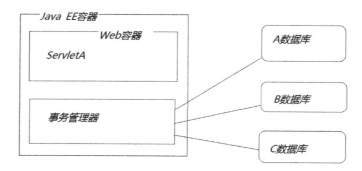

图 9-21 分布式事务管理器管理多个数据库的事务

事务管理器管理多个数据库的事务，在工作时采集各个数据库的操作状态，发现某一个数据库出现异常后，再告诉其他所有数据库进行事务回滚，也就保证了分布式事务的一致性。

9.5.3 接口 EntityManagerFactory

由 Persistence.java 根据持久化单元名称再创建出 EntityManagerFactory 接口。如果想实现一个持久化对象的操作，必须由 EntityManagerFactory 对象来根据持久化单元(Persistence Unit)的配置去生成 EntityManager 对象。EntityManager 对象再将实体类进行持久化，然后把这个新建实体转变成托管实体对象，并且放入 PersistenceContext 持久化上下文中，这就是典型 JPA 持久化实体在系统内部的交互过程。

从上面的过程中可以看到，EntityManagerFactory 对象根据持久化单元来创建 EntityManager 对象，然后由 EntityManager 对象创建持久化上下文环境。

EntityManagerFactory 接口只需要创建一次即可，通常都被声明成 static，它是线程安全的。

9.5.4 接口 EntityManager

接口 EntityManager 主要的功能就是通过管理实体类而转化对数据表中的数据记录进行增、删、改、查操作。

在程序运行时需要对接口 EntityManager 进行高频率的调用，所以各大 JPA 规范的实现厂商为了运行性能上的考虑，大多数都将 EntityManager 接口的实现类设计成了非线程安全的。EntityManager 常常结合 ThreadLocal 对象将自身变成线程安全的。

9.6 实体类的状态

实体类就是一个 Java 类，但又不是一个普通的 Java 类，它最基本的特性之一就是可以持久化，它是数据表中数据记录的一个载体。

EntityManager 类是处理实体类与数据表中记录之间操作的映射工具，实体类在 JPA 中主要有 4 种状态。

- 新建状态（瞬时状态）：新建的实体类对象，在内存中单独存在，示例代码如下。

```
            Userinfo u = new Userinfo();
            u.setUsername("myusername");
```

- 托管状态：由新建状态转变成托管状态主要的操作就是将新建状态的实体类保存进数据库，或者将数据表中的数据记录从数据库加载到内存，并且这个实体类已经由 EntityManager 的持久化上下文接管处理，这时的状态就是托管状态。实体类管理器内的托管实体的集合叫做持久化上下文（Persistence Context），可以将"持久化上下文"理解成是一个 List，这个 List 里面存储正在操作的实体类。
- 游离状态：实体的主键有值，但不在 EntityManager 对象的上下文接管中，典型的情况就是持久化上下文关闭时，托管的实体就变成游离状态。
- 删除状态：与 EntityManager 对象的上下文关联，并且将要从数据库中删除相对应的数据记录。

9.7 使用原生 JPA 的 API 实现 1 个添加记录的操作

借助于前面使用的 Web 项目，在其中创建测试用的 Java 类，示例代码如下。

```java
public class Test10 {
    public static void main(String[] args) {
        Userinfo userinfo = new Userinfo();
        userinfo.setUsername("美国");
        userinfo.setPassword("美国人");
        userinfo.setAge(100L);
        userinfo.setInsertdate(new Timestamp(new Date().getTime()));

        EntityManagerFactory entityManagerFactory = Persistence.createEntityManagerFactory("jpaTest");
        EntityManager entityManager = entityManagerFactory.createEntityManager();
        entityManager.getTransaction().begin();
        entityManager.persist(userinfo);
        entityManager.getTransaction().commit();
        entityManager.close();
    }
}
```

想要此程序的代码正确添加数据记录，则 JPA 事务要设置成 RESOURCE_LOCAL，示例代码如下。

```xml
<persistence-unit name="jpaTest" transaction-type="RESOURCE_LOCAL">
```

9.8 从零开始搭建 JPA 开发环境

前面章节使用 MyEclipse 工具来添加 JPA 开发及运行环境，此章节将使用手动的方式从 0 为起始，将一个最普通的 Java 项目变成支持 JPA 开发与运行的环境。

1. 新建 Java 项目 javaJPA。
2. 复制 Userinfo.java 实体类到 orm 包中。
3. 创建文件夹 META-INF，里面存放 persistence.xml 配置文件，配置文件代码如下。

```xml
<?xml version="1.0" encoding="UTF-8"?>
<persistence version="2.1"
```

```xml
        xmlns="http://xmlns.jcp.org/xml/ns/persistence"
xmlns:xsi="http://www.w3.org/2001/XMLSchema-instance"
        xsi:schemaLocation="http://xmlns.jcp.org/xml/ns/persistence
http://xmlns.jcp.org/xml/ns/persistence/persistence_2_1.xsd">
        <persistence-unit name="jpaTest" transaction-type="RESOURCE_LOCAL">
            <provider>org.hibernate.jpa.HibernatePersistenceProvider</provider>
            <class>orm.Userinfo</class><properties>
                <property name="hibernate.connection.driver_class" value="oracle.jdbc.OracleDriver" />
                <property name="hibernate.connection.url" value="jdbc:oracle:thin:@localhost:1521:orcl" />
                <property name="hibernate.connection.username" value="y2" />
                <property name="hibernate.connection.password" value="123" />
            </properties>
        </persistence-unit>
</persistence>
```

4. 添加 Hibernate 必备 jar 包与 Oracle 驱动程序，文件如图 9-22 所示。

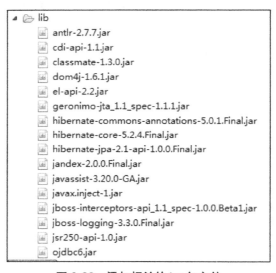

图 9-22　添加相关的 jar 包文件

5. 创建运行类 Test.java，代码如下。

```java
package test;

import java.sql.Timestamp;
import java.util.Date;

import javax.persistence.EntityManager;
import javax.persistence.EntityManagerFactory;
import javax.persistence.Persistence;

import orm.Userinfo;

public class Test {
    public static void main(String[] args) {
        Userinfo userinfo = new Userinfo();
        userinfo.setUsername("法国");
        userinfo.setPassword("法国人");
```

```
            userinfo.setAge(100L);
            userinfo.setInsertdate(new Timestamp(new Date().getTime()));

            EntityManagerFactory entityManagerFactory =
Persistence.createEntityManagerFactory("jpaTest");
            EntityManager entityManager = entityManagerFactory.createEntityManager();
            entityManager.getTransaction().begin();
            entityManager.persist(userinfo);
            entityManager.getTransaction().commit();
            entityManager.close();
    }
}
```

6. 运行 Test.java 类，成功在数据库中添加了一条记录。

9.9 EntityManager 核心方法的使用

使用 JPA 规范时，接口 EntityManager 的使用率是最高的。它提供了很多操作数据库的方法，在下面的章节将介绍这些常用方法的使用。

本章节的源代码还在项目 javaJPA 中。

9.9.1 方法 void persist(Object entity)保存一条记录

persist(Object)方法的功能将一个实体类对象中的数据变成数据表中的一条记录。如果执行flush()方法或将事务进行提交，则实体类对象中属性的值将作为数据表中字段的值插入到指定的数据表中。

使用 persist()方法可以将一个新建状态的实体对象转变成托管状态。

此方法在前面章节已经演示过，不再重复演示。

9.9.2 <T> T merge(T entity)方法和<T> T find(Class<T> entityClass, Object primaryKey)方法

方法 merge()是将实体对象与 EntityManager 环境的上下文进行关联，这样纳入到持久化容器中的实体类对象里最新版本的数据随着事务的提交就能正确并且自动化地同步到数据库中，也就是具有 update 更新的功能。

find()方法的功能是通过主键 id 值找到映射后的实体对象。

下面来实现一个根据 id 找到实体类并实现更新数据的示例。

示例代码如下。

```
public class Test1 {
    public static void main(String[] args) {
        EntityManagerFactory entityManagerFactory =
Persistence.createEntityManagerFactory("jpatest1");
        EntityManager entityManager = entityManagerFactory.createEntityManager();
        entityManager.getTransaction().begin();
        Userinfo userinfo = entityManager.find(Userinfo.class, 2082446L);
        userinfo.setUsername("xxxx");
        entityManager.merge(userinfo);
```

```
            entityManager.getTransaction().commit();
            entityManager.close();
        }
    }
```

9.9.3 方法 void remove(Object entity)

方法 remove()的功能是从数据表中删除记录。

示例代码如下。

```
    public class Test2 {
        public static void main(String[] args) {
            EntityManagerFactory entityManagerFactory =
Persistence.createEntityManagerFactory("jpatest1");
            EntityManager entityManager = entityManagerFactory.createEntityManager();
            entityManager.getTransaction().begin();
            Userinfo userinfo = entityManager.find(Userinfo.class, 2082446L);
            entityManager.remove(userinfo);
            entityManager.getTransaction().commit();
            entityManager.close();
        }
    }
```

9.9.4 getReference(Class<T>, Object)方法

方法 getReference()和 find()功能一样，都是从数据库中查询出指定 id 的实体类，但它们还是有具体的区别。

- 找不到数据时的情况。方法 find()会返回为 null。方法 getReference()返回 id 属性有值，其他属性为 null 的代理实体类对象。
- 如果找到对应数据表中记录的情况下，返回实体类的数据类型性质也不一样。方法 find()返回的是 1 个代理实体类。方法 getReference()返回的也是 1 个代理实体类，但具有延迟加载特性。
- 实体类对象属性值加载情况。方法 find()是立即加载数据。方法 getReference()是延迟加载数据。

被查找到的实体状态为托管状态。

下面先测试找不到数据时的情况，示例代码如下。

```
    public class Test3 {
        public static void main(String[] args) {
            EntityManagerFactory entityManagerFactory =
Persistence.createEntityManagerFactory("jpatest1");
            EntityManager entityManager = entityManagerFactory.createEntityManager();
            entityManager.getTransaction().begin();
            Userinfo userinfo1 = entityManager.find(Userinfo.class, 123456789L);
            Userinfo userinfo2 = entityManager.getReference(Userinfo.class, 123456789L);
            entityManager.getTransaction().commit();
            entityManager.close();
        }
    }
```

在 commit()方法处设置断点，以 debug 方式运行程序，如图 9-23 所示。

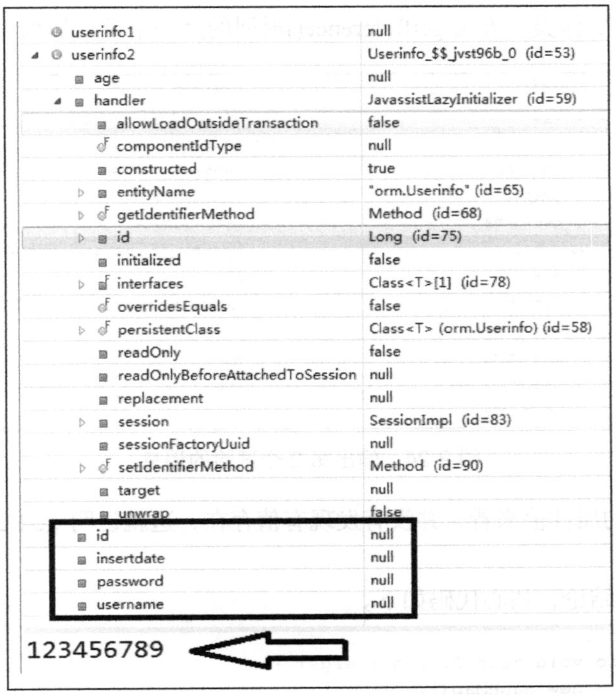

9.9 EntityManager 核心方法的使用

图 9-23 两个实体对象的值类型并不相同

第一个实验已经完毕，还得出下面的结论。

找不到数据时的情况：方法 find()会返回为 null。方法 getReference()返回实体对象，实体对象的 handler 属性的 id 属性有值，但实体对象的 id、username、password、insertdate 和 age 属性无值。

继续做第 2 个实验。

如果找到对应数据表中记录的情况下，返回实体类的数据类型也不一样。

创建实验用的代码如下。

```java
public class Test3 {
    public static void main(String[] args) {
        EntityManagerFactory entityManagerFactory =
Persistence.createEntityManagerFactory("jpatest1");
        EntityManager entityManager = entityManagerFactory.createEntityManager();
        entityManager.getTransaction().begin();
        Userinfo userinfo1 = entityManager.find(Userinfo.class, 2072497L);
        Userinfo userinfo2 = entityManager.getReference(Userinfo.class, 2072498L);
        entityManager.getTransaction().commit();
        entityManager.close();
    }
}
```

数据表中有 id 为 2072497 和 2072498 的记录。

在 commit()处设置断点，运行程序后的结果如图 9-24 所示。

上面是第 2 个实验的运行结果，结论如下所示。

如果找到对应数据表中记录的情况下，返回实体类的数据类型性质也不一样。方法 find()

返回的是一个真正的实体类。方法 getReference()返回的是一个延迟加载对象。

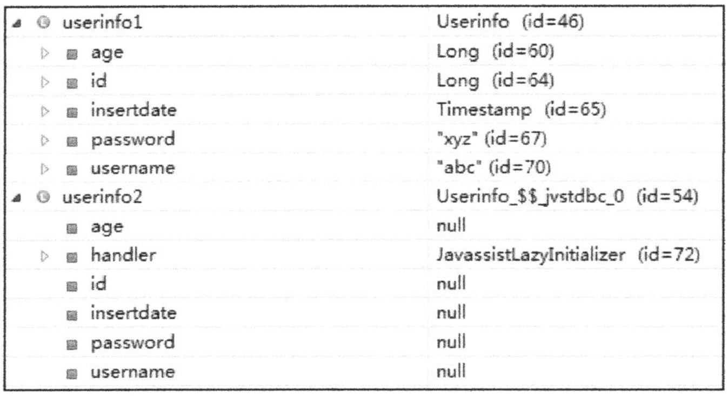

图 9-24　输出第 2 个试验的消息

从 userinfo2 中的属性值来看，并没有发现有值存在，这就说明 getReference()方法具有延迟加载特性。

继续测试第 3 个结论，核心代码如下。

```
public class Test5 {
    public static void main(String[] args) {
        Map map = new HashMap();
        map.put("hibernate.show_sql", "true");
        EntityManagerFactory entityManagerFactory = 
Persistence.createEntityManagerFactory("jpatest1", map);
        EntityManager entityManager = entityManagerFactory.createEntityManager();
        entityManager.getTransaction().begin();
        Userinfo userinfo1 = entityManager.find(Userinfo.class, 2072497L);
        System.out.println(
                "userinfo1的数据：" + userinfo1.getId() + " " + userinfo1.getUsername()
+ " " + userinfo1.getPassword());
        Userinfo userinfo2 = entityManager.getReference(Userinfo.class, 2072498L);
        System.out.println(
                "userinfo2的数据：" + userinfo2.getId() + " " + userinfo2.getUsername()
+ " " + userinfo2.getPassword());
        entityManager.getTransaction().commit();
        entityManager.close();
    }
}
```

在第 19、20、22 和 23 行设置断点，如图 9-25 所示。

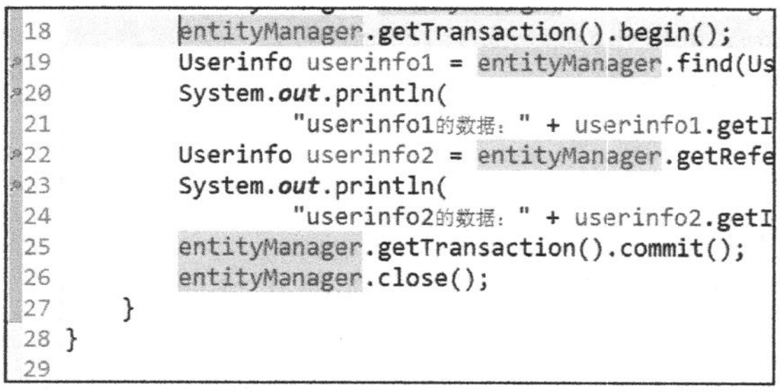

图 9-25　设置 4 处断点

通过调试可以发现，find()方法立即执行 SQL 语句，而 getReference()方法直到调用实体类的 getXXX()方法取得属性值时才发起 SQL 语句。

第 3 个实验的结论如下。

实体类对象属性值加载情况：方法 find()是立即加载数据；方法 getReference()是延迟加载数据。

9.9.5　createNativeQuery(string)方法

JPA 还支持使用 createNativeQuery()方法来执行原生的 SQL 语句。

（1）增加数据。

核心代码如下。

```java
public class Test7 {

    public static void main(String[] args) {
        Map map = new HashMap();
        map.put("hibernate.show_sql", "true");

        EntityManagerFactory entityManagerFactory =
Persistence.createEntityManagerFactory("jpatest4", map);

        EntityManager entityManager = entityManagerFactory.createEntityManager();
        entityManager.getTransaction().begin();

        Query query = entityManager.createNativeQuery(
                "insert into userinfo(id,username,password,age,insertdate) values(idauto.nextval,'中国','中国人',100,sysdate)");
        query.executeUpdate();

        entityManager.getTransaction().commit();
        entityManager.close();

    }
}
```

（2）更新数据。

核心代码如下。

```java
public class Test8 {

    public static void main(String[] args) {
        Map map = new HashMap();
        map.put("hibernate.show_sql", "true");

        EntityManagerFactory entityManagerFactory =
Persistence.createEntityManagerFactory("jpatest4", map);

        EntityManager entityManager = entityManagerFactory.createEntityManager();
        entityManager.getTransaction().begin();

        Query query = entityManager.createNativeQuery("update userinfo set username='zzzzzzzz' where id=2");
        query.executeUpdate();

        entityManager.getTransaction().commit();
        entityManager.close();
```

（3）删除数据。

核心代码如下。

```java
public class Test9 {

    public static void main(String[] args) {
        Map map = new HashMap();
        map.put("hibernate.show_sql", "true");

        EntityManagerFactory entityManagerFactory =
Persistence.createEntityManagerFactory("jpatest4", map);

        EntityManager entityManager = entityManagerFactory.createEntityManager();
        entityManager.getTransaction().begin();

        Query query = entityManager.createNativeQuery("delete from userinfo where id=2");
        query.executeUpdate();

        entityManager.getTransaction().commit();
        entityManager.close();

    }
}
```

（4）Object[]对象数组封装查询法。

核心代码如下。

```java
public class Test10 {

    public static void main(String[] args) {
        Map map = new HashMap();
        map.put("hibernate.show_sql", "true");

        EntityManagerFactory entityManagerFactory =
Persistence.createEntityManagerFactory("jpatest4", map);

        EntityManager entityManager = entityManagerFactory.createEntityManager();
        entityManager.getTransaction().begin();

        Query query = entityManager.createNativeQuery("select * from userinfo where id=1");
        List<Object[]> list = query.getResultList();
        for (int i = 0; i < list.size(); i++) {
            Object[] objectArray = list.get(i);
            System.out.println(objectArray[0] + " " + objectArray[1] + " " + objectArray[2]
+ " " + objectArray[3] + " "
                    + objectArray[4]);

        }
        System.out.println();
        System.out.println();
        query = entityManager.createNativeQuery("select * from userinfo order by id desc");
        list = query.getResultList();
        for (int i = 0; i < list.size(); i++) {
            Object[] objectArray = list.get(i);
```

```
                System.out.println(objectArray[0] + " " + objectArray[1] + " " + objectArray[2]
+ " " + objectArray[3] + " "
                        + objectArray[4]);

            }

            entityManager.getTransaction().commit();
            entityManager.close();

        }
    }
```

（5）实体对象封装查询法。

核心代码如下。

```
public class Test11 {

    public static void main(String[] args) {
        Map map = new HashMap();
        map.put("hibernate.show_sql", "true");

        EntityManagerFactory entityManagerFactory =
Persistence.createEntityManagerFactory("jpatest4", map);

        EntityManager entityManager = entityManagerFactory.createEntityManager();
        entityManager.getTransaction().begin();

        Query query = entityManager.createNativeQuery("select * from userinfo order by
id desc", Userinfo.class);
        List<Userinfo> listUserinfo = query.getResultList();
        for (int i = 0; i < listUserinfo.size(); i++) {
            Userinfo userinfo = listUserinfo.get(i);
            System.out.println(userinfo.getId() + " " + userinfo.getUsername() + " " +
userinfo.getPassword() + " "
                    + userinfo.getAge() + " " + userinfo.getInsertdate());
        }

        entityManager.getTransaction().commit();
        entityManager.close();

    }
}
```

9.9.6 clear()和 contains(Object)方法

方法 clear()是将持久化上下文中所有的实体类进行销毁以释放内存空间。

方法 contains()是判断当前的实体类是否在持久化上下文中受托管,如果存在就返回为 true,否则为 false。

测试代码如下。

```
public class Test6 {
    public static void main(String[] args) {
        EntityManagerFactory entityManagerFactory =
Persistence.createEntityManagerFactory("jpatest1");
        EntityManager entityManager = entityManagerFactory.createEntityManager();
        entityManager.getTransaction().begin();
```

```
            Userinfo userinfo1 = entityManager.find(Userinfo.class, 2072497L);
            System.out.println("A=" + entityManager.contains(userinfo1));
            entityManager.clear();
            System.out.println("B=" + entityManager.contains(userinfo1));
            entityManager.getTransaction().commit();
            entityManager.close();
        }
    }
```

运行结果如图 9-26 所示。

图 9-26 运行结果

方法 clear() 是删除一级缓存中的实体类，删除后当再次查询实体类时 Hibernate 会发起新的 SQL 语句进行查询，示例代码如下。

```
public class Test12_1____ {

    public static void main(String[] args) {
        Map map = new HashMap();
        map.put("hibernate.show_sql", "true");

        EntityManagerFactory entityManagerFactory =
Persistence.createEntityManagerFactory("jpatest4", map);

        EntityManager entityManager = entityManagerFactory.createEntityManager();
        entityManager.getTransaction().begin();

        Userinfo userinfo1 = (Userinfo) entityManager.find(Userinfo.class, 1L);
        Userinfo userinfo2 = (Userinfo) entityManager.find(Userinfo.class, 1L);
        System.out.println("userinfo1 hashCode=" + userinfo1.hashCode());
        System.out.println("userinfo2 hashCode=" + userinfo2.hashCode());
        System.out.println("A=" + entityManager.contains(userinfo1));
        System.out.println("B=" + entityManager.contains(userinfo2));
        entityManager.clear();// 从 1 级缓存中删除 userinfo 对象
        // 再发起新的 find() 查询时重新执行新的 SQL 语句进行查询数据
        Userinfo userinfo3 = (Userinfo) entityManager.find(Userinfo.class, 1L);
        entityManager.getTransaction().commit();
        entityManager.close();

    }
}
```

运行结果如图 9-27 所示。

```
Hibernate: select userinfo0_.ID as ID1_0_0_, userinfo0_.AGE as AGE
userinfo1 hashCode=1037475674
userinfo2 hashCode=1037475674
A=true
B=true
Hibernate: select userinfo0_.ID as ID1_0_0_, userinfo0_.AGE as AGE
```

图 9-27 发起两个 SQL 语句

9.9.7 createQuery(String)方法

ORM 框架 Hibernate 使用 HQL 语言来查询数据库中的记录,而在 JPA 中也有一种类似的语言,叫作 JPQL。JPQL 的使用和 HQL 的使用非常类似,本小节就来实现一个 JPQL 的示例。示例代码如下。

```java
public class Test13 {

    public static void main(String[] args) {
        Map map = new HashMap();
        map.put("hibernate.show_sql", "true");

        EntityManagerFactory entityManagerFactory = Persistence.createEntityManagerFactory("jpatest4", map);

        EntityManager entityManager = entityManagerFactory.createEntityManager();
        entityManager.getTransaction().begin();

        Query query = entityManager.createQuery("select u from Userinfo u order by u.id desc");
        List<Userinfo> listUserinfo = query.getResultList();
        for (int i = 0; i < listUserinfo.size(); i++) {
            Userinfo userinfo = listUserinfo.get(i);
            System.out.println(userinfo.getId() + " " + userinfo.getUsername() + " " + userinfo.getPassword() + " "
                    + userinfo.getAge() + " " + userinfo.getInsertdate());
        }

        entityManager.getTransaction().commit();
        entityManager.close();

    }
}
```

9.10 双向一对多的 CURD 实验

JPA 也支持双向一对多的关系映射。在数据库中创建 main 与 sub 物理主外关系表,然后对这两个表进行 JPA 逆向。

创建实验用的项目 jpaMainSub。

9.10.1 逆向 Main.java 和 Sub.java 实体类

查看一下 Main.java 文件的代码。

```java
package orm;

import java.util.HashSet;
import java.util.Set;
import javax.persistence.CascadeType;
import javax.persistence.Column;
import javax.persistence.Entity;
import javax.persistence.FetchType;
import javax.persistence.Id;
import javax.persistence.OneToMany;
```

```java
import javax.persistence.Table;

@Entity
@Table(name = "MAIN", schema = "Y2")
public class Main implements java.io.Serializable {

    private Long id;
    private String mainname;
    private Set<Sub> subs = new HashSet<Sub>(0);

    public Main() {
    }

    public Main(Long id) {
        this.id = id;
    }

    public Main(Long id, String mainname, Set<Sub> subs) {
        this.id = id;
        this.mainname = mainname;
        this.subs = subs;
    }

    @Id
    @Column(name = "ID", unique = true, nullable = false, precision = 18, scale = 0)
    @SequenceGenerator(name = "idautoRef", sequenceName = "idauto")
    @GeneratedValue(strategy = GenerationType.SEQUENCE, generator = "idautoRef")
    public Long getId() {
        return this.id;
    }

    public void setId(Long id) {
        this.id = id;
    }

    @Column(name = "MAINNAME", length = 50)
    public String getMainname() {
        return this.mainname;
    }

    public void setMainname(String mainname) {
        this.mainname = mainname;
    }

    @OneToMany(cascade = CascadeType.ALL, fetch = FetchType.LAZY, mappedBy = "main")
    public Set<Sub> getSubs() {
        return this.subs;
    }

    public void setSubs(Set<Sub> subs) {
        this.subs = subs;
    }

}
```

可以看到 Main.java 文件中使用了数据类型为 Set 的 subs 变量来存放 Sub.java 对象，是多个 Sub.java 对象的容器，这样一个主就可以对应多个子了。

在方法 public Set<Sub> getSubs()的上方使用了注解。
`@OneToMany(cascade = CascadeType.ALL, fetch = FetchType.LAZY, mappedBy = "main")`

其中 cascade 属性指的是级联关联，比如删除主表的同时是否将子表中的数据记录也一起删除。

而 fetch 属性定义加载子表数据的方式，有两种方式，分别为立即加载和延迟加载。如果是立即加载，取得 Main 对象的同时，数据类型是 Set 的 subs 变量中已经有多个 Sub.java 对象；如果是延迟加载方式，则直到需要取得 Sub 对象时才真正的从数据库中取出数据，这样的优点是有非常高的访问效率。自动生成的代码默认的加载方式是 FetchType.LAZY 延迟加载。

而 mappedBy 属性定义在 Sub.java 类中取得 Main.java 对象的属性名。

再来查看一下 Sub.java 文件的代码。

```java
package orm;

import javax.persistence.Column;
import javax.persistence.Entity;
import javax.persistence.FetchType;
import javax.persistence.Id;
import javax.persistence.JoinColumn;
import javax.persistence.ManyToOne;
import javax.persistence.Table;

@Entity
@Table(name = "SUB", schema = "Y2")
public class Sub implements java.io.Serializable {

    private Long id;
    private Main main;
    private String subname;

    public Sub() {
    }

    public Sub(Long id) {
        this.id = id;
    }

    public Sub(Long id, Main main, String subname) {
        this.id = id;
        this.main = main;
        this.subname = subname;
    }

    @Id
    @Column(name = "ID", unique = true, nullable = false, precision = 18, scale = 0)
    @SequenceGenerator(name = "idautoRef", sequenceName = "idauto")
    @GeneratedValue(strategy = GenerationType.SEQUENCE, generator = "idautoRef")
    public Long getId() {
        return this.id;
    }

    public void setId(Long id) {
        this.id = id;
    }

    @ManyToOne(fetch = FetchType.LAZY)
```

```
        @JoinColumn(name = "MAINID")
        public Main getMain() {
            return this.main;
        }

        public void setMain(Main main) {
            this.main = main;
        }

        @Column(name = "SUBNAME", length = 50)
        public String getSubname() {
            return this.subname;
        }

        public void setSubname(String subname) {
            this.subname = subname;
        }

}
```

在 public Main getMain()方法上使用了注解。

```
        @ManyToOne(fetch = FetchType.LAZY)
        @JoinColumn(name = "MAINID")
```

注解@ManyToOne 的属性 fetch 也是延迟加载策略，而注解@JoinColumn 的 name 属性的功能是使用 MAINID 字段引用主表的数据。

到这一步，基本的环境配置工作就完成了，下一步就开始实现功能了。

9.10.2 创建 Main

实验用的代码如下。

```
public class Test1 {

    public static void main(String[] args) {
        Main main = new Main();
        main.setMainname("山东省");

        Map map = new HashMap();
        map.put("hibernate.show_sql", "true");
        EntityManagerFactory entityManagerFactory = 
Persistence.createEntityManagerFactory("jpatest4", map);
        EntityManager entityManager = entityManagerFactory.createEntityManager();

        entityManager.persist(main);

        entityManager.getTransaction().begin();
        entityManager.getTransaction().commit();
        entityManager.close();
    }
}
```

9.10.3 创建 Sub

实验用的代码如下。

```java
public class Test2 {

    public static void main(String[] args) {
        Map map = new HashMap();
        map.put("hibernate.show_sql", "true");
        EntityManagerFactory entityManagerFactory =
Persistence.createEntityManagerFactory("jpatest4", map);
        EntityManager entityManager = entityManagerFactory.createEntityManager();

        Main main = entityManager.find(Main.class, 2102468L);

        Sub sub = new Sub();
        sub.setSubname("烟台市");
        sub.setMain(main);

        entityManager.persist(sub);

        entityManager.getTransaction().begin();
        entityManager.getTransaction().commit();
        entityManager.close();
    }
}
```

9.10.4 更新 Main

实验用的代码如下。

```java
public class Test3 {

    public static void main(String[] args) {
        Map map = new HashMap();
        map.put("hibernate.show_sql", "true");
        EntityManagerFactory entityManagerFactory =
Persistence.createEntityManagerFactory("jpatest4", map);
        EntityManager entityManager = entityManagerFactory.createEntityManager();

        Main main = entityManager.find(Main.class, 2102468L);
        main.setMainname("新的山东省");

        entityManager.merge(main);

        entityManager.getTransaction().begin();
        entityManager.getTransaction().commit();
        entityManager.close();
    }
}
```

9.10.5 更新 Sub

实验用的代码如下。

```java
public class Test4 {

    public static void main(String[] args) {
        Map map = new HashMap();
        map.put("hibernate.show_sql", "true");
```

```
            EntityManagerFactory entityManagerFactory =
Persistence.createEntityManagerFactory("jpatest4", map);
            EntityManager entityManager = entityManagerFactory.createEntityManager();

            Main main = entityManager.find(Main.class, 2102468L);

            Sub sub = (Sub) entityManager.find(Sub.class, 1445492L);
            sub.setSubname("青岛市");
            sub.setMain(main);

            entityManager.merge(sub);

            entityManager.getTransaction().begin();
            entityManager.getTransaction().commit();
            entityManager.close();
        }
    }
```

9.10.6 删除 Main 时默认将 Sub 也一同删除

实验用的代码如下。

```
public class Test6 {

    public static void main(String[] args) {
        Map map = new HashMap();
        map.put("hibernate.show_sql", "true");
        EntityManagerFactory entityManagerFactory =
Persistence.createEntityManagerFactory("jpatest4", map);
        EntityManager entityManager = entityManagerFactory.createEntityManager();

        // id 为 1445483 的 main 表记录在 sub 中有子表记录
        // 关联 1445483 这个主键值
        Main main = entityManager.find(Main.class, 1445483L);
        entityManager.remove(main);

        entityManager.getTransaction().begin();
        entityManager.getTransaction().commit();
        entityManager.close();
    }
}
```

在 Main.java 实体类中默认配置为级联删除 cascade = CascadeType.ALL，配置代码如下。

```
    @OneToMany(cascade = CascadeType.ALL, fetch = FetchType.LAZY, mappedBy = "main")
    public Set<Sub> getSubs() {
        return this.subs;
    }
```

9.10.7 从 Main 加载 Sub 时默认为延迟加载

实验用的代码如下。

```
public class Test7 {

    public static void main(String[] args) {
        Map map = new HashMap();
```

```
            map.put("hibernate.show_sql", "true");
            EntityManagerFactory entityManagerFactory =
Persistence.createEntityManagerFactory("jpatest4", map);
            EntityManager entityManager = entityManagerFactory.createEntityManager();
            entityManager.getTransaction().begin();
            // 此实验加载方式为"延迟加载"
            Main main = entityManager.find(Main.class, 2102468L);
            System.out.println(main.getId() + " " + main.getMainname());
            Set set = main.getSubs();
            Iterator<Sub> iterator = set.iterator();
            while (iterator.hasNext()) {
                Sub sub = iterator.next();
                System.out.println("  " + sub.getId() + " " + sub.getSubname());
            }

            entityManager.getTransaction().commit();
            entityManager.close();
        }
    }
```

程序运行后当执行代码：

```
Set set = main.getSubs();
```

才发起新的 SQL 查询语句。

9.11 JPQL 语言的使用

在 JPA 中使用的查询语言叫作 JPQL（Java Presistence Query Language）。JPQL 语言和 Hibernate 框架中的 HQL 语言一样，都是以面向对象的方式来进行数据的操作，所以 HQL 语言在 Hibernate 的使用中占据了很大部分，而 JPQL 也在 JPA 的使用中发挥了最大的作用。本章主要使用 JPQL 语言来进行数据的查询、检索操作。

9.11.1 参数索引式查询

索引式参数查询和 JDBC 中的"?"占位符功能相似，都是在查询语句中以传递参数值的方式来进行数据的查询。JPQL 也支持这种特性，另外使用参数式查询也能避免 SQL 注入。

创建实验用的项目 JPQLTest，核心代码如下。

```
public class Test1 {

    public static void main(String[] args) {
        Map map = new HashMap();
        map.put("hibernate.show_sql", "true");
        EntityManagerFactory entityManagerFactory =
Persistence.createEntityManagerFactory("jpatest4", map);
        EntityManager entityManager = entityManagerFactory.createEntityManager();

        Query query = entityManager
                .createQuery("select u from Userinfo u where u.username like ?1 and u.password like ?2");
        query.setParameter(1, "%b%");
```

```
        query.setParameter(2, "%b%");
        List<Userinfo> list = query.getResultList();
        for (int i = 0; i < list.size(); i++) {
            Userinfo userinfo = list.get(i);
            System.out.println(userinfo.getId() + " " + userinfo.getUsername() + " " + userinfo.getPassword());
        }
        entityManager.getTransaction().begin();
        entityManager.getTransaction().commit();
        entityManager.close();
    }
}
```

9.11.2 命名式参数查询

索引式参数查询虽然能解决一些不用"拼接 JPQL"字符串的问题，但在字段数量比较多的情况下，依然不能解决维护复杂的问题。JPQL 在支持命名式参数查询的方式上比索引式查询更加方便和易于维护。

核心代码如下。

```
public class Test2 {
    public static void main(String[] args) {
        Map map = new HashMap();
        map.put("hibernate.show_sql", "true");
        EntityManagerFactory entityManagerFactory = Persistence.createEntityManagerFactory("jpatest4", map);
        EntityManager entityManager = entityManagerFactory.createEntityManager();
        Query query = entityManager.createQuery("select u from Userinfo u where u.username like :usernameKey");
        query.setParameter("usernameKey", "%b%");
        List<Userinfo> list = query.getResultList();
        for (int i = 0; i < list.size(); i++) {
            Userinfo userinfo = list.get(i);
            System.out.println(userinfo.getId() + " " + userinfo.getUsername() + " " + userinfo.getPassword());
        }
        entityManager.getTransaction().begin();
        entityManager.getTransaction().commit();
        entityManager.close();
    }
}
```

9.11.3 JPQL 支持的运算符与聚合函数与排序

JPQL 与 SQL 语言一样，支持各种的常用运算符，包括+、-、*、/、=、>=、>、<、<=、<>、between（判断区间）、like（模糊查询）、in（指定类别）、is null（为空）and（并且）和 or（或者）关系。

JPQL 也支持 avg、count、max、min 和 sum 聚合函数。

JPQL 也支持对查询结果集进行排序，正序 asc 以及倒序 desc。

9.11.4　is null 为空运算符的使用

改变 userinfo 表数据，添加两条 username 字段为空的记录，结果如图 9-28 所示。

ID	USERNAME
1	b
2102498	
2102499	
2102500	中国

图 9-28　添加两条空记录

核心代码如下。

```java
public class Test3 {

    public static void main(String[] args) {
        Map map = new HashMap();
        map.put("hibernate.show_sql", "true");
        EntityManagerFactory entityManagerFactory = Persistence.createEntityManagerFactory("jpatest4", map);
        EntityManager entityManager = entityManagerFactory.createEntityManager();
        Query query = entityManager.createQuery("select u from Userinfo u where u.username is null");
        List<Userinfo> list = query.getResultList();
        for (int i = 0; i < list.size(); i++) {
            Userinfo userinfo = list.get(i);
            System.out.println(userinfo.getId() + " " + userinfo.getUsername() + " " + userinfo.getPassword());
        }
        entityManager.getTransaction().begin();
        entityManager.getTransaction().commit();
        entityManager.close();
    }
}
```

9.11.5　查询指定字段的示例

1. Object[]对象数组封装法

在前面的章节中查询结果的 List 中存储的都是实体类，但如果想查询指定的几个字段时，返回的结果 List 中存储的就不是实体类，而是 Object[]对象数组的形式。

核心代码如下。

```java
public class Test4 {

    public static void main(String[] args) {
        Map map = new HashMap();
        map.put("hibernate.show_sql", "true");
        EntityManagerFactory entityManagerFactory = Persistence.createEntityManagerFactory("jpatest4", map);
        EntityManager entityManager = entityManagerFactory.createEntityManager();
        Query query = entityManager.createQuery("select id,password,age from Userinfo");
        List<Object[]> list = query.getResultList();
        for (int i = 0; i < list.size(); i++) {
```

```
                Object[] eachObject = list.get(i);
                System.out.println(eachObject[0] + " " + eachObject[1] + " " + eachObject[2]);
            }
            entityManager.getTransaction().begin();
            entityManager.getTransaction().commit();
            entityManager.close();
        }
    }
```

2. 自定义实体对象封装法

虽然用 Object[]也实现了输出的效果,但用 index 索引取值在实际开发中也是非常不方便的,所以可以将取出来的字段数据封装进自定义的实体类中。在 entity 包中创建一个名为 NewUserinfo.java 类的代码如下。

```
package entity;

import java.util.Date;

public class NewUserinfo {

    private long id;
    private String password;
    private Date insertdate;

    public NewUserinfo() {
    }

    public long getId() {
        return id;
    }

    public void setId(long id) {
        this.id = id;
    }

    public NewUserinfo(long id, String password, Date insertdate) {
        super();
        this.id = id;
        this.password = password;
        this.insertdate = insertdate;
    }

    //省略 get 和 set 方法

}
```

运行类的核心代码如下。

```
public class Test5 {

    public static void main(String[] args) {
        Map map = new HashMap();
        map.put("hibernate.show_sql", "true");
        EntityManagerFactory entityManagerFactory =
Persistence.createEntityManagerFactory("jpatest4", map);
        EntityManager entityManager = entityManagerFactory.createEntityManager();
```

```
            Query query = entityManager.createQuery("select new
entity.NewUserinfo(id,password,insertdate) from Userinfo");
            List<NewUserinfo> list = query.getResultList();
            for (int i = 0; i < list.size(); i++) {
                NewUserinfo userinfo = list.get(i);
                System.out.println(userinfo.getId() + " " + userinfo.getPassword() + " " +
userinfo.getInsertdate());
            }
            entityManager.getTransaction().begin();
            entityManager.getTransaction().commit();
            entityManager.close();
        }
    }
```

9.11.6 JPQL 语言对日期的判断

准备多条实验用的数据记录，如图 9-29 所示。

ID	USERNAME	PASSWORD	AGE	INSERTDATE
1	b	bb	2	2016/11/25
2102498		中国人	100	2016/11/25
2102499		中国人	100	2016/11/25
2102500	中国	中国人	100	2016/11/25
2102501	中国	中国人	100	2016/11/25 15:04:14
2102502	b	中国人	100	2016/11/25 15:04:20
2102503	中国	中国人	100	2016/11/26 15:04:20

图 9-29 日期记录

下面测试一下日期来自于 String 字符串的示例，核心代码如下。

```
public class Test6 {

    public static void main(String[] args) throws ParseException {

        String dateString = "2016-11-25";
        String beginDateString = dateString + " 00:00:00";
        String endDateString = dateString + " 23:59:59";
        SimpleDateFormat format = new SimpleDateFormat("yyyy-MM-dd HH:mm:ss");
        Date beginDateObject = format.parse(beginDateString);
        Date endDateObject = format.parse(endDateString);

        Map map = new HashMap();
        map.put("hibernate.show_sql", "true");
        EntityManagerFactory entityManagerFactory =
Persistence.createEntityManagerFactory("jpatest4", map);
        EntityManager entityManager = entityManagerFactory.createEntityManager();
        Query query = entityManager.createQuery(
                "select u from Userinfo u where u.insertdate between :beginDate
and :endDate order by u.id asc");
        query.setParameter("beginDate", beginDateObject);
        query.setParameter("endDate", endDateObject);
        List<Userinfo> list = query.getResultList();
        for (int i = 0; i < list.size(); i++) {
            Userinfo userinfo = list.get(i);
            System.out.println(userinfo.getId() + " " + userinfo.getUsername() + " " +
```

```
userinfo.getInsertdate());
            }
                entityManager.getTransaction().begin();
                entityManager.getTransaction().commit();
                entityManager.close();
        }
    }
```

下面测试一下日期来自于 int 数字的示例，核心代码如下。

```
public class Test7 {

    public static void main(String[] args) throws ParseException {

        Calendar calendarBegin = Calendar.getInstance();
        calendarBegin.set(Calendar.YEAR, 2016);
        calendarBegin.set(Calendar.MONTH, 10);
        calendarBegin.set(Calendar.DAY_OF_MONTH, 25);
        calendarBegin.set(Calendar.HOUR_OF_DAY, 0);
        calendarBegin.set(Calendar.MINUTE, 0);
        calendarBegin.set(Calendar.SECOND, 0);
        calendarBegin.set(Calendar.MILLISECOND, 0);

        Calendar calendarEnd = Calendar.getInstance();
        calendarEnd.set(Calendar.YEAR, 2016);
        calendarEnd.set(Calendar.MONTH, 10);
        calendarEnd.set(Calendar.DAY_OF_MONTH, 25);
        calendarEnd.set(Calendar.HOUR_OF_DAY, 23);
        calendarEnd.set(Calendar.MINUTE, 59);
        calendarEnd.set(Calendar.SECOND, 59);
        calendarEnd.set(Calendar.MILLISECOND, 0);

        Date beginDateObject = calendarBegin.getTime();
        Date endDateObject = calendarEnd.getTime();

        Map map = new HashMap();
        map.put("hibernate.show_sql", "true");
        EntityManagerFactory entityManagerFactory =
Persistence.createEntityManagerFactory("jpatest4", map);
        EntityManager entityManager = entityManagerFactory.createEntityManager();
        Query query = entityManager.createQuery(
                "select u from Userinfo u where u.insertdate between :beginDate and :endDate order by u.id asc");
        query.setParameter("beginDate", beginDateObject);
        query.setParameter("endDate", endDateObject);
        List<Userinfo> list = query.getResultList();
        for (int i = 0; i < list.size(); i++) {
            Userinfo userinfo = list.get(i);
            System.out.println(userinfo.getId() + " " + userinfo.getUsername() + " " +
userinfo.getInsertdate());
        }

        entityManager.getTransaction().begin();
        entityManager.getTransaction().commit();
        entityManager.close();
    }
}
```

9.11.7 JPQL 语言中的分页功能

JPQL 语言中的分页功能和 Hibernate 中的分页非常相似，主要的核心代码就是使用 Query 对象的 setFirstResult()和 setMaxResults()方法，示例代码如下。

```java
public class Test8 {

    public static void main(String[] args) throws ParseException {

        Map map = new HashMap();
        map.put("hibernate.show_sql", "true");
        EntityManagerFactory entityManagerFactory = Persistence.createEntityManagerFactory("jpatest4", map);
        EntityManager entityManager = entityManagerFactory.createEntityManager();
        Query query = entityManager.createQuery("select u from Userinfo u order by id asc");
        query.setFirstResult(3);
        query.setMaxResults(4);
        List<Userinfo> list = query.getResultList();
        for (int i = 0; i < list.size(); i++) {
            Userinfo userinfo = list.get(i);
            System.out.println(userinfo.getId() + " " + userinfo.getUsername() + " " + userinfo.getInsertdate());
        }

        entityManager.getTransaction().begin();
        entityManager.getTransaction().commit();
        entityManager.close();
    }
}
```

第 10 章　Spring 4 的 DI 与 AOP

Spring 框架简化了 Java EE 开发的流程，是为了解决企业级应用开发复杂性而创建的。其强大之处在于对 Java EE 开发进行全方位地简化，对大部分常用的功能进行了封装，比如管理 JavaBean，包含创建及销毁，还提供了基于 Web 的 Spring MVC 分层框架，支持数据库操作，安全验证等功能。但这些功能的实现却要依赖于两个技术原理：DI 和 AOP。本章的目的就是学习并掌握 Spring 中的这两个核心技术，并能在实际的软件开发中得以运用。

本章应该着重掌握如下内容：
- 什么是注入；
- 什么是 DI 与 DI 容器；
- 为什么要使用面向切面编程；
- 什么是 AOP；
- 对属性注入常用的数据类型；
- 单例及多例模式的应用；
- 代理模式与 AOP 的关系；
- 面向切面编程的常用用法。

10.1　Spring 介绍

Spring 是一个开放源代码的 Java EE 框架，主要是为了解决企业应用程序维护复杂性而创建的。使用 Spring 简化了 Java EE 开发，提升 Java EE 软件项目的开发效率，提高开发效率的解决办法就是用模块架构，每个模块处理一个功能或者是业务，模块架构允许程序员选择使用哪一个模块参与开发，同时为 Java EE 应用程序开发提供集成的容器。在 Spring 框架中提供了一个 JavaBean 容器（可以暂时将容器理解成为一个 List），在该容器中存储不同数据类型的 JavaBean 对象，容器中可以将很多种不同功能的 JavaBean 进行整合，进行集成，来达到综合技术应用的目的。

本章节主要介绍 Spring 框架核心的原理：
- 依赖注入（DI，Dependency Injection）；
- 面向切面编程（AOP，Aspect Oriented Programming）。

使用依赖注入DI对不同数据类型的属性注入属性值,并且熟悉AOP在面向切面编程的知识。

10.2 依赖注入

在没有 Spring 框架的时候，如果在 A.java 类中使用 B.java 类，则必须在 A.java 类中实例化出 B.java 类的对象，这就造成了 A 和 B 类的紧耦合，A 类完全依赖于 B 类的功能实现，这样的情况就是典型的"侵入式开发"。随着软件业务复杂度的提升，当原有的 B 类不能满足 A 类的功能实现时，就需要创建更为高级的 BExt.java 类，结果就是把所有实例化 B.java 类的代码替换成 new BExt()代码。这就产生了源代码的改动，不利于软件运行的稳定性，并不符合商业软件的开发流程，Spring 的 DI 就可以解决这样的情况。其解决办法就是使用"反射"技术，来动态地对一个类中的属性进行反射赋值。这样的功能 Spring 形成了一个模块，模块的功能非常强大，并且 Spring 把这种机制进行命名，叫作依赖注入。

在介绍 DI 之前，先来了解一下什么是 IOC。IOC 的全称为控制反转（Inversion of Control），它想要达到的目的就是将调用者与被调用者进行分离，将类与类之间的关系进行解耦，是一种设计思想。

而 DI 则侧重于实现，A 类依赖于 B 类，B 类的对象由容器进行创建，容器再对 A 类中的 B 属性进行对象值的注入。DI 在 Java 中的底层技术原理就是"反射"，使用反射技术对某一个类中的属性值进行动态赋值，来达到 A 和 B 模块之间的解耦。

10.3 DI 容器

什么是 DI 容器？前面小节介绍过 Spring 的 DI 其实就是对 JavaBean 的属性使用反射技术进行赋值。当有很多的 JavaBean 需要这样的操作时，这些 JavaBean 的管理就成了问题。因为某些 JavaBean 之间是需要关联，而某些 JavaBean 之间并不需要关联，但所有这些 JavaBean 的创建、销毁都要统一的调度，由 Spring 框架处理它们的生命周期。为了方便这种管理，Spring 框架提供了 DI 容器，对 JavaBean 进行统一组织，便于后期代码的维护。曾经在任意的位置进行 new 实例化任何类对象的情况一去不复返了，所有 new 实例化的任务都要交给 DI 容器进行实现，这时，对 JavaBean 的管理就更加规范了。

另外 Spring 的 DI 容器完全脱离了平台，可以在任何支持 Java 语言的环境中进行运行，具有极好的移植性，其用最简单的接口与实现分离的原理，对组件的调配提供很好的支持。DI 容器就是去管理 JavaBean，创建 JavaBean 的一个内存区，在这个内存区中可以将操作 JavaBean 的代码以面向接口的方式进行开发与管理，这样将接口的多态性与 DI 技术进行结合，使程序结构的分层就更加灵活化了，维护和扩展也很方便。

DI 概念从编程技术上来讲就是使用反射技术将接口和实现相分离。DI 容器就是管理 JavaBean 的生命周期，以及管理多个 JavaBean 之间的注入关系。

通过前面的解释，读者已经大概了解 DI 与 DI 容器的作用与使用场景了。程序员随意地在任何位置创建任何类的对象在 Spring 框架中是不规范的。Spring 框架对 JavaBean 的管理更加具有规划性，比如创建、销毁，还可以动态地对一个属性注入值。通过使用 Spring 的 DI 容器，软件项目对 JavaBean 的管理更加统一和方便。

10.4　AOP 的介绍

什么是 AOP 呢？AOP 全称是面向切面编程（Aspect Oriented Programming）。

在没有 AOP 技术时，如果想对软件项目进行日志记录，则必须要在关键的业务点写上记录日志的程序代码，日志的信息包含"开始时间""结束时间""执行人"以及"角色"等信息。随着软件项目越来越稳定，曾经的日志代码需要进行删除，因为输出日志会影响程序运行的效率，这时就要在 Java 源代码中删除记录日志的程序代码，造成代码的改动。另外在未来有可能还需要记录日志时，还要重新加入日志程序代码，造成源代码反复更改，不利于软件运行的稳定性。使用 Spring 的 AOP 技术就可以解决这些"通用性"的问题。Spring 的 AOP 功能模块具有可插拔性，所以几乎不需要大幅的更改代码即可完成前面想要实现的功能。

AOP 的原理像 DI 一样也具有基础性。AOP 使用的技术原理就是"动态代理"，动态代理是 23 个标准设计模式中的一个，动态代理解决的问题是在不改变原有代码的基础上，对原有的模块进行功能上的加强并进行扩展，使扩展的功能与被扩展的模块充分解耦，利于软件项目模块化设计。应用 AOP 的场景比如可以在不改变 Servlet 代码的基础上加入日志的功能，在不改变 Struts 2 框架的 Action 代码的基础上加入数据库事务的功能等，所以 AOP 主要实现对功能模块进行扩展与模块间的解耦合。

10.5　Spring 的架构

Spring 框架就是实现了 AOP 功能的 DI 容器，在 DI 容器的基础上加入 AOP 就可以不仅做到松耦合开发，还具有面向切面编程的功能。

Spring 框架发展多年，现在已经是一个初具规模的 Java EE 开发平台。Spring 4 版本中主要的模块如图 10-1 所示。

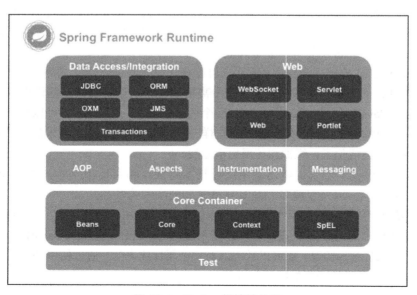

图 10-1　Spring 模块结构图

Spring 的设计一直推崇模块化，所以 Spring 中的每一个模块都可以单独进行使用，或者与其他一个或多个模块联合使用。

- Test 模块：Test 模块支持将 Spring 的组件在 JUnit 或 TestNG 框架中进行测试。
- Core Container 核心容器模块：主要的作用就是对 JavaBean 进行管理，负责 Context 核心上下文环境的创建与维护。
- AOP 和 Aspects 模块：支持面向切面编程，使类之间充分解耦。
- Instrumentation 模块：提供在应用服务器中使用的工具类以及类加载器的实现。
- Messaging 模块：提供系统内各个节点之间的交互。
- Data Access/Integration 模块：Spring 对 JDBC 访问数据库的操作进行了封装，还具有自己的 ORM 框架，支持 JMS 消息通讯技术，对数据库事务 Transactions 可以自动进行处理。OXM 是 Object-to-XML-Mapping 的缩写，它主要作用是将 Java 对象转换成 XML 数据，或者将 XML 数据转换成 Java 对象。
- Web 模块：提供了与 Web 有关技术的支持，包括 WebSocket、Servlet 等，最重要的是 Spring 提供了一个 Spring MVC 模块，使用它可以对 Web 开发进行分层设计。

Spring 框架的功能可以用在任何 Java EE 服务器中，大多数功能也适用于不受管理的环境。Spring 的核心要点是：支持不绑定到特定 Java EE 服务容器的可重用业务和数据访问对象。毫无疑问，这样的对象可以在不同的 Java EE 运行环境和测试环境之间进行重用。

10.6 一个使用传统方式保存数据功能的测试

前面使用了大段文字，一直在讨论一个话题，DI——依赖注入，那么这个 DI 是如何在 Spring 框架中实现接口与实现相分离的呢？在继续 DI 话题之前，先看一下使用传统的方法实现一个数据保存功能的弊端。

创建测试用的项目 firstSaveTest，创建业务类代码如下。

```java
package saveServiceTest;

public class SaveDB {
    public void save() {
        System.out.println("将数据保存进数据库！");
    }
}
```

运行类代码如下。

```java
package test;

import saveServiceTest.SaveDB;

public class Test {

    private SaveDB saveDB = new SaveDB();

    public SaveDB getSaveDB() {
        return saveDB;
    }
```

```java
    public void setSaveDB(SaveDB saveDB) {
        this.saveDB = saveDB;
    }

    public static void main(String[] args) {
        Test test = new Test();
        test.getSaveDB().save();
    }

}
```

程序运行结果如图 10-2 所示。

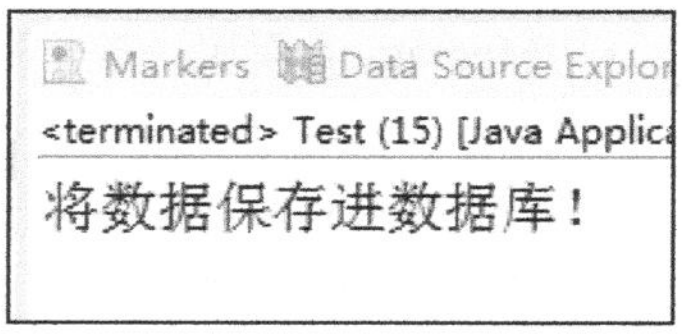

图中输出正确的结果，在大多数的软件项目中都在使用这样的代码结构，而且有些已经应用到商业的项目中。虽然输出的结果是正确的，但从项目的整合设计结构上来看很明显是不合理的，举例如下。

图 10-2　运行结果

- 源代码反复被修改：本项目会将数据保存进数据库，如果换成保存进 XML 文件中，那么就不得不更改程序，将 new SaveDB()的程序改成保存进 XML 的文件代码，这就造成了程序的更动，不利于项目的运行稳定性。
- 出现紧耦合：Test.java 和 SaveDB.java 类产生了紧耦合，不利于软件功能的扩展、测试与复用。
- 无法保证单例性：SaveDB.java 无法在单实例的情况下被重用，因为它的声明是在 Test.java 类中。也就是在 Test.java 类中可以随意地创建出很多 SaveDB.java 类的实例，无法保证该类实例的单例性。
- 无法保证资源被正确地释放：如果从 SaveDB.java 类中获取一些资源时，比如数据库的 Connection 连接、数据库的 JNDI、输入输出 Stream 流等，那么 Test.java 不得不维护这些资源的开启 open 和关闭 close。如果忘记 close，则资源不能有效地释放，造成资源占用，影响项目运行的稳定性。

典型的 4 个缺点就造成了这个设计的失败，那么根据面向对象 3 大特性中的"多态"技术，可以把 SaveDB.java 类改成接口与实现类的模式吗？也就是 Test.java 类中声明接口，然后再实例化实现这个接口的实现类（测试代码在 firstSave2 项目中）。这样的设计的确比上面的示例要灵活一些，也符合面向接口编程的方式，但仅仅是将接口和实现进行分离，也没有完全解决业务变更后源代码还要被更改的问题。那么如果在项目中经常有这样耦合的结构，该如何解决这样的问题呢？Spring 是如何解决的呢？

在知道 Spring 是如何解决紧耦合的问题之前，先来学习一下 Spring 框架的基本使用，然后才能掌握解决紧耦合的知识点。

10.7　在 Spring 中创建 JavaBean

在 Spring 中创建 JavaBean 的实例并不使用传统 new Object()实例化的方式，而是使用其他多种途径来创建出类的对象。

如果使用 new Object()方式创建出对象，那么创建类的对象的目的是达成了，但对象的管理却非常不方便，以至于非常的不工程化，所以在 Spring 框架中将创建出来的 JavaBean 对象放入 DI 容器中，在容器中统一管理这些对象。

下面的章节演示在 Spring 框架中常用的两种方式，分别是 xml 声明法以及 Annotation 注解法创建出类的对象。

10.7.1 使用 xml 声明法创建对象

xml 声明法是最原始，最有效的创建对象的方法。它使用 DOM4J 和反射技术来生成对象，在学习 Spring 创建 JavaBean 以及注入时，从 xml 声明法入手是快速掌握此技术的捷径。

1. 使用 xml 声明法创建对象

创建 Web 项目 useXMLCreateObject，添加 Spring 框架需要的 jar 包，在 src 路径中创建名称为 applicationContext.xml 配置文件，代码如下。

```xml
<?xml version="1.0" encoding="UTF-8"?>
<beans xmlns="http://www.springframework.org/schema/beans"
    xmlns:xsi="http://www.w3.org/2001/XMLSchema-instance"
    xsi:schemaLocation="http://www.springframework.org/schema/beans
http://www.springframework.org/schema/beans/spring-beans.xsd">
    <bean id="userinfo1" class="entity.Userinfo"></bean>
</beans>
```

在 Spring 框架中，对象是存放在 DI 容器中，所以要在 applicationContext.xml 文件中进行声明。声明的作用就是告诉 Spring 创建对象，使用<bean>声明对象的配置代码如下。

```xml
<bean id="userinfo1" class="entity.Userinfo"></bean>
```

上面代码的作用就是创建出 entity 包中 Userinfo 类的对象，该对象在容器中的 id 是 userinfo1
创建 Userinfo.java 类，代码如下。

```java
package entity;

public class Userinfo {
    public Userinfo() {
        System.out.println("类 Userinfo 被实例化=" + this);
    }
}
```

创建 Test1.java 类，代码如下。

```java
package test;

import org.springframework.context.ApplicationContext;
import org.springframework.context.support.ClassPathXmlApplicationContext;

public class Test1 {
    public static void main(String[] args) {
        ApplicationContext context = new ClassPathXmlApplicationContext("applicationContext.xml");
    }
}
```

程序运行后在控制台输出信息如图 10-3 所示。

```
类Userinfo被实例化=entity.Userinfo@464bee09
```

图 10-3　控制台打印结果

2. 使用 xml 声明法创建对象并获取

DI 容器中的对象可以使用 getBean()方法进行获取。

配置文件 applicationContext.xml 的代码如下。

```xml
<?xml version="1.0" encoding="UTF-8"?>
<beans xmlns="http://www.springframework.org/schema/beans"
    xmlns:xsi="http://www.w3.org/2001/XMLSchema-instance"
    xsi:schemaLocation="http://www.springframework.org/schema/beans
http://www.springframework.org/schema/beans/spring-beans.xsd">
    <bean id="userinfo1" class="entity.Userinfo"></bean>
</beans>
```

创建 Test1_1.java 类，代码如下。

```java
package test;

import org.springframework.context.ApplicationContext;
import org.springframework.context.support.ClassPathXmlApplicationContext;

import entity.Userinfo;

public class Test1_1 {
    public static void main(String[] args) {
        ApplicationContext context = new ClassPathXmlApplicationContext("applicationContext.xml");
        Userinfo userinfo = (Userinfo) context.getBean(Userinfo.class);
        System.out.println(userinfo);
    }
}
```

程序运行后控制台输出信息如图 10-4 所示。

```
类Userinfo被实例化=entity.Userinfo@464bee09
entity.Userinfo@464bee09
```

图 10-4　控制台输出结果

从控制台输出的信息来看，创建的 Userinfo 对象和获取的 Userinfo 对象是同一个。

3. 使用 xml 声明法出现 NoUniqueBeanDefinitionException 异常及解决办法

在使用 getBean(Class)方法获取对象时，如果有多个对象属于同一个类型，则在获取对象时会出现 NoUniqueBeanDefinitionException 异常。

配置文件 applicationContext.xml 的代码如下。

```xml
<?xml version="1.0" encoding="UTF-8"?>
<beans xmlns="http://www.springframework.org/schema/beans"
    xmlns:xsi="http://www.w3.org/2001/XMLSchema-instance"
    xsi:schemaLocation="http://www.springframework.org/schema/beans
http://www.springframework.org/schema/beans/spring-beans.xsd">
```

```
        <bean id="userinfo1" class="entity.Userinfo"></bean>
        <bean id="userinfo2" class="entity.Userinfo"></bean>
</beans>
```

创建 Test1_2.java 类，代码如下。

```
package test;

import org.springframework.context.ApplicationContext;
import org.springframework.context.support.ClassPathXmlApplicationContext;

import entity.Userinfo;

public class Test1_2 {
    public static void main(String[] args) {
        ApplicationContext context = new ClassPathXmlApplicationContext("applicationContext.xml");
        Userinfo userinfo = (Userinfo) context.getBean(Userinfo.class);
        System.out.println(userinfo);
    }
}
```

程序运行后，控制台输出信息如图 10-5 所示。

```
类Userinfo被实例化=entity.Userinfo@464bee09
类Userinfo被实例化=entity.Userinfo@10bdf5e5
Exception in thread "main" org.springframework.beans.factory.NoUniqueBeanDefinitionException: No qualifying bean of type 'entity.Userinfo' a
    at org.springframework.beans.factory.support.DefaultListableBeanFactory.resolveNamedBean(DefaultListableBeanFactory.java:1034)
    at org.springframework.beans.factory.support.DefaultListableBeanFactory.getBean(DefaultListableBeanFactory.java:340)
    at org.springframework.beans.factory.support.DefaultListableBeanFactory.getBean(DefaultListableBeanFactory.java:335)
    at org.springframework.context.support.AbstractApplicationContext.getBean(AbstractApplicationContext.java:1093)
    at test.Test1_2.main(Test1_2.java:11)
```

图 10-5　控制台输出结果

详细异常信息如下所示。

```
Exception in thread "main" org.springframework.beans.factory.NoUniqueBeanDefinitionException:
No qualifying bean of type 'entity.Userinfo' available: expected single matching bean but found
2: userinfo1,userinfo2
```

控制台输出的信息提示，Spring 找到两个对象，分别是 userinfo1 和 userinfo2，两者都属于 Userinfo.class 类的对象。Spring 并不知道应该获取哪一个，所以出现了异常。

使用 getBean(Class) 如果找到相同类型的对象时则出现异常，解决的办法是使用重载方法 getBean(String)，代码如下。

```
package test;

import org.springframework.context.ApplicationContext;
import org.springframework.context.support.ClassPathXmlApplicationContext;

import entity.Userinfo;

public class Test1_3 {
    public static void main(String[] args) {
        ApplicationContext context = new ClassPathXmlApplicationContext("applicationContext.xml");
        Userinfo userinfo1 = (Userinfo) context.getBean("userinfo1");
        Userinfo userinfo2 = (Userinfo) context.getBean("userinfo2");
        System.out.println("Userinfo userinfo1=" + userinfo1);
        System.out.println("Userinfo userinfo2=" + userinfo2);
    }
}
```

根据前面实验的结果可以总结出 getBean(Userinfo.class)和 getBean("userinfo1")具有如下使用场景。
- getBean(Userinfo.class)：如果 DI 容器中只有 1 个 Userinfo 类的实例时可以使用此方法。
- getBean("userinfo1")：如果在 DI 容器中出现多个相同数据类型的对象时，就要使用 getBean(String)的写法来根据对象的 id 找到指定的对象了。

10.7.2 使用 Annotation 注解法创建对象

使用 XML 声明法创建对象时容易造成 applicationContext.xml 文件中<bean>声明的配置代码过多，对后期项目代码维护比较不利。可以使用 Spring 新版本里提供的 Annotation 注解法来解决这个问题。

1. 使用<context:component-scan base-package="">创建对象

配置代码<context:component-scan base-package="">的作用是在指定的包中扫描符合创建对象的类，如果某些类需要被 Spring 实例化，则 class 类的上方必须使用@Component 注解。

创建测试用的项目 annotationCreateBean，配置文件 applicationContext.xml 的代码如下。

```xml
<?xml version="1.0" encoding="UTF-8"?>
<beans xmlns="http://www.springframework.org/schema/beans"
    xmlns:context="http://www.springframework.org/schema/context"
    xmlns:xsi="http://www.w3.org/2001/XMLSchema-instance"
    xsi:schemaLocation="http://www.springframework.org/schema/beans
http://www.springframework.org/schema/beans/spring-beans.xsd
http://www.springframework.org/schema/context
http://www.springframework.org/schema/context/spring-context.xsd
">
    <context:component-scan base-package="entity"></context:component-scan>
</beans>
```

实体类 Userinfo.java 的代码如下。

```java
package entity;

import org.springframework.stereotype.Component;

@Component
public class Userinfo {
    public Userinfo() {
        System.out.println("Userinfo 构造方法执行了：" + this);
    }
}
```

注解@Component 的作用就是标识 Userinfo.java 类是一个组件，能被<context:component-scan base-package="entity">扫描器所识别并进行自动实例化，最后将 Userinfo 对象放入 DI 容器中。另外也可以不使用@Component 注解，转而使用注解@Repository 来进行声明，程序运行的效果是没有变化的，但两者还是有一些区别。

- @Repository 主要用来声明 DAO 层。
- @Component 主要用来声明一些通用性的组件。

运行类代码如下。

```java
package test;

import org.springframework.context.ApplicationContext;
import org.springframework.context.support.ClassPathXmlApplicationContext;

public class Test {
    public static void main(String[] args) {
        ApplicationContext context = new ClassPathXmlApplicationContext("applicationContext.xml");
    }
}
```

程序运行后的结果如图 10-6 所示。

> Userinfo构造方法执行了：entity.Userinfo@5579bb86

图 10-6　程序运行结果

使用注解法成功进行了 Userinfo.java 类的实例化。

2. 使用<context:component-scan base-package="">创建对象并获取

使用<context:component-scan base-package="">扫描器创建对象后，获取对象还是使用 getBean()方法，示例代码如下。

```java
package test;

import org.springframework.context.ApplicationContext;
import org.springframework.context.support.ClassPathXmlApplicationContext;

import entity.Userinfo;

public class Test2 {
    public static void main(String[] args) {
        ApplicationContext context = new ClassPathXmlApplicationContext("applicationContext.xml");
        Userinfo userinfo = (Userinfo) context.getBean(Userinfo.class);
        System.out.println("main run " + userinfo);
    }
}
```

程序运行结果如图 10-7 所示。

> Userinfo构造方法执行了：entity.Userinfo@5579bb86
> main run entity.Userinfo@5579bb86

图 10-7　程序运行结果

3. 使用"全注解"创建出对象

本小节将实现使用"全注解"的方式创建出对象，不再使用 applicationContext.xml 配置文件。"全注解"配置法也称为 JavaConfig 配置法。

配置文件 applicationContext.xml 的根节点是<beans>，它起到全局 Configuration 配置定义的作用，它的子节点<bean>包含了创建 JavaBean 的细节信息。

在使用"全注解"法创建对象时，与<beans>标记相同作用的注解就是@Configuration，与<bean>标记相同功能的注解是@Bean。

创建测试用的项目 allAnnotationCreateObject1，示例代码如下。

```java
package tools;

import org.springframework.context.annotation.Bean;
import org.springframework.context.annotation.Configuration;

import entity.Userinfo;

@Configuration
public class CreateBean {
    @Bean
    public Userinfo getUserinfo() {
        Userinfo userinfo1 = new Userinfo();
        System.out.println("创建 userinfo1=" + userinfo1);
        return userinfo1;
    }

    @Bean
    public Userinfo createUserinfo() {
        Userinfo userinfo2 = new Userinfo();
        System.out.println("创建 userinfo2=" + userinfo2);
        return userinfo2;
    }
}
```

使用@Bean 注解声明的创建对象的方法名称可以是任意的，但必须要有返回值，不然会出现没有声明返回值的异常。

实体类代码如下。

```java
package entity;

public class Userinfo {
    public Userinfo() {
        System.out.println("Userinfo 构造方法执行了：" + this);
    }
}
```

运行类代码如下。

```java
package test;

import org.springframework.context.annotation.AnnotationConfigApplicationContext;

public class Test1 {
    public static void main(String[] args) {
        AnnotationConfigApplicationContext context = new AnnotationConfigApplicationContext("tools");
    }
}
```

实例化 new AnnotationConfigApplicationContext()类时传入参数"tools"，代表在 tools 包中寻找哪个类带有@Configuration 注解，再根据该类中的信息创建出 JavaBean 对象。

10.7 在 Spring 中创建 JavaBean

程序运行结果如图 10-8 所示。

```
Userinfo构造方法执行了：entity.Userinfo@5a4aa2f2
创建userinfo1=entity.Userinfo@5a4aa2f2
Userinfo构造方法执行了：entity.Userinfo@6591f517
创建userinfo2=entity.Userinfo@6591f517
```

图 10-8　程序运行结果

成功使用"全注解"法创建了两个 Userinfo.java 类的对象。

4．使用"全注解"获取对象出现 NoUniqueBeanDefinitionException 异常及解决办法

使用"全注解"法获取对象时也会出现获取相同类型对象的情况，并且也会出现 NoUniqueBeanDefinitionException 异常。

JavaBean 工厂类代码如下。

```java
package tools;

import org.springframework.context.annotation.Bean;
import org.springframework.context.annotation.Configuration;

import entity.Userinfo;

@Configuration
public class CreateBean {
    @Bean
    public Userinfo getUserinfo() {
        Userinfo userinfo1 = new Userinfo();
        System.out.println("创建userinfo1=" + userinfo1);
        return userinfo1;
    }

    @Bean
    public Userinfo createUserinfo() {
        Userinfo userinfo2 = new Userinfo();
        System.out.println("创建userinfo2=" + userinfo2);
        return userinfo2;
    }
}
```

运行类代码如下。

```java
package test;

import org.springframework.context.annotation.AnnotationConfigApplicationContext;

import entity.Userinfo;

public class Test2 {
    public static void main(String[] args) {
        AnnotationConfigApplicationContext context = new AnnotationConfigApplicationContext("tools");
        context.getBean(Userinfo.class);
    }
}
```

程序运行后出现 org.springframework.beans.factory.NoSuchBeanDefinitionException 异常，说明有多个对象的数据类型是相同的。解决这个问题的办法是对相同类型的对象的 id 进行命名即可，示例代码如下。

```java
package tools;

import org.springframework.context.annotation.Bean;
import org.springframework.context.annotation.Configuration;

import entity.Userinfo;

@Configuration
public class CreateBean {
    @Bean(name = "userinfo1")
    public Userinfo getUserinfo() {
        Userinfo userinfo1 = new Userinfo();
        System.out.println("创建userinfo1=" + userinfo1);
        return userinfo1;
    }

    @Bean(name = "userinfo2")
    public Userinfo createUserinfo() {
        Userinfo userinfo2 = new Userinfo();
        System.out.println("创建userinfo2=" + userinfo2);
        return userinfo2;
    }
}
```

对 @Bean 中的 name 属性赋值，设置每个 Userinfo.java 类的对象拥有不同的 id 值，并且还要结合 getBean(String)方法，运行类代码如下。

```java
package test;

import org.springframework.context.annotation.AnnotationConfigApplicationContext;
import entity.Userinfo;

public class Test3 {
    public static void main(String[] args) {
        AnnotationConfigApplicationContext context = new AnnotationConfigApplicationContext("tools");
        Userinfo userinfo1 = (Userinfo) context.getBean("userinfo1");
        Userinfo userinfo2 = (Userinfo) context.getBean("userinfo2");
        System.out.println("main userinfo1=" + userinfo1);
        System.out.println("main userinfo2=" + userinfo2);
    }
}
```

程序运行后的结果如图 10-9 所示。

```
Userinfo构造方法执行了：entity.Userinfo@5a4aa2f2
创建userinfo1=entity.Userinfo@5a4aa2f2
Userinfo构造方法执行了：entity.Userinfo@6591f517
创建userinfo2=entity.Userinfo@6591f517
main userinfo1=entity.Userinfo@5a4aa2f2
main userinfo2=entity.Userinfo@6591f517
```

图 10-9　程序运行结果

分别获取不同 Userinfo.java 类的对象。

5. 使用@ComponentScan(basePackages = "")创建对象并获取

与命名空间<context:component-scan base-package="entity"></context:component-scan>作用相同的注解是@ComponentScan(basePackages = "")，它也可以进行类扫描并实例化。

创建测试用的项目 ComponentScanTest，创建工厂类的代码如下。

```
package tools;

import java.util.Date;

import org.springframework.context.annotation.Bean;
import org.springframework.context.annotation.Configuration;

@Configuration
public class CreateBean {
    @Bean
    public Date createDate() {
        Date nowDate = new Date();
        System.out.println("createDate " + nowDate.getTime());
        return nowDate;
    }
}
```

实体类代码如下。

```
package entity;

import org.springframework.stereotype.Repository;

@Component
public class Userinfo {
    public Userinfo() {
        System.out.println("Userinfo 构造方法执行了：" + this);
    }
}
```

运行类代码如下。

```
package test;

import org.springframework.context.annotation.AnnotationConfigApplicationContext;

public class Test4 {
    public static void main(String[] args) {
        AnnotationConfigApplicationContext context = new AnnotationConfigApplicationContext("tools");
    }
}
```

程序运行结果如图 10-10 所示。

```
createDate 1480264382888
```

图 10-10 并没有将 Userinfo 类实例化

从控制台输出的信息来看，只将 Date 对象进行了实例化，并没有创建出 Userinfo.java 类的对象。如果想将其他包中的类进行实例化，那么就需要使用如下注解。

```
@ComponentScan(basePackages = "entity")
```

更改工厂类代码如下。

```java
package tools;

import java.util.Date;

import org.springframework.context.annotation.Bean;
import org.springframework.context.annotation.ComponentScan;
import org.springframework.context.annotation.Configuration;

@Configuration
@ComponentScan(basePackages = "entity")
public class CreateBean {
    @Bean
    public Date createDate() {
        Date nowDate = new Date();
        System.out.println("createDate " + nowDate.getTime());
        return nowDate;
    }
}
```

程序运行结果如图 10-11 所示。

```
Userinfo构造方法执行了：entity.Userinfo@62bd765
createDate 1480264432441
```

图 10-11　成功将 Userinfo 实例化

6．使用@ComponentScan(basePackages = "")扫描多个包

注解@ComponentScan 支持对多个包进行扫描。

创建测试用的项目 ComponentScanMorePackage，示例代码如图 10-12 所示。

```java
package entity1;

import org.springframework.stereotype.Repository;

@Repository
public class Bookinfo {
    public Bookinfo() {
        System.out.println("Bookinfo构造方法执行了：" + this);
    }
}
```

```java
package entity2;

import org.springframework.stereotype.Repository;

@Repository
public class Userinfo {
    public Userinfo() {
        System.out.println("Userinfo构造方法执行了：" + this);
    }
}
```

```java
package tools;

import java.util.Date;

@Configuration
@ComponentScan(
    basePackages = { "entity1", "entity2" }
)
public class CreateBean {
    @Bean
    public Date createDate() {
        Date nowDate = new Date();
        System.out.println("createDate " + nowDate.getTime());
        return nowDate;
    }
}
```

图 10-12　在包 entity1 和 entity2 中分别有类

运行类代码如下。

```
package test;

import org.springframework.context.annotation.AnnotationConfigApplicationContext;

import tools.CreateBean;

public class Test1 {
    public static void main(String[] args) {
        AnnotationConfigApplicationContext context = new AnnotationConfigApplicationContext(CreateBean.class);
    }
}
```

程序运行后的结果如图 10-13 所示。

```
Bookinfo构造方法执行了：entity1.Bookinfo@78a2da20
Userinfo构造方法执行了：entity2.Userinfo@dd3b207
createDate 1480311084512
```

图 10-13　成功扫描到不同包中的类

7. 使用@ComponentScan 的 basePackageClasses 属性进行扫描

注解@ComponentScan 的 basePackageClasses 属性的作用是扫描指定*.class 文件所在的包路径，然后再创建出该包下以及子孙包下类的对象。

创建测试用的项目 basePackageClassesTest，项目结构如图 10-14 所示。

图 10-14　项目结构

在 entity1.entity2.entity3 包中有 Bookinfo.java 和 Userinfo.java 类，在 entity1.entity2 包中有 Bookinfo22222222.java 类。下面要使用 basePackageClasses 属性扫描 Userinfo.java 类所在的包中所有的组件，工厂类 CreateBean.java 代码如下。

```
package tools;

import java.util.Date;

import org.springframework.context.annotation.Bean;
```

```
import org.springframework.context.annotation.ComponentScan;
import org.springframework.context.annotation.Configuration;

import entity1.entity2.entity3.Userinfo;

@Configuration
// 设置 Userinfo.class 类所在的包为基础扫描包
@ComponentScan(basePackageClasses = { Userinfo.class })
public class CreateBean {
    @Bean
    public Date createDate() {
        Date nowDate = new Date();
        System.out.println("createDate " + nowDate.getTime());
        return nowDate;
    }
}
```

运行类代码如下。

```
package test;

import org.springframework.context.annotation.AnnotationConfigApplicationContext;

import tools.CreateBean;

public class Test1 {
    public static void main(String[] args) {
        AnnotationConfigApplicationContext context = new AnnotationConfigApplicationContext(CreateBean.class);
    }
}
```

程序运行结果如图 10-15 所示。

```
Bookinfo构造方法执行了：entity1.Bookinfo@dd3b207
Userinfo构造方法执行了：entity1.Userinfo@551bdc27
createDate 1480311363120
```

图 10-15 程序运行结果

类 Bookinfo22222222.java 并没有被实例化，只是将 Bookinfo.java 和 Userinfo.java 类实例化了，说明属性 basePackageClasses 实例化的对象为同级以及子孙。

注解@ComponentScan 还可以指定多个 class 文件所在的路径进行扫描，示例代码如下。

```
@ComponentScan(basePackageClasses = { Userinfo.class, Userinfo2222222222222.class })
```

注解@ComponentScan 允许拆分多行，下面的写法是有效的。

```
@ComponentScan(basePackageClasses = { Userinfo.class })
@ComponentScan(basePackageClasses = { Userinfo2222222222222.class })
```

8. 使用@ComponentScan 不使用 basePackages 属性的效果

如果单独使用@ComponentScan 注解而不使用任何的属性，@ComponentScan 注解默认扫描的是使用@Configuration 注解的配置类所在的包路径下的所有组件，包含子包中的组件。

10.7 在 Spring 中创建 JavaBean

创建项目 ComponentScanTest2，项目结构如图 10-16 所示。

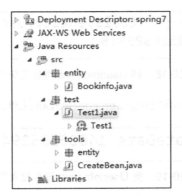

图 10-16 项目结构

注意：在 tools.entity 包下有 Userinfo.java 类。

主要代码如图 10-17 所示。

```
Bookinfo.java
1  package entity;
2
3  import org.springframework.stereotype.Repository;
4
5  @Repository
6  public class Bookinfo {
7      public Bookinfo() {
8          System.out.println("Bookinfo构造方法执行了：" + this);
9      }
10 }
11
```

```
Userinfo.java
1  package tools.entity;
2
3  import org.springframework.stereotype.Repository;
4
5  @Repository
6  public class Userinfo {
7      public Userinfo() {
8          System.out.println("Userinfo构造方法执行了：" + this);
9      }
10 }
11
```

```
CreateBean.java
1  package tools;
2
3  import java.util.Date;
10
11 @Configuration
12 @ComponentScan
13 public class CreateBean {
14     @Bean
15     public Date createDate() {
16         Date nowDate = new Date();
17         System.out.println("createDate " +
18         nowDate.getTime());
19         return nowDate;
20     }
21 }
22
```

图 10-17 主要代码

运行类代码如下。

```
package test;

import org.springframework.context.annotation.AnnotationConfigApplicationContext;

import tools.CreateBean;

public class Test1 {
    public static void main(String[] args) {
        AnnotationConfigApplicationContext context = new AnnotationConfigApplicationContext(CreateBean.class);
    }
}
```

程序运行后的结果如图 10-18 所示。

```
Userinfo构造方法执行了：tools.entity.Userinfo@551bdc27
createDate 1480310305318
```

图 10-18　将 Userinfo.java 类也实例化了

如果在 CreateBean.java 类中不使用@ComponentScan 注解，则输出的结果如图 10-19 所示。

```
createDate 1480310529428
```

图 10-19　类 Userinfo.java 没有被实例化

10.8　DI 的使用

DI 会产生类与类之间的关联，产生类之间关联的行为叫作装配（wiring）。本节将使用多种方式来装配 JavaBean 之间的关系。

在 Spring 框架中，一个 JavaBean 不需要创建另外一个 JavaBean，JavaBean 之间的关系全部由 DI 容器进行管理。也就是在 A 类中虽然声明了 B 类的对象，但是不需要实例化 B 类，B 类的对象由容器进行创建，容器还能对 A 类中的 B 对象进行赋值，这一切都是由 Spring 框架的 DI 容器来完成。

把原来 A 类实例化 B 类的写法交由 DI 容器来处理，这样就使 A 和 B 类产生了松耦合，程序代码利于扩展与后期维护。

10.8.1　使用 xml 声明法注入对象

使用 xml 声明法能实现注入对象的功能，这也是 Spring 最先采用的注入方式。

创建测试用的项目 diTest1，创建 applicationContext.xml 配置文件，代码如下。

```xml
<?xml version="1.0" encoding="UTF-8"?>
<beans xmlns="http://www.springframework.org/schema/beans"
    xmlns:xsi="http://www.w3.org/2001/XMLSchema-instance"
    xsi:schemaLocation="http://www.springframework.org/schema/beans
http://www.springframework.org/schema/beans/spring-beans.xsd">
    <bean id="userinfo1" class="entity.Userinfo"></bean>
    <bean id="test1" class="test.Test1">
        <property name="userinfo" ref="userinfo1"></property>
    </bean>
</beans>
```

使用配置代码：

```xml
<bean id="test1" class="test.Test1">
    <property name="userinfo" ref="userinfo1"></property>
</bean>
```

对 test 包中 Test1.java 类中的 userinfo 属性注入 userinfo1 对象，引用关系如图 10-20 所示。

```
<bean id="userinfo1" class="entity.Userinfo"></bean>
<bean id="test1" class="test.Test1">
    <property name="userinfo" ref="userinfo1"></property>
</bean>
```

图 10-20 注入的引用关系

实体类代码如下。

```java
package entity;

public class Userinfo {
    public Userinfo() {
        System.out.println("类 Userinfo 被实例化=" + this);
    }
}
```

运行类代码如下。

```java
package test;

import org.springframework.context.ApplicationContext;
import org.springframework.context.support.ClassPathXmlApplicationContext;

import entity.Userinfo;

public class Test1 {

    public Test1() {
        System.out.println("类 Test1 被实例化=" + this);
    }

    private Userinfo userinfo;

    public Userinfo getUserinfo() {
        return userinfo;
    }

    public void setUserinfo(Userinfo userinfo) {
        this.userinfo = userinfo;
        System.out.println("setUserinfo userinfo=" + userinfo);
    }

    public static void main(String[] args) {
        ApplicationContext context = new ClassPathXmlApplicationContext("applicationContext.xml");
        Test1 test1 = (Test1) context.getBean("test1");
        System.out.println("main userinfo=" + test1.getUserinfo());
    }
}
```

程序运行后结果如图 10-21 所示。

```
类Userinfo被实例化=entity.Userinfo@3abbfa04
类Test1被实例化=test.Test1@1e127982
setUserinfo userinfo=entity.Userinfo@3abbfa04
main userinfo=entity.Userinfo@3abbfa04
```

图 10-21 运行结果

使用反射技术动态调用 setUserinfo(Userinfo userinfo)方法传入 Userinfo.java 类的对象，最后在 main()方法中输出 Userinfo.java 类的对象信息。

使用 xml 注入还可以使用如下的写法。

```xml
<bean id="test1" class="test.Test1">
    <property name="userinfo">
        <ref bean="userinfo1" />
    </property>
</bean>
```

10.8.2　使用注解声明法注入对象

使用注解声明法注入对象时还是需要使用<context:component-scan base-package=*"entity"*></context:component-scan>配置代码来发现 JavaBean 并实例化成类的对象，再使用@Autowired 注解对变量进行值的注入。

创建测试用的项目 diTest2，配置文件 applicationContext.xml 代码如下。

```xml
<?xml version="1.0" encoding="UTF-8"?>
<beans xmlns="http://www.springframework.org/schema/beans"
    xmlns:context="http://www.springframework.org/schema/context"
    xmlns:xsi="http://www.w3.org/2001/XMLSchema-instance"
    xsi:schemaLocation="http://www.springframework.org/schema/beans
http://www.springframework.org/schema/beans/spring-beans.xsd
http://www.springframework.org/schema/context
http://www.springframework.org/schema/context/spring-context.xsd
">
    <context:component-scan base-package="entity"></context:component-scan>
    <context:component-scan base-package="test"></context:component-scan>
</beans>
```

实体类代码如下。

```java
package entity;

import org.springframework.stereotype.Repository;

@Component
public class Userinfo {
    public Userinfo() {
        System.out.println("类 Userinfo 被实例化=" + this);
    }
}
```

运行类代码如下。

```java
package test;

import org.springframework.beans.factory.annotation.Autowired;
import org.springframework.context.ApplicationContext;
import org.springframework.context.support.ClassPathXmlApplicationContext;
import org.springframework.stereotype.Repository;

import entity.Userinfo;

@Component
public class Test {
    public Test() {
```

```java
        System.out.println("类 Test 被实例化=" + this);
    }

    @Autowired
    private Userinfo userinfo;

    public Userinfo getUserinfo() {
        return userinfo;
    }

    public void setUserinfo(Userinfo userinfo) {
        this.userinfo = userinfo;
        System.out.println("setUserinfo userinfo=" + userinfo);
    }

    public static void main(String[] args) {
        ApplicationContext context = new ClassPathXmlApplicationContext("applicationContext.xml");
        Test test = (Test) context.getBean(Test.class);
        System.out.println("main userinfo=" + test.getUserinfo());
    }
}
```

程序运行结果如图 10-22 所示。

```
类Userinfo被实例化=entity.Userinfo@5204062d
类Test被实例化=test.Test@17baae6e
main userinfo=entity.Userinfo@5204062d
```

图 10-22 使用注解成功注入

方法 setUserinfo(Userinfo)并没有执行,使用反射技术直接对 userinfo 变量进行赋值。

10.8.3 多实现类的歧义性

如果对接口类型的变量注入,当 DI 容器发现有多个实现类时,在注入前会抛出异常,Spring 并不知道应该把哪个实现类对接口进行注入。

创建测试用的项目 diTest3。

接口示例代码如下。

```java
package service;

public interface IUserinfoService {
    public void save();
}
```

实现类 A 的代码如下。

```java
package service;

import org.springframework.stereotype.Repository;

@Component
public class UserinfoServiceA implements IUserinfoService {
    @Override
```

```java
    public void save() {
        System.out.println("UserinfoServiceA save");
    }
}
```

实现类 B 的代码如下。

```java
package service;

import org.springframework.stereotype.Repository;

@Component
public class UserinfoServiceB implements IUserinfoService {
    @Override
    public void save() {
        System.out.println("UserinfoServiceB save");
    }
}
```

配置文件 applicationContext.xml 的代码如下。

```xml
<?xml version="1.0" encoding="UTF-8"?>
<beans xmlns="http://www.springframework.org/schema/beans"
    xmlns:context="http://www.springframework.org/schema/context"
    xmlns:xsi="http://www.w3.org/2001/XMLSchema-instance"
    xsi:schemaLocation="http://www.springframework.org/schema/beans
http://www.springframework.org/schema/beans/spring-beans.xsd
http://www.springframework.org/schema/context
http://www.springframework.org/schema/context/spring-context.xsd
">
    <context:component-scan base-package="service"></context:component-scan>
    <context:component-scan base-package="test"></context:component-scan>
</beans>
```

运行类代码如下。

```java
package test;

import org.springframework.beans.factory.annotation.Autowired;
import org.springframework.context.ApplicationContext;
import org.springframework.context.support.ClassPathXmlApplicationContext;
import org.springframework.stereotype.Repository;

import service.IUserinfoService;

@Component
public class Test {
    public Test() {
        System.out.println("类 Test 被实例化=" + this);
    }

    @Autowired
    private IUserinfoService userinfoService;

    public IUserinfoService getUserinfoService() {
        return userinfoService;
    }

    public void setUserinfoService(IUserinfoService userinfoService) {
```

```java
        this.userinfoService = userinfoService;
    }

    public static void main(String[] args) {
        ApplicationContext context = new ClassPathXmlApplicationContext("applicationContext.xml");
        Test test = (Test) context.getBean(Test.class);
        test.getUserinfoService().save();
    }
}
```

程序运行后出现异常信息如下。

```
Caused by: org.springframework.beans.factory.NoUniqueBeanDefinitionException: No qualifying bean of type 'service.IUserinfoService' available: expected single matching bean but found 2: userinfoServiceA,userinfoServiceB
```

再次出现 NoUniqueBeanDefinitionException 异常，Spring 框架不能确定具体注入的是哪个实现类对象，所以出现异常。解决的办法就是对服务层的 id 进行命名，根据 id 进行指定服务层的注入，修改后的 A 服务类代码如下。

```java
@Component(value = "userinfoServiceA")
public class UserinfoServiceA implements IUserinfoService {
    @Override
    public void save() {
        System.out.println("UserinfoServiceA save");
    }
}
```

修改后的 B 服务类代码如下。

```java
@Component(value = "userinfoServiceB")
public class UserinfoServiceB implements IUserinfoService {
    @Override
    public void save() {
        System.out.println("UserinfoServiceB save");
    }
}
```

运行类代码如下。

```java
package test;

import javax.annotation.Resource;

import org.springframework.beans.factory.annotation.Autowired;
import org.springframework.context.ApplicationContext;
import org.springframework.context.support.ClassPathXmlApplicationContext;
import org.springframework.stereotype.Repository;

import service.IUserinfoService;

@Component
public class Test {
    public Test() {
        System.out.println("类 Test 被实例化=" + this);
    }

    @Resource(name = "userinfoServiceB")
    private IUserinfoService userinfoService;
```

```java
    public IUserinfoService getUserinfoService() {
        return userinfoService;
    }

    public void setUserinfoService(IUserinfoService userinfoService) {
        this.userinfoService = userinfoService;
    }

    public static void main(String[] args) {
        ApplicationContext context = new ClassPathXmlApplicationContext("applicationContext.xml");
        Test test = (Test) context.getBean(Test.class);
        test.getUserinfoService().save();
    }
}
```

程序运行结果如图 10-23 所示。

```
类Test被实例化=test.Test@5f71c76a
UserinfoServiceB save
```

图 10-23　将服务类 B 进行注入

以上代码在 diTest4 项目中。

10.8.4　使用@Autowired 注解向构造方法参数注入

使用注解@Autowired 可以向构造方法的参数进行注入。

创建测试用的项目 diTest5。

实体类代码如下。

```java
package entity1;

import org.springframework.stereotype.Component;

@Component
public class Userinfo {
    public Userinfo() {
        System.out.println("创建出了 Bookinfo 对象：" + this);
    }
}
```

实体类代码如下。

```java
package entity1;

import org.springframework.beans.factory.annotation.Autowired;
import org.springframework.stereotype.Component;

@Component
public class Bookinfo {
    @Autowired
    public Bookinfo(Userinfo userinfo) {
        System.out.println("执行了 public Bookinfo(Userinfo userinfo)有参构造，参数是：" + userinfo);
```

配置类代码如下。

```
package spring7;

import org.springframework.context.annotation.ComponentScan;
import org.springframework.context.annotation.Configuration;

@Configuration
@ComponentScan(basePackages = { "entity1" })
public class MyContext {
}
```

运行类代码如下。

```
package test;

import org.springframework.context.annotation.AnnotationConfigApplicationContext;

public class Test {
    public static void main(String[] args) {
        new AnnotationConfigApplicationContext("spring7");
    }
}
```

程序运行结果如图 10-24 所示。

```
创建出了Bookinfo对象：entity1.Userinfo@67b467e9
执行了public Bookinfo(Userinfo userinfo)有参构造，参数是：entity1.Userinfo@67b467e9
```

图 10-24 成功向构造方法的参数进行注入

10.8.5 在 set 方法中使用 @Autowired 注解

可以向 set 方法中的参数进行注入。

创建测试用的项目 diTest6，实体类示例代码如下。

```
package entity1;

import org.springframework.beans.factory.annotation.Autowired;
import org.springframework.stereotype.Component;

@Component
public class Bookinfo {
    private Userinfo userinfo;

    @Autowired
    public void setUserinfo(Userinfo userinfo) {
        this.userinfo = userinfo;
        System.out.println("执行了public void setUserinfo(Userinfo userinfo)方法，参数是：" + userinfo);
    }
}
```

通过前面的实验可以得知，使用注解@Autowired 可以对字段、方法的参数以及构造方法的参数进行注入。

10.8.6 使用@Bean 向工厂方法的参数传参

前面章节介绍过向 set 方法注入参数、向构造方法注入参数，以及向 Filed 字段注入值，本小节将测试对工厂方法进行注入参数。

创建测试用的项目 diTest7。

示例代码如下。

```
package spring7;

import org.springframework.context.annotation.Bean;
import org.springframework.context.annotation.ComponentScan;
import org.springframework.context.annotation.Configuration;

import entity1.Bookinfo;
import entity1.Userinfo;

@Configuration
@ComponentScan(basePackages = { "entity1" })
public class MyContext {
    @Bean
    public Bookinfo createBookinfo(Userinfo userinfo) {
        System.out.println("执行了 public Bookinfo createBookinfo(Userinfo userinfo)方法，参数为：" + userinfo);
        return new Bookinfo();
    }
}
```

10.8.7 使用@Autowired(required = false)的写法

在注入时，如果找不到符合条件的 JavaBean 对象，控制台会出现 NoSuchBeanDefinitionException 异常。

创建项目 diTest8，项目文件结构如图 10-25 所示。

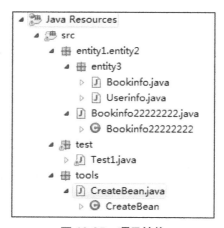

图 10-25 项目结构

10.8 DI 的使用

配置类代码如下。

```java
package tools;

import java.util.Date;

import org.springframework.context.annotation.Bean;
import org.springframework.context.annotation.ComponentScan;
import org.springframework.context.annotation.Configuration;

@Configuration
@ComponentScan(basePackages = { "test" })
public class CreateBean {
    @Bean
    public Date createDate() {
        Date nowDate = new Date();
        System.out.println("createDate " + nowDate.getTime());
        return nowDate;
    }
}
```

运行类代码如下。

```java
package test;

import org.springframework.beans.factory.annotation.Autowired;
import org.springframework.context.annotation.AnnotationConfigApplicationContext;
import org.springframework.stereotype.Repository;

import entity1.entity2.entity3.Userinfo;
import tools.CreateBean;

@Component
public class Test1 {

    @Autowired
    private Userinfo userinfo;

    public Userinfo getUserinfo() {
        return userinfo;
    }

    public void setUserinfo(Userinfo userinfo) {
        this.userinfo = userinfo;
    }

    public static void main(String[] args) {
        AnnotationConfigApplicationContext context = new AnnotationConfigApplicationContext(CreateBean.class);
        Test1 test1 = (Test1) context.getBean(Test1.class);
        System.out.println(test1.getUserinfo());
    }
}
```

程序运行后出现异常信息如下。

```
Caused by: org.springframework.beans.factory.NoSuchBeanDefinitionException: No qualifying bean of type 'entity1.entity2.entity3.Userinfo' available: expected at least 1 bean which qualifies as autowire candidate. Dependency annotations:
```

```
{@org.springframework.beans.factory.annotation.Autowired(required=true)}
```

异常信息提示没有找到符合条件的 JavaBean 对象。这时为了避免发生异常,可以加入 required = false 属性,更改后的代码如下。

```
@Autowired(required = false)
private Userinfo userinfo;
```

程序运行后并没有出现异常,由于没有找到符合条件的记录,所以对象的值为 null,结果如图 10-26 所示。

```
createDate 1480312145123
null
```

图 10-26 空值 null

10.8.8 使用 @Bean 注入多个相同类型的对象时出现异常

创建测试用的项目 diTest9。
配置类代码如下。

```java
package tools;

import java.util.Date;

import org.springframework.context.annotation.Bean;
import org.springframework.context.annotation.ComponentScan;
import org.springframework.context.annotation.Configuration;

@Configuration
@ComponentScan(basePackages = { "test" })
public class CreateBean {
    @Bean
    public Date createDate() {
        Date nowDate = new Date();
        System.out.println("createDate " + nowDate.getTime());
        return nowDate;
    }

    @Bean
    public Date createDateOther() {
        Date nowDate = new Date();
        System.out.println("createDate " + nowDate.getTime());
        return nowDate;
    }
}
```

运行类代码如下。

```java
package test;

import java.util.Date;

import org.springframework.beans.factory.annotation.Autowired;
import org.springframework.context.annotation.AnnotationConfigApplicationContext;
import org.springframework.stereotype.Repository;
```

```java
import tools.CreateBean;

@Component
public class Test1 {

    @Autowired
    private Date date;

    public Date getDate() {
        return date;
    }

    public void setDate(Date date) {
        this.date = date;
    }

    public static void main(String[] args) {
        AnnotationConfigApplicationContext context = new AnnotationConfigApplicationContext(CreateBean.class);
    }
}
```

程序运行后出现异常信息如下。

```
Caused by: org.springframework.beans.factory.NoUniqueBeanDefinitionException: No qualifying bean of type 'java.util.Date' available: expected single matching bean but found 2: createDate,createDateOther
```

DI 容器一共创建出两个 Date 对象，到底注入哪个 Date 对象呢？Spring 并不能确定，所以出现了异常。

使用@Bean 声明 JavaBean 时，JavaBean 的 id 值默认是方法名称，所以异常信息中出现了 createDate 和 createDateOther。

解决这个问题的办法就是注入指定的 JavaBean 对象，使用 @Resource(name = "createDateOther")来引用 id 为 createDateOther 的 Date 对象，示例代码如下。

```java
@Resource(name = "createDateOther")
private Date date;
```

再次运行程序，就不再出现异常了。

10.8.9　使用@Bean 对 JavaBean 的 id 重命名

默认情况下，使用@Bean 创建的 JavaBean 的 id 值就是方法名称，id 值是可以自定义的。

创建测试用的项目 diTest10。

示例代码如下。

```java
@Bean(name = "zzzzzzzzzzzz")
public Date createDateOther() {
    Date nowDate = new Date();
    System.out.println("createDate2 " + nowDate.getTime());
    return nowDate;
}
```

在运行类中注入指定的 JavaBean 对象，代码如下。

```
@Resource(name = "zzzzzzzzzzzzz")
private Date date;
```

10.8.10 对构造方法进行注入

使用@Autowired 可以对构造方法进行注入，在 XML 配置文件中也可以对构造方法进行注入。

1．使用<constructor-arg>对构造方法注入基本类型

创建测试用的项目 diTest11。

创建具有两个构造方法的测试类，代码如下。

```
package test;

public class test {
    public test(String str_ref, int int_ref) {
        System.out.println("在public test(String str_ref,int int_ref)构造方法中打印");
        System.out.println(str_ref + " " + int_ref);
    }

    public test(String str1_ref, String str2_ref) {
        System.out
                .println("在public test(String str1_ref,String str2_ref)构造方法中打印");
        System.out.println(str2_ref + " " + str2_ref);
    }
}
```

在程序代码中声明了两个重载的构造方法。

```
public test(String str_ref, int int_ref)
public test(String str1_ref, String str2_ref)
```

配置文件 applicationContext.xml 文件内容如下。

```xml
<?xml version="1.0" encoding="UTF-8"?>
<beans xmlns="http://www.springframework.org/schema/beans"
    xmlns:xsi="http://www.w3.org/2001/XMLSchema-instance"
xmlns:p="http://www.springframework.org/schema/p"
    xsi:schemaLocation="http://www.springframework.org/schema/beans
http://www.springframework.org/schema/beans/spring-beans-3.1.xsd">

    <bean id="test_ref" class="test.test">
        <constructor-arg type="java.lang.String" value="hello"></constructor-arg>
        <constructor-arg type="java.lang.String" value="world"></constructor-arg>
    </bean>

    <bean id="runit_ref" class="runit.runit">
        <property name="test_ref" ref="test_ref"></property>
    </bean>

</beans>
```

使用<constructor-arg type="java.lang.String" value="hello"></constructor-arg>配置代码向构造方法的参数注入简单数据，简单数据类型使用 value 属性。

运行类代码如下。

```
package runit;

import org.springframework.context.ApplicationContext;
import org.springframework.context.support.ClassPathXmlApplicationContext;

import test.test;

public class runit {

    test test_ref;

    public test getTest_ref() {
        return test_ref;
    }

    public void setTest_ref(test test_ref) {
        this.test_ref = test_ref;
    }

    public static void main(String[] args) {
//取得应用程序上下文接口
        ApplicationContext context = new ClassPathXmlApplicationContext(
                "applicationContext.xml");
//从 DI 容器中取出指定的 BEAN 对象
        runit runit = (runit) context.getBean("runit_ref");
    }

}
```

通过 getBean 方法来取得 DI 容器中的对象，再执行对象中的方法。

运行结果如图 10-27 所示。

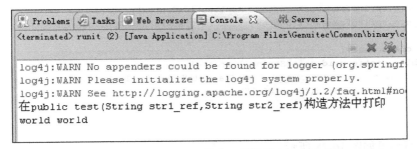

图 10-27　程序运行结果

2．使用<constructor-arg>对构造方法注入复杂类型

使用<constructor-arg>标签可以对构造方法注入复杂数据类型。

创建测试用的项目 diTest12。

实体类代码如下。

```
package entity1;

public class Bookinfo {
    public Bookinfo(Userinfo userinfo) {
        System.out.println("public Bookinfo(Userinfo userinfo) userinfo=" + userinfo);
```

        }
    }

实体类代码如下。

```
package entity1;

public class Userinfo {
    public Userinfo() {
        System.out.println("创建出了 Userinfo 对象: " + this);
    }
}
```

配置文件代码如下。

```xml
<?xml version="1.0" encoding="UTF-8"?>
<beans xmlns="http://www.springframework.org/schema/beans"
    xmlns:xsi="http://www.w3.org/2001/XMLSchema-instance"
    xmlns:p="http://www.springframework.org/schema/p"
    xsi:schemaLocation="http://www.springframework.org/schema/beans
http://www.springframework.org/schema/beans/spring-beans-4.1.xsd">

    <bean id="bookinfo1" class="entity1.Bookinfo">
        <constructor-arg ref="userinfo1"></constructor-arg>
    </bean>
    <bean id="userinfo1" class="entity1.Userinfo"></bean>


</beans>
```

运行类代码如下。

```java
package test;

import org.springframework.context.ApplicationContext;
import org.springframework.context.support.ClassPathXmlApplicationContext;

public class Test {
    public static void main(String[] args) {
        ApplicationContext context = new ClassPathXmlApplicationContext("applicationContext.xml");
        context.getBean("bookinfo1");
    }
}
```

程序运行结果如图 10-28 所示。

```
创建出了Userinfo对象: entity1.Userinfo@1794d431
public Bookinfo(Userinfo userinfo) userinfo=entity1.Userinfo@1794d431
```

图 10-28　运行结果

如果构造方法有两个参数，则使用如下的配置进行注入。

```xml
<bean id="bookinfo1" class="entity1.Bookinfo">
    <constructor-arg ref="userinfo1"></constructor-arg>
    <constructor-arg ref="userinfo2"></constructor-arg>
</bean>
<bean id="userinfo1" class="entity1.Userinfo"></bean>
```

```xml
<bean id="userinfo2" class="entity1.Userinfo"></bean>
```

3. 使用 C 命名空间对构造方法进行注入

前面章节使用配置代码`<constructor-arg ref="`*userinfo1*`"></constructor-arg>`对构造方法进行注入，配置代码比较冗长，可以使用 C 命名空间进行简化。

创建测试用的项目 diTest13。

配置文件代码如下。

```xml
<?xml version="1.0" encoding="UTF-8"?>
<beans xmlns="http://www.springframework.org/schema/beans"
    xmlns:xsi="http://www.w3.org/2001/XMLSchema-instance"
    xmlns:p="http://www.springframework.org/schema/p"
    xmlns:c="http://www.springframework.org/schema/c"
    xsi:schemaLocation="
    http://www.springframework.org/schema/beans
http://www.springframework.org/schema/beans/spring-beans-4.1.xsd
    http://www.springframework.org/schema/c
http://www.springframework.org/schema/c/spring-c-4.1.xsd
">
    <bean id="bookinfo1" class="entity1.Bookinfo" c:userinfo1-ref="userinfo1"
        c:userinfo2-ref="userinfo2">
    </bean>
    <bean id="userinfo1" class="entity1.Userinfo"></bean>
    <bean id="userinfo2" class="entity1.Userinfo"></bean>
</beans>
```

如果构造方法只有一个参数时可以使用如下的配置。

```xml
<?xml version="1.0" encoding="UTF-8"?>
<beans xmlns="http://www.springframework.org/schema/beans"
    xmlns:xsi="http://www.w3.org/2001/XMLSchema-instance"
    xmlns:p="http://www.springframework.org/schema/p"
    xmlns:c="http://www.springframework.org/schema/c"
    xsi:schemaLocation="
    http://www.springframework.org/schema/beans
http://www.springframework.org/schema/beans/spring-beans-4.1.xsd
    http://www.springframework.org/schema/c
http://www.springframework.org/schema/c/spring-c-4.1.xsd
">
    <bean id="bookinfo1" class="entity1.Bookinfo" c:_-ref="userinfo1">
    </bean>
    <bean id="userinfo1" class="entity1.Userinfo"></bean>
    <bean id="userinfo2" class="entity1.Userinfo"></bean>
</beans>
```

4. 使用 C 命名空间的索引功能对构造方法进行注入

可以根据构造方法参数的索引位置进行注入。

创建测试用的项目 diTest14。

配置文件代码如下。

```xml
<?xml version="1.0" encoding="UTF-8"?>
<beans xmlns="http://www.springframework.org/schema/beans"
    xmlns:xsi="http://www.w3.org/2001/XMLSchema-instance"
    xmlns:p="http://www.springframework.org/schema/p"
    xmlns:c="http://www.springframework.org/schema/c"
```

```xml
        xsi:schemaLocation="
    http://www.springframework.org/schema/beans
http://www.springframework.org/schema/beans/spring-beans-4.1.xsd
    http://www.springframework.org/schema/c
http://www.springframework.org/schema/c/spring-c-4.1.xsd
    ">
        <bean id="bookinfo1" class="entity1.Bookinfo" c:_0-ref="userinfo1"
            c:_1-ref="userinfo2">
        </bean>
        <bean id="userinfo1" class="entity1.Userinfo"></bean>
        <bean id="userinfo2" class="entity1.Userinfo"></bean>
</beans>
```

5. 使用 C 命名空间对构造方法注入简单数据类型

使用 C 命名空间可以对构造方法注入简单的数据类型。

创建测试用的项目 diTest15。

实体类代码如下。

```java
package entity1;

public class Bookinfo {
    public Bookinfo(String username) {
        System.out.println("public Bookinfo(String username) username=" + username);
    }

    public Bookinfo(String username, int age) {
        System.out.println("public Bookinfo(String username, int age) username=" + username + " age=" + age);
    }
}
```

配置文件代码如下。

```xml
<?xml version="1.0" encoding="UTF-8"?>
<beans xmlns="http://www.springframework.org/schema/beans"
    xmlns:xsi="http://www.w3.org/2001/XMLSchema-instance"
    xmlns:p="http://www.springframework.org/schema/p"
    xmlns:c="http://www.springframework.org/schema/c"
    xsi:schemaLocation="
    http://www.springframework.org/schema/beans
http://www.springframework.org/schema/beans/spring-beans-4.1.xsd
    http://www.springframework.org/schema/c
http://www.springframework.org/schema/c/spring-c-4.1.xsd
    ">
        <bean id="bookinfo1" class="entity1.Bookinfo" c:username="中国1">
        </bean>
        <bean id="bookinfo2" class="entity1.Bookinfo" c:_="中国2">
        </bean>
        <bean id="bookinfo3" class="entity1.Bookinfo" c:username="中国3"
            c:age="100">
        </bean>
        <bean id="bookinfo4" class="entity1.Bookinfo" c:_0="中国4" c:_1="200">
        </bean>
</beans>
```

6. 使用<constructor-arg>对构造方法注入 List/Set/Map

可以使用<constructor-arg>对构造方法注入 List/Set/Map 数据类型。

创建测试用的项目 diTest16。

实体类代码如下。

```java
package entity1;

import java.util.Date;
import java.util.List;

public class Bookinfo {
    public Bookinfo(List list) {
        for (int i = 0; i < list.size(); i++) {
            System.out.println("bookinfo1 " + list.get(i));
        }
    }

    public Bookinfo(String username, List<String> list) {
        for (int i = 0; i < list.size(); i++) {
            System.out.println("bookinfo2 " + " " + username + " " + list.get(i));
        }
    }

    public Bookinfo(int age, List<Date> list) {
        for (int i = 0; i < list.size(); i++) {
            System.out.println("bookinfo2 " + " " + age + " " + list.get(i).getTime());
        }
    }
}
```

配置文件代码如下。

```xml
<?xml version="1.0" encoding="UTF-8"?>
<beans xmlns="http://www.springframework.org/schema/beans"
    xmlns:xsi="http://www.w3.org/2001/XMLSchema-instance"
    xmlns:p="http://www.springframework.org/schema/p"
    xmlns:c="http://www.springframework.org/schema/c"
    xsi:schemaLocation="
http://www.springframework.org/schema/beans
http://www.springframework.org/schema/beans/spring-beans-4.1.xsd
    http://www.springframework.org/schema/c
http://www.springframework.org/schema/c/spring-c-4.1.xsd
    ">

    <bean id="date1" class="java.util.Date"></bean>
    <bean id="date2" class="java.util.Date"></bean>
    <bean id="date3" class="java.util.Date"></bean>

    <bean id="bookinfo1" class="entity1.Bookinfo">
        <constructor-arg>
            <list>
                <value>中国 1</value>
                <value>中国 2</value>
                <value>中国 3</value>
                <value>中国 4</value>
```

```xml
            </list>
        </constructor-arg>
    </bean>

    <bean id="bookinfo2" class="entity1.Bookinfo">
        <constructor-arg name="username" value="美国">
        </constructor-arg>
        <constructor-arg type="java.util.List">
            <list>
                <value>中国 1</value>
                <value>中国 2</value>
                <value>中国 3</value>
                <value>中国 4</value>
            </list>
        </constructor-arg>
    </bean>

    <bean id="bookinfo3" class="entity1.Bookinfo">
        <constructor-arg type="int" value="123">
        </constructor-arg>
        <constructor-arg>
            <list>
                <ref bean="date1" />
                <ref bean="date2" />
                <ref bean="date3" />
            </list>
        </constructor-arg>
    </bean>
</beans>
```

前面的配置是注入 List 数据类型，还可以注入 Map 数据类型，示例配置代码如下。

```xml
<constructor-arg>
    <map>
        <entry key="a1" value-ref="nowDate1"></entry>
        <entry key="a2" value-ref="nowDate2"></entry>
        <entry key="a3" value-ref="nowDate3"></entry>
    </map>
</constructor-arg>
```

10.8.11 使用 p 命名空间对属性进行注入

前面章节使用<bean>的子标签<property>来对属性值进行注入，还可以使用 p 命名空间进行值的注入。

创建测试用的项目 diTest17。

实体类代码如下。

```java
package entity1;

import java.util.Date;
import java.util.List;

public class Bookinfo {
    private String username;
    private Date nowDate;
    private List<String> listString;
    private List<Date> listDate;
```

```
        //set 和 get 方法
    }
```

配置文件代码如下。

```xml
<?xml version="1.0" encoding="UTF-8"?>
<beans xmlns="http://www.springframework.org/schema/beans"
    xmlns:xsi="http://www.w3.org/2001/XMLSchema-instance"
xmlns:p="http://www.springframework.org/schema/p"
    xmlns:c="http://www.springframework.org/schema/c"
xmlns:util="http://www.springframework.org/schema/util"
    xsi:schemaLocation="
    http://www.springframework.org/schema/beans
http://www.springframework.org/schema/beans/spring-beans-4.1.xsd
    http://www.springframework.org/schema/c
http://www.springframework.org/schema/c/spring-c-4.1.xsd
    http://www.springframework.org/schema/util
http://www.springframework.org/schema/util/spring-util-4.1.xsd
    ">

    <bean id="date1" class="java.util.Date"></bean>
    <bean id="date2" class="java.util.Date"></bean>
    <bean id="date3" class="java.util.Date"></bean>

    <bean id="bookinfo1" class="entity1.Bookinfo">
        <property name="username" value="中国"></property>
    </bean>

    <bean id="bookinfo2" class="entity1.Bookinfo">
        <property name="nowDate" ref="date1"></property>
    </bean>

    <bean id="bookinfo3" class="entity1.Bookinfo">
        <property name="listString">
            <list>
                <value>大中国 1</value>
                <value>大中国 2</value>
                <value>大中国 3</value>
            </list>
        </property>
    </bean>

    <bean id="bookinfo4" class="entity1.Bookinfo">
        <property name="listDate">
            <list>
                <ref bean="date1" />
                <ref bean="date2" />
                <ref bean="date3" />
            </list>
        </property>
    </bean>

    <!-- ++++++++++++++++++++++++++++++++++ -->

    <bean id="bookinfo5" class="entity1.Bookinfo" p:username="中国">
    </bean>
```

```xml
    <bean id="bookinfo6" class="entity1.Bookinfo" p:nowDate-ref="date1">
    </bean>

    <util:list id="myList1">
        <value>美国 1</value>
        <value>美国 2</value>
        <value>美国 3</value>
        <value>美国 4</value>
    </util:list>

    <bean id="bookinfo7" class="entity1.Bookinfo" p:listString-ref="myList1">
    </bean>

    <util:list id="myList2">
        <ref bean="date1" />
        <ref bean="date2" />
        <ref bean="date3" />
    </util:list>

    <bean id="bookinfo8" class="entity1.Bookinfo" p:listDate-ref="myList2">
    </bean>
</beans>
```

10.8.12　Spring 上下文环境的相关知识

Spring 的上下文环境可以理解成为 Spring 运行的环境，可以创建多个 Spring 上下文环境。默认的情况下，在不同上下文环境中的 JavaBean 对象不可以共享。

1．创建多个 Spring 上下文环境

创建多个 Spring 上下文环境就是创建多个 ApplicationContext 对象。

创建测试用的项目 springTest1。

实体类代码如下。

```java
package entity1;

public class Bookinfo {
    public Bookinfo() {
        System.out.println("创建出了 Bookinfo 对象：" + this);
    }
}
```

实体类代码如下。

```java
package entity1;

public class Userinfo {
    public Userinfo() {
        System.out.println("创建出了 Userinfo 对象：" + this);
    }
}
```

配置类代码如下。

```java
package spring7;

import org.springframework.context.annotation.Bean;
```

```
import org.springframework.context.annotation.Configuration;

import entity1.Bookinfo;

@Configuration
public class MyContext1 {
    @Bean
    public Bookinfo createBookinfo() {
        System.out.println("MyContext1 public Bookinfo createBookinfo()");
        return new Bookinfo();
    }
}
```

配置类代码如下。

```
package spring7;

import org.springframework.context.annotation.Bean;
import org.springframework.context.annotation.Configuration;

import entity1.Userinfo;

@Configuration
public class MyContext2 {
    @Bean
    public Userinfo createUserinfo() {
        System.out.println("MyContext2 public Userinfo createUserinfo()");
        return new Userinfo();
    }
}
```

运行类代码如下。

```
package test;

import org.springframework.context.ApplicationContext;
import org.springframework.context.annotation.AnnotationConfigApplicationContext;

import entity1.Bookinfo;
import entity1.Userinfo;
import spring7.MyContext1;
import spring7.MyContext2;

public class Test1 {
    public static void main(String[] args) {
        ApplicationContext context1 = new AnnotationConfigApplicationContext(MyContext1.class);
        System.out.println("A=" + context1.getBean(Bookinfo.class));
        ApplicationContext context2 = new AnnotationConfigApplicationContext(MyContext2.class);
        System.out.println("B=" + context2.getBean(Userinfo.class));
    }
}
```

程序运行后的结果如图 10-29 所示。

```
MyContext1 public Bookinfo createBookinfo()
创建出了Bookinfo对象：entity1.Bookinfo@1786f9d5
A=entity1.Bookinfo@1786f9d5
MyContext2 public Userinfo createUserinfo()
创建出了Userinfo对象：entity1.Userinfo@df27fae
B=entity1.Userinfo@df27fae
```

图 10-29　程序运行结果

成功创建出不同的 Spring 上下文环境，在自己的上下文环境中可以取得到对应的 JavaBean。

2. 多个 Spring 上下文环境各有自己的 JavaBean 的测试

本小节要测试在不同的 Spring 上下文环境中的 JavaBean 对象是不共享的。

创建测试用的项目 springTest2。

运行类代码如下。

```java
package test;

import org.springframework.context.ApplicationContext;
import org.springframework.context.annotation.AnnotationConfigApplicationContext;

import entity1.Userinfo;
import spring7.MyContext1;

public class Test2 {
    public static void main(String[] args) {
        ApplicationContext context1 = new AnnotationConfigApplicationContext(MyContext1.class);
        System.out.println("A=" + context1.getBean(Userinfo.class));
    }
}
```

程序运行后出现如下异常。

```
Exception in thread "main" org.springframework.beans.factory.NoSuchBeanDefinitionException: No qualifying bean of type [entity1.Userinfo] is defined
```

提示 Userinfo.java 类并没有被找到。

3. 让多个配置类互相通信

那么如何实现使用配置类创建出不同的上下文环境中的 JavaBean 是可以互相共享呢？

创建测试用的项目 springTest3。

配置类代码如下。

```java
package spring7;

import org.springframework.context.annotation.Bean;
import org.springframework.context.annotation.Configuration;
import org.springframework.context.annotation.Import;

import entity1.Bookinfo;

@Configuration
```

```java
@Import(MyContext2.class)
public class MyContext1 {
    @Bean
    public Bookinfo createBookinfo() {
        System.out.println("MyContext1 public Bookinfo createBookinfo()");
        return new Bookinfo();
    }
}
```

使用注解:

```java
@Import(MyContext2.class)
```

导入其他的上下文环境中的 JavaBean, 还可以使用如下的注解写法一次性导入多个配置类, 代码如下。

```java
@Import({ MyContext2.class, MyContext3.class })
```

注意: 在不同配置类中创建 JavaBean 工厂方法的名称不能一样。因为在默认的情况下, JavaBean 的 id 和方法名称一样。如果多个工厂方法名称一样, 则不会创建其他的 JavaBean, 导致找不到对象的异常发生, 此实验的源代码在项目 springTest4 中进行了测试。

4. 创建 AllContext.java 全局配置类

对多个 Spring 上下文共享 JavaBean 的写法比较好的代码组织方式是创建一个全局的配置类, 然后在这个类中使用注解@Import 导入其他的配置类。

创建测试用的项目 springTest5。

全局配置代码如下。

```java
package spring7;

import org.springframework.context.annotation.Configuration;
import org.springframework.context.annotation.Import;

@Configuration
@Import({ MyContext1.class, MyContext2.class })
public class AllContext {
}
```

```java
package test;

import org.springframework.context.ApplicationContext;
import org.springframework.context.annotation.AnnotationConfigApplicationContext;

import entity1.Bookinfo;
import entity1.Userinfo;
import spring7.AllContext;

public class Test3 {
    public static void main(String[] args) {
        ApplicationContext context1 = new AnnotationConfigApplicationContext(AllContext.class);
        System.out.println("A=" + context1.getBean(Bookinfo.class));
        System.out.println("B=" + context1.getBean(Userinfo.class));
```

        }
    }

5. 使用@ImportResource 导入 xml 文件中的配置

如果使用 XML 文件的方式创建出另外一个 Spring 上下文环境，则在配置类中如何引用这个上下文环境呢？

创建测试用的项目 springTest6。

配置文件 XML 代码如下。

```xml
<?xml version="1.0" encoding="UTF-8"?>
<beans xmlns="http://www.springframework.org/schema/beans"
    xmlns:xsi="http://www.w3.org/2001/XMLSchema-instance"
xmlns:p="http://www.springframework.org/schema/p"
    xmlns:c="http://www.springframework.org/schema/c"
xmlns:util="http://www.springframework.org/schema/util"
    xsi:schemaLocation="
http://www.springframework.org/schema/beans
http://www.springframework.org/schema/beans/spring-beans-4.1.xsd
    http://www.springframework.org/schema/c
http://www.springframework.org/schema/c/spring-c-4.1.xsd
    http://www.springframework.org/schema/util
http://www.springframework.org/schema/util/spring-util-4.1.xsd
">
    <bean id="userinfo1" class="entity1.Userinfo">
    </bean>
</beans>
```

配置类代码如下。

```java
package spring7;

import org.springframework.context.annotation.Bean;
import org.springframework.context.annotation.Configuration;
import org.springframework.context.annotation.ImportResource;

import entity1.Bookinfo;

@Configuration
@ImportResource({ "applicationContext.xml" })
public class MyContext1 {
    @Bean
    public Bookinfo createBookinfo() {
        System.out.println("MyContext1 public Bookinfo createBookinfo()");
        return new Bookinfo();
    }
}
```

在配置类中使用注解@ImportResource 导入 XML 配置文件中的 JavaBean。

此实验也证明可以使用不同的途径来创建出 Spring 上下文环境，不同 Spring 上下文环境中的 JavaBean 是可以互相共享的。

6. 使用 xml 文件创建多个上下文环境的总结

前面章节都是使用配置类的方式来创建多个 Spring 上下文，其实使用多个 xml 配置文件也可以创建多个 Spring 上下文环境。

在 xml 文件中多个 Spring 上下文环境的总结。
- 使用 xml 文件可以创建多个 Spring 上下文，但每个 Spring 上下文环境都有自己私有的 JavaBean 对象，在默认的情况下互相不可以共享这些 JavaBean 对象。测试的源代码在项目 springTest7 中。
- 让多个使用 xml 文件创建的 Spring 上下文中的 JavaBean 能共享，要在 xml 配置文件中使用<import resource="ac.xml" />标签导入其他的 xml 配置文件，也就是导入其他的 Spring 上下文环境。测试的源代码在项目 springTest8 中。
- 可以创建一个全局的 allContext.xml 配置文件，然后使用多个<import resource="ac.xml" />标签导入其他的 xml 配置文件。测试的源代码在项目 springTest9 中。
- 在 xml 中导入 JavaConfig 配置类 Spring 上下文环境的用法是在 xml 文件中使用下述代码来将配置类导入到当前上下文环境中。测试的源代码在项目 springTest10 中。

```
<context:annotation-config />
<bean class="config.MyContext1">
</bean>
```

10.8.13 使用 Spring 的 DI 方式保存数据功能的测试

在本章最开始有一个保存数据的案例，项目中的 SaveDB.java 类的控制权是在 Test.java 类中，Test.java 类完全掌握 SaveDB.java 类的创建和使用，这两个类之间产生了非常紧的耦合性，分离不开，独立不了。即使使用接口 interface 也依然会在代码中出现 new 创建一个实现类并且赋值给接口的情况。

在 DI 容器的帮助下可以实现松耦合，并且模块之间是分离的，彼此可以互相访问。使用 DI 技术加上 DI 容器之后，创建对象的操作是由 DI 容器来进行控制的，并且也完全基于接口 interface 和实现类的分离开发。这样将 SaveDB.java 类的控制权原来在 Test.java 类中转变成控制权在 DI 容器中，interface 接口的实现类是依赖 DI 容器进行注入赋值的。

下面就使用 Spring 框架来解决这两个类之间的紧耦合问题，把这个问题划上一个句号。

创建测试用的项目 springTest11。

（1）创建一个数据保存的 interface 接口，代码如下。

```
package service;

public interface ISave {
    void save();
}
```

（2）创建一个将数据保存进数据库的实现类，代码如下。

```
package service;

import org.springframework.stereotype.Component;

@Component
public class Save_DB implements ISave {
    @Override
    public void save() {
```

```java
        System.out.println("将数据保存进 DB 数据库！");
    }
}
```

（3）配置文件代码如下。

```xml
<?xml version="1.0" encoding="UTF-8"?>
<beans xmlns="http://www.springframework.org/schema/beans"
    xmlns:xsi="http://www.w3.org/2001/XMLSchema-instance"
    xmlns:p="http://www.springframework.org/schema/p"
    xmlns:context="http://www.springframework.org/schema/context"
    xmlns:aop="http://www.springframework.org/schema/aop"
    xmlns:util="http://www.springframework.org/schema/util"
    xsi:schemaLocation="http://www.springframework.org/schema/beans
    http://www.springframework.org/schema/beans/spring-beans-4.1.xsd
        http://www.springframework.org/schema/context
    http://www.springframework.org/schema/context/spring-context-4.1.xsd
        http://www.springframework.org/schema/aop
    http://www.springframework.org/schema/aop/spring-aop-4.1.xsd
        http://www.springframework.org/schema/util
    http://www.springframework.org/schema/util/spring-util-4.1.xsd">
    <context:component-scan base-package="service"></context:component-scan>
    <context:component-scan base-package="test"></context:component-scan>
</beans>
```

（4）创建一个运行类 Run.java，代码如下。

```java
package test;

import org.springframework.beans.factory.annotation.Autowired;
import org.springframework.context.ApplicationContext;
import org.springframework.context.support.ClassPathXmlApplicationContext;
import org.springframework.stereotype.Component;

import service.ISave;

@Component
public class Test {

    @Autowired
    private ISave save;

    public ISave getSave() {
        return save;
    }

    public void setSave(ISave save) {
        this.save = save;
    }

    public static void main(String[] args) {
        ApplicationContext context = new ClassPathXmlApplicationContext("applicationContext.xml");
        Test test = (Test) context.getBean("test");
        test.getSave().save();
    }
}
```

程序执行结果如图 10-30 所示。

> 将数据保存进DB数据库！

图 10-30　程序执行结果

通过使用 Spring 框架，如果业务变更，需要用 XML 文件来保存数据的情况下，只需要创建一个数据保存进 XML 文件的实现类，并且对变量进行注入，这样就可以非常灵活地处理数据，将其保存进数据库或 XML 文件中了，这也符合"开闭原则"的设计思想。

另外通过使用 DI 容器来对创建的 JavaBean 对象进行管理，完全可以非常独立地与各个模块进行 DI 注入，这样的使用所带来的结果就是软件的 JavaBean 完全重用了。在面对复杂的业务逻辑、灵活变化的模块关系、烦恼的软件后期维护及扩展等难题时，Spring 提供了很大的帮助，这也是 Spring 的哲学思想，用最简单的技术原理"反射"来解决复杂的问题。

10.8.14　BeanFactory 与 ApplicationContext

其实 Spring 的 DI 容器就是一个实现了 BeanFactory 接口的实现类，因为 ApplicationContext 接口继承自 BeanFactory 接口。通过工厂模式来取得相对应 JavaBean 对象的引用。

看一下 Spring 的 API DOC，如图 10-31 所示。

```
org.springframework.beans.factory
Interface BeanFactory

All Known Subinterfaces:
ApplicationContext, AutowireCapableBeanFactory, ConfigurableApplicationContext, ConfigurableBeanFactory, ConfigurableListableBeanFactory, ConfigurablePortletApplicationContext,
ConfigurableWebApplicationContext, HierarchicalBeanFactory, ListableBeanFactory, WebApplicationContext

All Known Implementing Classes:
AbstractApplicationContext, AbstractAutowireCapableBeanFactory, AbstractBeanFactory, AbstractRefreshableApplicationContext, AbstractRefreshableConfigApplicationContext,
AbstractRefreshablePortletApplicationContext, AbstractRefreshableWebApplicationContext, AbstractXmlApplicationContext, AnnotationConfigApplicationContext,
AnnotationConfigWebApplicationContext, ClassPathXmlApplicationContext, DefaultListableBeanFactory, FileSystemXmlApplicationContext, GenericApplicationContext,
GenericGroovyApplicationContext, GenericWebApplicationContext, GenericXmlApplicationContext, GroovyWebApplicationContext, ResourceAdapterApplicationContext, SimpleJndiBeanFactory,
StaticApplicationContext, StaticListableBeanFactory, StaticPortletApplicationContext, StaticWebApplicationContext, XmlBeanFactory, XmlPortletApplicationContext,
XmlWebApplicationContext
```

图 10-31　BeanFactory 接口结构

从图中可以发现 ApplicationContext 是 BeanFactory 的子接口，BeanFactory 接口提供了最基本的对象管理的功能，而子接口 ApplicationContext 提供了更多附加功能，比如与 Web 整合、支持国际化、事件发布和通知等功能。

10.8.15　注入 null 类型

创建测试用的项目 springTest12。

最为关键的配置文件 applicationContext.xml 代码如下。

```xml
<bean id="null_string_ref" class="test.test">
    <property name="null_string">
        <null />
    </property>
</bean>
```

10.8.16 注入 Properties 类型

创建测试用的项目 springTest13。

工具类代码如下。

```java
package test;

import java.util.Iterator;
import java.util.Properties;

public class test {
    Properties properties;//声明一个属性类对象Properties properties
    public void println_ALL() {
//打印属性类对象properties的内容
        Iterator iterator = properties.keySet().iterator();
        while (iterator.hasNext()) {
            Object key = iterator.next();
            Object value = properties.get(key);
            System.out.println("key=" + key + " value=" + value);
        }
    }

    public Properties getProperties() {//get 方法
        return properties;
    }

    public void setProperties(Properties properties) {//set 方法
        this.properties = properties;
    }
}
```

配置文件 applicationContext.xml 文件的内容如下。

```xml
<?xml version="1.0" encoding="UTF-8"?>
<beans xmlns="http://www.springframework.org/schema/beans"
    xmlns:xsi="http://www.w3.org/2001/XMLSchema-instance"
    xmlns:p="http://www.springframework.org/schema/p"
    xsi:schemaLocation="http://www.springframework.org/schema/beans
http://www.springframework.org/schema/beans/spring-beans-3.1.xsd">

    <bean id="test_ref" class="test.test">
        <property name="properties">
            <props>
                <prop key="1">11</prop>
                <prop key="2">22</prop>
                <prop key="3">22</prop>
            </props>
        </property>
    </bean>

    <bean id="runit_ref" class="runit.runit">
        <property name="test_ref" ref="test_ref"></property>
    </bean>
</beans>
```

运行类代码如下。

```
package runit;

import org.springframework.context.ApplicationContext;
import org.springframework.context.support.ClassPathXmlApplicationContext;

import test.test;

public class runit {

    test test_ref;

    public test getTest_ref() {
        return test_ref;
    }

    public void setTest_ref(test test_ref) {
        this.test_ref = test_ref;
    }

    public static void main(String[] args) {
//取得应用程序上下文接口
        ApplicationContext context = new ClassPathXmlApplicationContext(
            "applicationContext.xml");
//在DI容器中取出指定的BEAN对象
        runit runit = (runit) context.getBean("runit_ref");
//运行BEAN的println_ALL()方法
        runit.getTest_ref().println_ALL();
    }

}
```

通过 getBean 方法来获取 DI 容器中的对象，再执行对象中的方法，运行结果如图 10-32 所示。

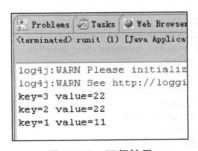

图 10-32　运行结果

10.8.17　在 DI 容器中创建 Singleton 单例和 Prototype 多例的 JavaBean 对象

DI 容器对 JavaBean 的管理也有作用域，在 Spring 中一共有 7 种作用域：Singleton、Prototype、request、session、globalSession、application 和 websocket。

Singleton 作用域代表在 DI 容器中只有一个 JavaBean 的对象，而 Prototype 代表当使用 getBean 方法取得一个 JavaBean 时，DI 容器新建一个指定 JavaBean 的实例并且返回给程序员。它们的区别仅是 Singleton 永远是一个实例，而 Prototype 是多实例。

Singleton 一定要注意 JavaBean 的线程安全问题。非线程安全是指多个线程访问同一个对象的同一个实例变量，此变量值有可能被覆盖。

做一个实例来具体看一下这两种作用域的区别！

创建测试用的项目 springTest14。

配置文件 applicationContext.xml 文件的内容如下。

```xml
<?xml version="1.0" encoding="UTF-8"?>
<beans xmlns="http://www.springframework.org/schema/beans"
    xmlns:xsi="http://www.w3.org/2001/XMLSchema-instance"
xmlns:p="http://www.springframework.org/schema/p"
    xsi:schemaLocation="http://www.springframework.org/schema/beans
http://www.springframework.org/schema/beans/spring-beans-3.1.xsd">
    <bean id="my_Date" class="java.util.Date" scope="prototype" />
</beans>
```

运行类代码如下。

```java
package runit;

import java.util.Date;

import org.springframework.context.ApplicationContext;
import org.springframework.context.support.ClassPathXmlApplicationContext;

public class runit {

    public static void main(String[] args) {
        ApplicationContext context = new ClassPathXmlApplicationContext(
                "applicationContext.xml");
        try {
//反复的调用getBean()来查看日期时间
            Date date = (Date) context.getBean("my_Date");
            date = (Date) context.getBean("my_Date");
//sleep 休息 3 秒钟
            Thread.sleep(3000);
            System.out.println(date);
            date = (Date) context.getBean("my_Date");
            Thread.sleep(3000);
            System.out.println(date);
            date = (Date) context.getBean("my_Date");
            Thread.sleep(3000);
            System.out.println(date);
        } catch (Exception e) {
            // TODO: handle exception
        }
    }

}
```

在运行类中使用 getBean 方法来获取 java.util.Date 类的实例。由于在配置文件中使用了 prototype 作用域，即每次执行 getBean 时 DI 容器都要新创建一个 java.util.Date 类的实例，代码如下。

```xml
<bean id="my_Date" class="java.util.Date" scope="prototype" />
```

这样在控制台输出的时候，肯定是不同的时间，运行结果如图 10-33 所示。

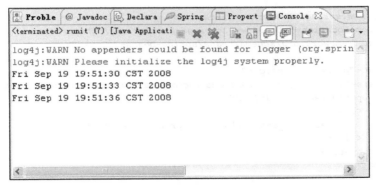

图 10-33　prototype 作用域的打印结果

将代码<bean id="my_Date" class="java.util.Date" scope="prototype" />改成<bean id="my_Date" class="java.util.Date" scope="singleton" />。根据上面介绍的知识，Singleton 是单实例，所以取得的 JavaBean 永远是一个，那么 Date 的日期也是一样，输出的结果如图 10-34 所示。

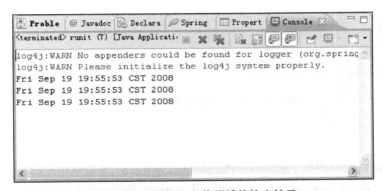

图 10-34　Singleton 作用域的输出结果

当有多个线程对同一个对象的同一个实例变量进行写操作时，这时要避免出现"非线程安全"问题，所以要使用 Prototype 多例，Struts 2 的 Action 必须使用多例，因为 Action 中存在实例变量。

当没有出现多个线程对同一个对象的同一个实例变量进行写操作时，为了减少内存的使用率，可以使用 Singleton 单例模式。在开发中 Service 类和 DAO 类都使用单例模式，因为这两个类中的实例变量在大多数的情况下都是只读的。

10.8.18　Spring 中注入外部属性文件的属性值

在开发针对数据库软件的项目中，大多数将数据库的连接信息放入属性文件中。但如何通过 Spring 将属性文件中的属性值取出来并且注入到类中的属性呢？

Spring 提供了 PropertyPlaceholderConfigurer 类来操作属性文件。

创建测试用的项目 springTest15。

applicationContext.xml 配置文件的内容如下。

```xml
<?xml version="1.0" encoding="UTF-8"?>
<beans xmlns="http://www.springframework.org/schema/beans"
    xmlns:xsi="http://www.w3.org/2001/XMLSchema-instance"
```

```xml
xmlns:p="http://www.springframework.org/schema/p"
    xsi:schemaLocation="http://www.springframework.org/schema/beans
http://www.springframework.org/schema/beans/spring-beans-3.1.xsd">

    <bean id="property_configer"
        class="org.springframework.beans.factory.config.PreferencesPlaceholderConfigurer">
        <property name="locations">
            <list>
                <value>db_connection_info.properties</value>
            </list>
        </property>
    </bean>

    <bean id="show_db_info" class="runit.runit">
        <property name="username" value="${username}" />
        <property name="password" value="${password}" />
        <property name="db_type" value="${db_type}" />
        <property name="db_url" value="${db_url}" />
    </bean>
</beans>
```

代码段如下。

```xml
<bean id="property_configer"
    class="org.springframework.beans.factory.config.PreferencesPlaceholderConfigurer">
    <property name="locations">
        <list>
            <value>db_connection_info.properties</value>
        </list>
    </property>
</bean>
```

使用 Spring 框架中的类 PreferencesPlaceholderConfigurer 来装载指定的属性文件。在本示例中将属性文件 db_connection_info.properties 从 claspath 路径中装载进内存，然后通过指定属性文件中的 key 来将属性文件中的 value 注入到类的属性中，代码如下。

```xml
<bean id="show_db_info" class="runit.runit">
    <property name="username"
        value="${username}" />
    <property name="password"
        value="${password}" />
    <property name="db_type"
        value="${db_type}" />
    <property name="db_url"
        value="${db_url}" />
</bean>
```

通过使用类似 EL 表达式的形式 ${}，指定属性文件中的 key 来将对应的 value 注入类的属性中。运行类代码如下。

```java
package runit;

import java.util.Date;

import org.springframework.context.ApplicationContext;
import org.springframework.context.support.ClassPathXmlApplicationContext;

public class runit {
```

```
        String username;
        String password;
        String db_type;
        String db_url;

        //省略 get 和 set 方法

        public static void main(String[] args) {
//取得应用程序上下文接口
            ApplicationContext context = new ClassPathXmlApplicationContext(
                    "applicationContext.xml");
//在 DI 容器中取得指定的 BEAN 对象
            runit runit = (runit) context.getBean("show_db_info");
//打印 BEAN 对象的属性值
            System.out.println("username=" + runit.username);
            System.out.println("password=" + runit.password);
            System.out.println("db_type=" + runit.db_type);
            System.out.println("db_url=" + runit.db_url);
        }

}
```

程序输出结果如图 10-35 所示。

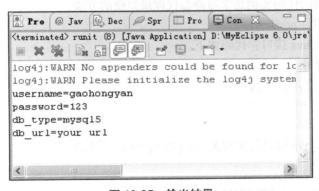

图 10-35 输出结果

10.9 面向切面编程 AOP 的使用

在以往的程序设计中，当处理记录日志这样的功能时，是需要将记录日志相关的程序代码嵌入到业务代码中，这样就造成了模块之间严重的紧耦合，不利于代码的后期维护。这种情况下可以使用 AOP 面向切面编程进行分离。

AOP 面向切面编程给程序员最直观的感受就是它可以在不改动程序代码的基础上做功能上的增强。比如在原来代码的基础上加入数据库事务处理和记录日志的功能，这种增强的方式是通过 Proxy 代理进行实现的。

那么代理 Proxy 具体是什么概念呢？先来看看生活中的示例。比如你在当当网买书，当当网只需要将书放在邮递包中，并且写上你的目的地址就可以了，剩下的任务由快递公司来完成。在这个过程中，快递公司就相当于当当网送书服务的代理 Proxy。当当网并不负责送书，送书的任务由快递公司进行代理，这就是代理模式最白话的解释。所以如果想对某一个类进行功能上的增强而又不改变原始代码，只需要将这个类让代理类进行增强就可以了。增强什么，在哪方面增强由代理类进行决定。

10.9.1　AOP 的原理之代理设计模式

在学习 AOP 技术之前，一定要先了解代理设计模式。Spring 的 AOP 技术原理就是基于代理设计模式的，下面来看一下代理模式的定义：为其他对象提供一种代理以控制对这个对象的访问。在某些情况下，一个对象不适合或者不能直接引用另一个对象，而代理对象可以在客户端和目标对象之间起到中介的作用。

代理设计模式可以在不改变原始代码的基础上，进行功能上的增强，使原始对象中的代码与增强代码进行充分解耦。代理模式分为静态代理与动态代理，为了更好地理解 AOP 技术，那就先来学习一下代理设计模式吧。

1. 静态代理的实现

在静态代理中，代理对象与被代理的对象必须实现同一个接口，完整保留被代理对象的接口样式，并且一直保持接口不变的原则。

看一个静态代理的代码实例是学习静态代理最简便的方法。

创建测试用的项目 aopTest1。

新建 ISendBook.java 接口，代码如下。

```java
package service;

public interface ISendBook {
    public void sendBook();
}
```

新建 ISendBook.java 接口的实现类 DangDang.java，代码如下。

```java
package service;

public class DangDang implements ISendBook {
    @Override
    public void sendBook() {
        System.out.println("当当网-书籍部门，将书送到指定地址的客户手中！");
    }
}
```

新建顺风快递送书代理类 SFSendBookProxy.java，代码如下。

```java
package proxy;

import service.ISendBook;

public class SFSendBookProxy implements ISendBook {

    private ISendBook sendBook;

    public SFSendBookProxy(ISendBook sendBook) {
        super();
        this.sendBook = sendBook;
    }
```

```java
    @Override
    public void sendBook() {
        System.out.println("顺风接收书籍");//增强算法
        sendBook.sendBook();
        System.out.println("顺风书籍已送达");//增强算法
    }
}
```

代理类 SFSendBookProxy.java 也要实现 ISendBook.java 接口，以达到行为的一致性。

新建 Test.java 运行类，代码如下。

```java
package run;

import proxy.SFSendBookProxy;
import service.DangDang;
import service.ISendBook;

public class Test {
    public static void main(String[] args) {
        ISendBook sendBook = new SFSendBookProxy(new DangDang());
        sendBook.sendBook();
    }
}
```

程序运行结果如图 10-36 所示。

```
顺风接收书籍
当当网-书籍部门，将书送到指定地址的客户手中！
顺风书籍已送达
```

图 10-36　静态代理输出结果

在不改变原有类 DangDang.java 代码的基础上进行功能的增强，在控制台输出的"当当网-书籍部门，将书送到指定地址的客户手中！"信息之前和之后分别输出"顺风接收书籍"和"顺风书籍已送达"的信息，这就是一个典型的日志或事务的功能增强模型。不改变原始的代码实现了事务处理，不改变原始的代码实现了日志处理。

但从上面这个示例代码中可以发现静态代理类有自身的缺点，即扩展性不好。因为 SFSendBookProxy.java 类绑死了 ISendBook.java 接口，如果顺风快递想代理更多的送货任务，则需要创建更多的代理类，举例如下。

- 送鼠标代理类：public class SFSendMouseProxy implements ISendMouse。
- 送电视代理类：public class SFSendTVProxy implements ISendTV。
- 送电话代理类：public class SFSendPhoneProxy implements ISendPhone。
- 许多许多代理类：……

静态代理这个最致命的缺点造成了静态代理不会被应用到实际的软件项目中，那有没有一个更好的解决办法不再使代理类绑死接口呢？有！这就是动态代理。

2．动态代理的实现

静态代理的缺点是代理类绑死了固定的接口，不利于扩展。动态代理则不然，通过动态代理，可以对任何实现某一接口的类进行功能上的增强，不会出现代理类绑死接口的情况发生。

在 Java 中动态代理类的对象由 Proxy.java 类的 newProxyInstance()方法进行创建，这就说明 Java 实现动态代理中的代理类并不像静态代理一样是由程序员自己创建的。动态代理中的代理类对象而是由 JVM 创建出来的，增强的算法需要由 InvocationHandler 接口来进行实现。

创建测试用的项目 aopTest2。

新建增强算法类 SendBookInvocationHandler.java，代码如下。

```java
package myandler;

import java.lang.reflect.InvocationHandler;
import java.lang.reflect.Method;

public class SendBookInvocationHandler implements InvocationHandler {

    private Object object;

    public SendBookInvocationHandler(Object object) {
        super();
        this.object = object;
    }

    @Override
    public Object invoke(Object proxy, Method method, Object[] args) throws Throwable {
        System.out.println("顺风接收书籍");
        Object returnValue = method.invoke(object, args);
        System.out.println("顺风书籍已送达");
        return returnValue;
    }

}
```

运行类代码如下。

```java
package run;

import java.lang.reflect.Proxy;

import myandler.SendBookInvocationHandler;
import service.DangDang;
import service.ISendBook;

public class Test {
    public static void main(String[] args) {
        DangDang dangdang = new DangDang();
        ISendBook sendBook = (ISendBook) Proxy.newProxyInstance(Test.class.getClassLoader(),
                dangdang.getClass().getInterfaces(), new SendBookInvocationHandler(dangdang));
        sendBook.sendBook();
    }
}
```

程序运行结果如图 10-37 所示。

程序正确运行，有以下 3 点结论。

- 代理类由 JVM 创建，程序员不需要自己创建代理类，代理类*.java 文件的数量急剧下降。
- 代理类不再绑死固定的接口，达到代理类与接口的解耦。

- 在 InvocationHandler 中通过反射技术能更灵活地处理增强算法。

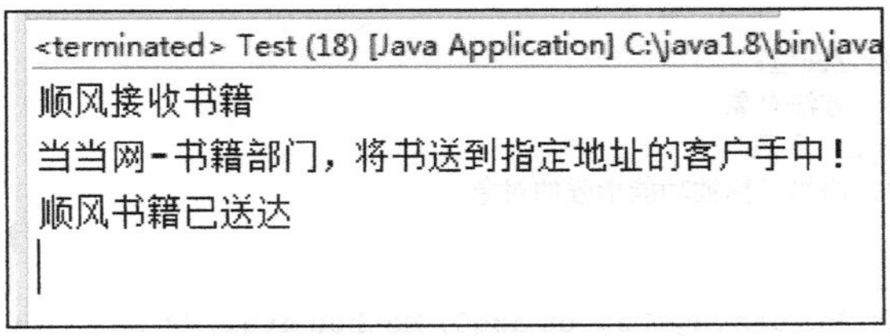

```
<terminated> Test (18) [Java Application] C:\java1.8\bin\java
顺风接收书籍
当当网-书籍部门，将书送到指定地址的客户手中！
顺风书籍已送达
```

图 10-37　动态代理的输出结果

静态代理和动态代理都是针对 public void sendBook() 方法进行增强。在动态代理中虽然使用了反射技术，但 Spring 只支持对 Method 方法进行增强，不支持 Field 字段级的增强。Spring 认为那样做是违反了面向对象编程 OOP 的思想，所以支持 Method 方法的增强是最合适不过的，而且与 Spring 的其他模块进行整合开发时更有标准性。

3. 在 Spring 中对动态代理 Proxy 的封装

虽然使用 JDK 中原始的动态代理能实现功能，但还是不太方便，并且每个程序员的写法会不同，不利于代码风格的一致性与后期维护。在 Spring 框架中对动态代理进行了封装。

- 方法前通知要实现 org.springframework.aop.MethodBeforeAdvice 接口。
- 方法后通知要实现 org.springframework.aop.AfterReturningAdvice 接口。
- 方法环绕通知要实现 org.aopalliance.intercept.MethodInterceptor 接口。
- 异常处理通知要实现 org.springframework.aop.ThrowsAdvice 接口。

创建测试用的项目 aopTest3。

创建方法执行前通知类 My_MethodBeforeAdvice.java 的代码如下。

```java
package myadvice;

import java.lang.reflect.Method;

import org.springframework.aop.MethodBeforeAdvice;

public class My_MethodBeforeAdvice implements MethodBeforeAdvice {
    // Method arg0：方法对象
    // Object[] arg1：参数
    // Object arg2：原始目标被功能增强的对象
    @Override
    public void before(Method arg0, Object[] arg1, Object arg2) throws Throwable {
        System.out.println("MethodBeforeAdvice 信息，方法名称=" + arg0.getName() + " 参数个数：" + arg1.length + " 原始对象：" + arg2);
    }

}
```

创建方法执行后通知类 My_AfterReturningAdvice.java 的代码如下。

```java
package myadvice;

import java.lang.reflect.Method;

import org.springframework.aop.AfterReturningAdvice;
```

```java
public class My_AfterReturningAdvice implements AfterReturningAdvice {

    // Object arg0:返回值:
    // Method arg1：方法对象
    // Object[] arg2：参数
    // Object arg3：原始目标被功能增强的对象

    @Override
    public void afterReturning(Object arg0, Method arg1, Object[] arg2, Object arg3) throws Throwable {
        System.out.println("AfterReturningAdvice 信息，返回值：" + arg0 + " 方法名称=" + arg1.getName() + " 参数个数：" + arg2.length
                + " 原始对象：" + arg3);
    }

}
```

创建方法环绕通知类 My_MethodInterceptor.java 的代码如下。

```java
package myadvice;

import org.aopalliance.intercept.MethodInterceptor;
import org.aopalliance.intercept.MethodInvocation;

public class My_MethodInterceptor implements MethodInterceptor {

    // MethodInvocation arg0：MethodInvocation 类的对象
    // 可以通过此对象
    // arg0.getArguments()：获取参数
    // arg0.getMethod()：获取方法对象
    // arg0.getThis()：原始目标被功能增强的对象
    // arg0.proceed()：调用原始目标对象的指定 Method 方法
    // 返回值就是调用 Method 方法的返回值，类型为 Object

    @Override
    public Object invoke(MethodInvocation arg0) throws Throwable {
        System.out.println("MethodInterceptor begin 信息，方法名称=" + arg0.getMethod().getName() + " 参数个数："
                + arg0.getArguments().length + " 原始对象：" + arg0.getThis());
        Object returnValue = arg0.proceed();
        System.out.println("MethodInterceptor   end 信息，方法名称=" + arg0.getMethod().getName() + " 参数个数：" 
                + arg0.getArguments().length + " 原始对象：" + arg0.getThis() + " 返回值：" + returnValue);
        return returnValue;
    }

}
```

创建异常通知类 My_ThrowsAdvice.java 的代码如下。

```java
package myadvice;

import java.lang.reflect.Method;

import org.springframework.aop.ThrowsAdvice;

public class My_ThrowsAdvice implements ThrowsAdvice {
```

```java
        // Exception ex: 异常对象
        public void afterThrowing(Method method, Object[] args, Object target, Exception ex) {
            System.out.println("ThrowsAdvice 信息,方法名称=" + method.getName() + " 参数个数: " + args.length + " 原始对象: " + target
                    + " 异常信息: " + ex.getMessage());
        }
    }
```

业务接口 ISendBook.java 的代码如下。

```java
package service;

public interface ISendBook {
    public String sendBook();
    public String sendBookError();
}
```

业务接口 ISendBook.java 的实现类 DangDang.java 的代码如下。

```java
package service;

public class DangDang implements ISendBook {
    @Override
    public String sendBook() {
        System.out.println("sendBookMethod 当当网-书籍部门,将书送到指定地址的客户手中!");
        return "返回值A";
    }

    @Override
    public String sendBookError() {
        System.out.println("sendBookError 当当网-书籍部门,将书送到指定地址的客户手中!");
        Integer.parseInt("a");
        return "返回值B";
    }
}
```

配置文件 applicationContext.xml 的代码如下。

```xml
<?xml version="1.0" encoding="UTF-8"?>
<beans xmlns="http://www.springframework.org/schema/beans"
    xmlns:xsi="http://www.w3.org/2001/XMLSchema-instance"
    xsi:schemaLocation="http://www.springframework.org/schema/beans
http://www.springframework.org/schema/beans/spring-beans.xsd">

    <bean id="after" class="myadvice.My_AfterReturningAdvice"></bean>
    <bean id="before" class="myadvice.My_MethodBeforeAdvice"></bean>
    <bean id="around" class="myadvice.My_MethodInterceptor"></bean>
    <bean id="throwError" class="myadvice.My_ThrowsAdvice"></bean>



    <bean id="dangdang" class="service.DangDang">
    </bean>

    <bean id="proxy" class="org.springframework.aop.framework.ProxyFactoryBean">
        <property name="interfaces">
            <list>
                <value>service.ISendBook</value>
            </list>
        </property>
```

```xml
            <property name="target" ref="dangdang"></property>
            <property name="interceptorNames">
                <list>
                    <value>after</value>
                    <value>before</value>
                    <value>around</value>
                    <value>throwError</value>
                </list>
            </property>
    </bean>

</beans>
```

运行类 Test1.java 的代码如下。

```java
package test;

import org.springframework.context.ApplicationContext;
import org.springframework.context.support.ClassPathXmlApplicationContext;

import service.ISendBook;

public class Test1 {
    public static void main(String[] args) {
        ApplicationContext context = new ClassPathXmlApplicationContext("applicationContext.xml");
        ISendBook sendBook = (ISendBook) context.getBean("proxy");
        sendBook.sendBook();

        System.out.println();
        System.out.println();

        sendBook.sendBookError();
    }
}
```

程序运行结果如图 10-38 所示。

```
MethodBeforeAdvice信息,方法名称=sendBook 参数个数: 0 原始对象: service.DangDang@6121c9d6
MethodInterceptor begin 信息,方法名称=sendBook 参数个数: 0 原始对象: service.DangDang@6121c9d6
sendBookMethod 当当网-书籍部门,将书送到指定地址的客户手中!
MethodInterceptor   end 信息,方法名称=sendBook 参数个数: 0 原始对象: service.DangDang@6121c9d6 返回值: 返回值A
AfterReturningAdvice信息,返回值: 返回值A 方法名称=sendBook 参数个数: 0 原始对象: service.DangDang@6121c9d6

MethodBeforeAdvice信息,方法名称=sendBookError 参数个数: 0 原始对象: service.DangDang@6121c9d6
MethodInterceptor begin 信息,方法名称=sendBookError 参数个数: 0 原始对象: service.DangDang@6121c9d6
sendBookError 当当网-书籍部门,将书送到指定地址的客户手中!
ThrowsAdvice信息,方法名称=sendBookError 参数个数: 0 原始对象: service.DangDang@6121c9d6 异常信息: For input string: "a"
Exception in thread "main" java.lang.NumberFormatException: For input string: "a"
    at java.lang.NumberFormatException.forInputString(NumberFormatException.java:65)
    at java.lang.Integer.parseInt(Integer.java:580)
    at java.lang.Integer.parseInt(Integer.java:615)
    at service.DangDang.sendBookError(DangDang.java:13)
    at sun.reflect.NativeMethodAccessorImpl.invoke0(Native Method)
    at sun.reflect.NativeMethodAccessorImpl.invoke(NativeMethodAccessorImpl.java:62)
    at sun.reflect.DelegatingMethodAccessorImpl.invoke(DelegatingMethodAccessorImpl.java:43)
    at java.lang.reflect.Method.invoke(Method.java:498)
    at org.springframework.aop.support.AopUtils.invokeJoinpointUsingReflection(AopUtils.java:333)
    at org.springframework.aop.framework.ReflectiveMethodInvocation.invokeJoinpoint(ReflectiveMethodInvocation.java:190)
    at org.springframework.aop.framework.ReflectiveMethodInvocation.proceed(ReflectiveMethodInvocation.java:157)
    at org.springframework.aop.framework.adapter.ThrowsAdviceInterceptor.invoke(ThrowsAdviceInterceptor.java:125)
    at org.springframework.aop.framework.ReflectiveMethodInvocation.proceed(ReflectiveMethodInvocation.java:179)
    at myadvice.My_MethodInterceptor.invoke(My_MethodInterceptor.java:21)
    at org.springframework.aop.framework.ReflectiveMethodInvocation.proceed(ReflectiveMethodInvocation.java:179)
    at org.springframework.aop.framework.adapter.MethodBeforeAdviceInterceptor.invoke(MethodBeforeAdviceInterceptor.java:52)
    at org.springframework.aop.framework.ReflectiveMethodInvocation.proceed(ReflectiveMethodInvocation.java:179)
    at org.springframework.aop.framework.adapter.AfterReturningAdviceInterceptor.invoke(AfterReturningAdviceInterceptor.java:52)
    at org.springframework.aop.framework.ReflectiveMethodInvocation.proceed(ReflectiveMethodInvocation.java:179)
    at org.springframework.aop.framework.JdkDynamicAopProxy.invoke(JdkDynamicAopProxy.java:213)
    at com.sun.proxy.$Proxy2.sendBookError(Unknown Source)
    at test.Test1.main(Test1.java:17)
```

图 10-38　运行结果

在控制台中输出了全部通知类中的功能增强方法,所有通知都参与了动态代理的算法增强。

10.9.2 与 AOP 相关的必备概念

前面章节介绍过 Spring 框架的 AOP 技术的原理就是基于动态代理的,掌握动态代理就是掌握 AOP 技术的前奏。AOP 现在已经成为一个独立的技术,有自己一些独有的术语,学习这些术语并掌握理解有助于学习 AOP 的知识,这些术语是学习 AOP 技术的必备知识。

1. 什么是横切关注点

在开发软件项目时,需要关注一些"通用性"的功能,比如需要计算出 Method 方法的执行时间,有没有访问资源的权限,做一些常规的日志,日志内容包括哪位用户正在登录,在什么时间做了哪些操作,操作的结果是正常还是出现异常等需要记录的信息,另外还有分散在系统中多处的数据库 Connection 连接的开启与关闭等功能。这些"通用性"功能的代码大多数是交织在 Service 业务对象中,代码结构如图 10-39 所示。

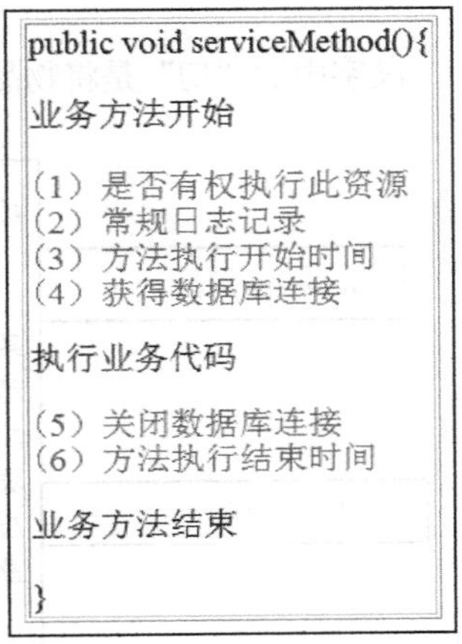

在大多数 Service 业务类的代码中都会出现这些"通用性"的功能代码,这些代码都与业务代码进行混合,两者之间产生了密不可分的紧耦合,不利于软件的后期维护与扩展。为了解决这个问题,程序员就要关注这些"通用性"的功能。它们在 Spring 框架的 AOP 中被称为"横切关注点"(Cross-cutting Concerns)。把"通用性"的功能代码从 Service 业务代码中进行分离正是 AOP 所要达到的核心目的:解耦合。

图 10-39 交织在一起的面条式代码

DI 中的"解耦"主要是对象与对象之间进行解耦,而 AOP 中的"解耦"主要是将"横切关注点"相关的代码与"业务代码"进行分离,进行解耦。虽然都是解耦,但性质是不一样的。

常用的横切关注点如图 10-40 所示。

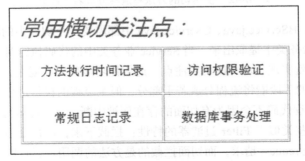

图 10-40 常用的横切关注点

横切关注点就是通用性的功能。

2．什么是切面（Aspect）以及 AOP

在研究什么是切面之前，先来研究一下汉字中的"切"字，百度百科解释如图 10-41 所示。

图 10-41　百度百科解释

汉字中的"切"是将物品分成若干部分，这个特性和 AOP 是一致的，如图 10-42 所示。

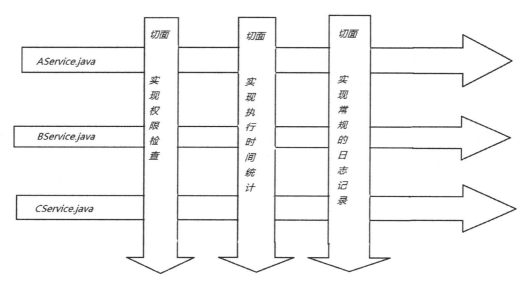

图 10-42　使用切的方法将流程分成若干部分

在 AService.java、BService.java、CService.java 中都需要 3 个切面中的功能，所以要把这些功能从 Service.java 服务类中提取出来。当 Service 业务类中的代码执行时动态地对切面中的功能代码进行调用，也就实现了对"横切关注点"通用性功能代码的复用，达到了解耦的目的。

虽然使用"切"的手段能将流程分成若干部分，但主流程的执行过程是不会被中断的，比如 AService.java 的流程代码不会因为有切面的存在而被中断，会一直运行后最后。在这个过程中，切面的存在就有些类似于 Filter 过滤器的特性：拦截下来，进行处理，然后放行。只不过 Filter 过滤器拦截的是 request 请求，而切面拦截的是方法的调用。运行的特性基本一致，但针对的目的不一样，一个是 request 请求，另一个是方法的调用。

汉字中的"切"以及在 Spring 中如何应用"切"去对执行的方法进行拦截已经解释完毕，那么什么是"切面"呢？"切面"就是对"横切关注点"的模块化，也就是将"横切关注点"

的功能代码提取出来放入一个单独的类中进行统一处理,这个类就是"切面类",也可称为切面 Aspect。在 AOP 编程中,主要就是针对"切面"进行设计代码,所以 AOP 的全称就是面向切面编程——Aspect Oriented Programming。

"切面"对"横切关注点"模块化的示例代码如图 10-43 所示。

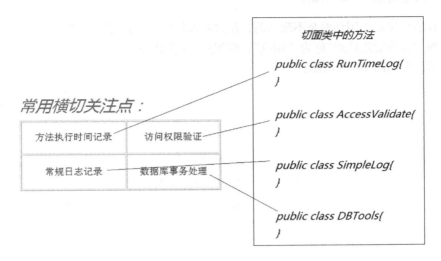

图 10-43　对横切关注点的代码模块化

将横切关注点中的功能代码放入切面类中的方法里,达到了横切关注点的模块化,切面中的方法可以被很多 Java 类以共享的方式进行访问。

3．什么是连接点（Join Point）

前面介绍了切面,那什么是连接点呢?连接点是在软件执行过程中能够插入切面的一个点。连接点可以是在调用方法前、调用方法后、方法抛出异常时、方法返回了值之后等这样的时机。在这些连接点中可以插入切面中定义的通用性功能而添加新的软件行为。

连接点示意图如图 10-44 所示。

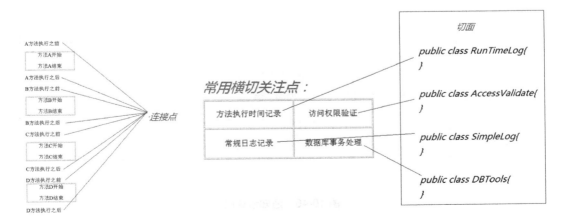

图 10-44　连接点示意图

在软件系统中存在很多个连接点，在这些连接点处可以插入切面，比如在执行 B 方法之前插入 SimpleLog()日志的功能，在执行 C 方法之前和之后插入 RunTimeLog()记录执行时间的功能等。

4．什么是切点（Pointcut）

对所有的连接点应用切面是不现实的，在大多数的情况下只想针对"部分的连接点"应用切面，这些"部分的连接点"称为"切点"，在切点上应用切面。

切点示意图如图 10-45 所示。

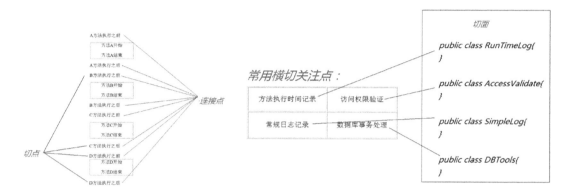

图 10-45　切点示意图

切点是缩小连接点数量的范围，只针对某几个切点进行切面的参与，切点就是精确地定义在什么位置放置切面。

5．什么是通知（Advice）

切点是定义了应用切面的精准位置，但什么时候应用切面就由通知来决定了。比如可以在执行方法之前、执行方法之后、出现异常等情况应用切面，这个时机可称为通知。

通知示意图如图 10-46 所示。

图 10-46　通知示意图

在 Spring 的 AOP 中，通知分为 5 种。

- 前置通知（Before）：方法被调用之前。
- 后置通知（After）：方法被调用之后。
- 环绕通知（Around）：方法被调用之前与之后。
- 返回通知（After-returning）：方法返回了值。
- 异常通知（After-throwing）：方法出现了异常。

在这 5 种通知 Advice 类型中都可以应用切面中的功能代码，这 5 种通知的使用在后面的章节有专门介绍。

切面包含了通知和切点，是两者的结合，通知和切点是切面最基本的元素。

- 通知：定义了在什么时机进行切面的参与。
- 切点：定义了可以在哪些连接点上放置切面。

6. 什么是织入（Weaving）

织入是把切面应用到指定的对象中，在 Spring 的 AOP 中织入的原理是由 JVM 创建出代理对象，在代理对象中调用原始对象中的方法，再结合增强算法实现的。所以 Spring 的 AOP 技术的原理就是代理设计模式，由于 Spring 的 AOP 技术是基于动态代理的，所以 Spring 中的 AOP 只支持方法的连接点，不支持字段级的连接点。

织入示意图如图 10-47 所示。

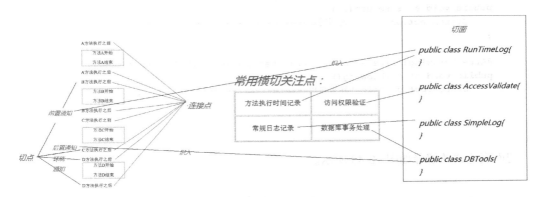

图 10-47　织入的示意图

织入使切面应用到目标对象中。

10.9.3　面向切面编程 AOP 核心案例

前面用大量篇幅介绍了 AOP 技术的概念与术语的解释，下面开始通过若干代码示例来学习一下 AOP 的程序设计吧。

1. 使用注解实现前置通知—后置通知—返回通知—异常通知

本章节使用注解的方式来实现前置通知、后置通知、返回通知、异常通知的应用。从运行的结果来看，和代理设计模式是非常一致的。

创建测试用的项目 aopTest4。

创建切面类 AspectObject.java 代码如下。

```java
package myaspect;

import org.aspectj.lang.annotation.After;
import org.aspectj.lang.annotation.AfterReturning;
import org.aspectj.lang.annotation.AfterThrowing;
import org.aspectj.lang.annotation.Aspect;
import org.aspectj.lang.annotation.Before;
import org.springframework.stereotype.Component;

@Aspect
@Component
public class AspectObject {

    @Before(value = "execution(* service.UserinfoService.*(..))")
    public void beginRun() {
        System.out.println("前置通知--------------------------");
    }

    @After(value = "execution(* service.UserinfoService.*(..))")
    public void endRun() {
        System.out.println("后置通知--------------------------");
    }

    @AfterReturning(value = "execution(* service.UserinfoService.*(..))")
    public void afterReturn() {
        System.out.println("返回通知--------------------------");
    }

    @AfterThrowing(value = "execution(* service.UserinfoService.*(..))")
    public void afterThrowing() {
        System.out.println("异常通知--------------------------");
    }
}
```

切面表达式的解释如图 10-48 所示。

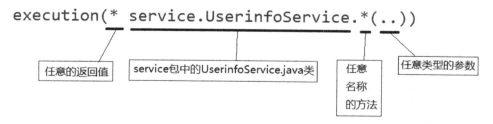

图 10-48 切面表达式的解释

切面类中的功能代码就是"横切关注点"的模块化，把这些通用性的功能代码统一管理起来，便于复用。

创建业务类 UserinfoService.java，代码如下。

```java
package service;

import org.springframework.stereotype.Service;
```

```java
@Service
public class UserinfoService {
    public void save() {
        System.out.println("执行了: userinfoService save()方法");
    }

    public String getUsername() {
        System.out.println("执行了: userinfoService getUsername()方法");
        return "我是返回值A";
    }

    public String hasErrorMethod() {
        System.out.println("执行了: userinfoService hasErrorMethod()方法");
        Integer.parseInt("a");
        return "我是返回值B";
    }
}
```

Service 服务类中的方法有些有返回值，有些没有返回值，有些会出现异常的情况。
创建配置类 MyContext.java 代码如下。

```java
package tools;

import org.springframework.context.annotation.ComponentScan;
import org.springframework.context.annotation.Configuration;
import org.springframework.context.annotation.EnableAspectJAutoProxy;

@Configuration
@ComponentScan(basePackages = { "myaspect", "service" })
// 不写@EnableAspectJAutoProxy 就没有对 UserinfoServie.java 产生 AOP 的功能
// 只是将 AspectObject.java 做为一个普通的 JavaBean 对象，而不是一个 AOP 切面
@EnableAspectJAutoProxy
public class MyContext {
}
```

创建运行类 Test1.java 代码如下。

```java
package test;

import org.springframework.context.annotation.AnnotationConfigApplicationContext;

import service.UserinfoService;
import tools.MyContext;

public class Test1 {
    public static void main(String[] args) {
        AnnotationConfigApplicationContext context = new AnnotationConfigApplicationContext(MyContext.class);
        UserinfoService service = (UserinfoService) context.getBean(UserinfoService.class);
        service.save();
        System.out.println();
        System.out.println(service.getUsername());
        System.out.println();
        service.hasErrorMethod();
    }
}
```

程序运行结果如图 10-49 所示。

```
前置通知------------------------
执行了：userinfoService save()方法
后置通知------------------------
返回通知------------------------

前置通知------------------------
执行了：userinfoService getUsername()方法
后置通知------------------------
返回通知------------------------
我是返回值A

前置通知------------------------
执行了：userinfoService hasErrorMethod()方法
后置通知------------------------
异常通知------------------------
Exception in thread "main" java.lang.NumberFormatException: For input string: "a"
        at java.lang.NumberFormatException.forInputString(NumberFormatException.java:65)
        at java.lang.Integer.parseInt(Integer.java:580)
        at java.lang.Integer.parseInt(Integer.java:615)
```

图 10-49 运行结果

在无异常的情况下，运行顺序如下所示。

1. 前置通知。
2. 业务类中的方法。
3. 后置通知。
4. 返回通知。
5. 返回值 A。

在有异常的情况下，运行顺序如下所示。

1. 前置通知。
2. 业务类中的方法。
3. 后置通知。
4. 出现异常通知。

在有异常出现时，返回通知的信息没有在控制台中进行输出，说明方法内部出现了程序运行流程的中断，程序不再继续执行，所以没有"返回通知"。

后置通知@After 和返回通知@AfterReturning 在输出效果上非常类似，这两者本质上的区别如下所示。

- @After 着重点在于方法执行完毕。
- @AfterReturning 着重点在于方法执行完毕后进行返回并继续执行后面的方法，在返回的同时可以使用@AfterReturning 取得返回值。这个功能@After 是不具有的，使用@AfterReturning 取得返回值的示例代码如下。

```
@Aspect
public class AfterReturningExample {
    @AfterReturning(
       value = "execution(* service.UserinfoService.*(..))", returning="retVal")
    public void doAccessCheck(Object retVal) {
        // ...
    }
}
```

此示例在后面的章节有介绍。

2. 使用 xml 实现前置通知—后置通知—返回通知—异常通知

使用 xml 方式来实现 AOP 就不需要任何的注解了,在 Java 类中删除全部的注解。

创建测试用的项目 aopTest5。

创建切面类 AspectObject.java 代码如下。

```java
package myaspect;

public class AspectObject {

    public void beginRun() {
        System.out.println("前置通知--------------------------");
    }

    public void endRun() {
        System.out.println("后置通知--------------------------");
    }

    public void afterReturn() {
        System.out.println("返回通知--------------------------");
    }

    public void afterThrowing() {
        System.out.println("异常通知--------------------------");
    }

}
```

创建业务类 UserinfoService.java 的代码如下。

```java
package service;

public class UserinfoService {
    public void save() {
        System.out.println("执行了: userinfoService save()方法");
    }

    public String getUsername() {
        System.out.println("执行了: userinfoService getUsername()方法");
        return "我是返回值A";
    }

    public String hasErrorMethod() {
        System.out.println("执行了: userinfoService hasErrorMethod()方法");
        Integer.parseInt("a");
        return "我是返回值B";
    }
}
```

创建配置文件 applicationContext.xml 的代码如下。

```xml
<?xml version="1.0" encoding="UTF-8"?>
<beans xmlns="http://www.springframework.org/schema/beans"
    xmlns:context="http://www.springframework.org/schema/context"
    xmlns:aop="http://www.springframework.org/schema/aop"
```

```xml
xmlns:xsi="http://www.w3.org/2001/XMLSchema-instance"
    xsi:schemaLocation="http://www.springframework.org/schema/beans
    http://www.springframework.org/schema/beans/spring-beans.xsd
    http://www.springframework.org/schema/aop
    http://www.springframework.org/schema/aop/spring-aop.xsd
    http://www.springframework.org/schema/context
    http://www.springframework.org/schema/context/spring-context.xsd
    ">

    <aop:aspectj-autoproxy></aop:aspectj-autoproxy>

    <bean id="userinfoService" class="service.UserinfoService"></bean>

    <bean id="myaspect" class="myaspect.AspectObject"></bean>

    <aop:config>
        <aop:aspect ref="myaspect">
            <aop:before method="beginRun"
                pointcut="execution(* service.UserinfoService.*(..))" />
            <aop:after method="endRun"
                pointcut="execution(* service.UserinfoService.*(..))" />
            <aop:after-returning method="afterReturn"
                pointcut="execution(* service.UserinfoService.*(..))" />
            <aop:after-throwing method="afterThrowing"
                pointcut="execution(* service.UserinfoService.*(..))" />
        </aop:aspect>
    </aop:config>

</beans>
```

配置代码的作用如下。

- <aop:before>：方法执行前通知。
- <aop:after>：方法执行后通知。
- <aop:after-returning>：方法返回通知。
- <aop:after-throwing>：方法异常通知。

创建运行类 Test1.java 的代码如下。

```java
package test;

import org.springframework.context.ApplicationContext;
import org.springframework.context.support.ClassPathXmlApplicationContext;

import service.UserinfoService;

public class Test1 {
    public static void main(String[] args) {
        ApplicationContext context = new ClassPathXmlApplicationContext("applicationContext.xml");
        UserinfoService service = (UserinfoService) context.getBean("userinfoService");
        service.save();
        System.out.println();
        System.out.println(service.getUsername());
        System.out.println();
        service.hasErrorMethod();
    }
}
```

程序运行结果如图 10-50 所示。

```
前置通知------------------------
执行了：userinfoService save()方法
后置通知------------------------
返回通知------------------------

前置通知------------------------
执行了：userinfoService getUsername()方法
后置通知------------------------
返回通知------------------------
我是返回值A

前置通知------------------------
执行了：userinfoService hasErrorMethod()方法
后置通知------------------------
异常通知------------------------
Exception in thread "main" java.lang.NumberFormatException: For input string: "a"
    at java.lang.NumberFormatException.forInputString(NumberFormatException.java:65)
    at java.lang.Integer.parseInt(Integer.java:580)
    at java.lang.Integer.parseInt(Integer.java:615)
```

图 10-50　运行结果

使用 xml 方式实现 AOP 与使用注解的方式实现 AOP 在运行结果上是一样的。

3．使用注解实现环绕通知

所谓的环绕通知是在执行方法之前和执行方法之后都有切面的参与。

创建测试用的项目 aopTest6。

创建切面类 AspectObject.java 代码如下。

```java
package test8;

import org.aspectj.lang.ProceedingJoinPoint;
import org.aspectj.lang.annotation.Around;
import org.aspectj.lang.annotation.Aspect;
import org.springframework.stereotype.Component;

@Aspect
@Component
public class AOPObject {

    @Around(value = "execution(* test8.UserinfoService.*(..))")
    public Object aroundMethod(ProceedingJoinPoint point) {
        Object returnValue = null;
        try {
            System.out.println("前置");
            returnValue = point.proceed();// 执行业务和 invoke 相似
            System.out.println("后置");
        } catch (Throwable e) {
            e.printStackTrace();
        }
        return returnValue;
    }

}
```

方法的参数 ProceedingJoinPoint point 必须要有，数据类型不能写错。

创建业务类 UserinfoService.java 的代码如下。

```java
package test8;

import org.springframework.stereotype.Service;

@Service
public class UserinfoService {

    public void save() {
        System.out.println("public void save()");
    }

    public String getUserinfo() {
        System.out.println("public String getUserinfo()");
        return "返回值A";
    }

    public String hasError() {
        System.out.println("public String hasError()");
        Integer.parseInt("a");
        return "返回值B";
    }

}
```

创建配置类 MyContext.java 的代码如下。

```java
package test8;

import org.springframework.context.annotation.ComponentScan;
import org.springframework.context.annotation.Configuration;
import org.springframework.context.annotation.EnableAspectJAutoProxy;

@Configuration
@EnableAspectJAutoProxy
@ComponentScan(basePackages = { "test8" })
public class MyContext {
}
```

创建运行类 Test1.java 的代码如下。

```java
package test8;

import org.springframework.context.annotation.AnnotationConfigApplicationContext;

public class Test {

    public static void main(String[] args) {
        AnnotationConfigApplicationContext context = new AnnotationConfigApplicationContext(MyContext.class);
        UserinfoService service = (UserinfoService) context.getBean(UserinfoService.class);
        service.save();
        System.out.println();
        System.out.println();
        System.out.println(service.getUserinfo());
        System.out.println();
        System.out.println();
        service.hasError();
```

        }
    }

程序运行结果如图 10-51 所示。

```
前置
public void save()
后置


前置
public String getUserinfo()
后置
返回值A


前置
public String hasError()
java.lang.NumberFormatException: For input string: "a"
        at java.lang.NumberFormatException.forInputString(NumberFormatE
        at java.lang.Integer.parseInt(Integer.java:580)
```

图 10-51 运行结果

4. 使用 xml 实现环绕通知

创建测试用的项目 aopTest7。

创建切面类 AspectObject.java 的代码如下。

```java
package test8;

import org.aspectj.lang.ProceedingJoinPoint;

public class AOPObject {

    public Object aroundMethod(ProceedingJoinPoint point) {
        Object returnValue = null;
        try {
            System.out.println("前置");
            returnValue = point.proceed();// 执行业务和 invoke 相似
            System.out.println("后置");
        } catch (Throwable e) {
            e.printStackTrace();
        }
        return returnValue;
    }

}
```

创建业务类 UserinfoService.java 的代码如下。

```java
package test8;

public class UserinfoService {

    public void save() {
        System.out.println("public void save()");
    }
```

```java
    public String getUserinfo() {
        System.out.println("public String getUserinfo()");
        return "返回值 A";
    }

    public String hasError() {
        System.out.println("public String hasError()");
        Integer.parseInt("a");
        return "返回值 B";
    }

}
```

创建配置文件 applicationContext.xml 的代码如下。

```xml
<?xml version="1.0" encoding="UTF-8"?>
<beans xmlns="http://www.springframework.org/schema/beans"
    xmlns:xsi="http://www.w3.org/2001/XMLSchema-instance"
    xmlns:p="http://www.springframework.org/schema/p"
    xmlns:aop="http://www.springframework.org/schema/aop"
    xsi:schemaLocation="http://www.springframework.org/schema/beans
        http://www.springframework.org/schema/beans/spring-beans-4.1.xsd
        http://www.springframework.org/schema/aop
        http://www.springframework.org/schema/aop/spring-aop-4.1.xsd">


    <aop:aspectj-autoproxy></aop:aspectj-autoproxy>

    <bean id="userinfoService" class="test8.UserinfoService"></bean>
    <bean id="aopObject" class="test8.AOPObject"></bean>

    <aop:config>
        <aop:aspect ref="aopObject">
            <aop:around method="aroundMethod"
                pointcut="execution(* test8.UserinfoService.*(..))" />
        </aop:aspect>
    </aop:config>


</beans>
```

创建运行类 Test1.java 的代码如下。

```java
package test8;

import org.springframework.context.ApplicationContext;
import org.springframework.context.support.ClassPathXmlApplicationContext;

public class Test {

    public static void main(String[] args) {
        ApplicationContext context = new ClassPathXmlApplicationContext("applicationContext.xml");
        UserinfoService service = (UserinfoService) context.getBean(UserinfoService.class);
        service.save();
        System.out.println();
        System.out.println();
```

```
            System.out.println(service.getUserinfo());
            System.out.println();
            System.out.println();
            service.hasError();
        }
    }
```

程序运行结果如图 10-52 所示。

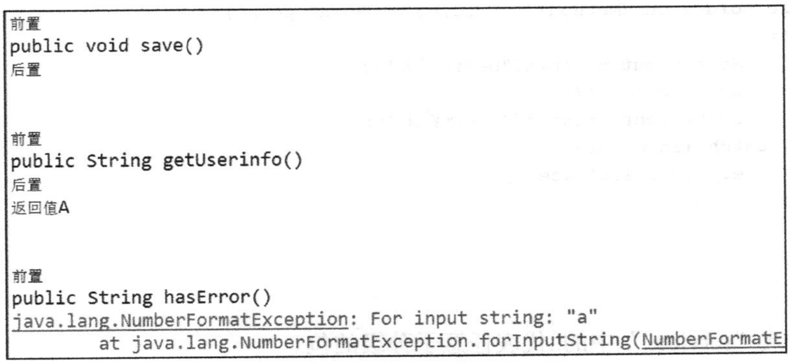

```
前置
public void save()
后置


前置
public String getUserinfo()
后置
返回值A


前置
public String hasError()
java.lang.NumberFormatException: For input string: "a"
          at java.lang.NumberFormatException.forInputString(NumberFormatE
```

图 10-52　运行结果

需要注意的是，@Before 和@After 并不真正是一对，在经过作者测试后，如果@Before 和 @After 和@Around 联合使用时，打印的顺序如下：

```
aroundMethod begin
beforeMethod
a run !
aroundMethod    end
afterMethod
```

并不是想象中的：

```
aroundMethod begin
beforeMethod
a run !
afterMethod
aroundMethod    end
```

另外，如果在 XML 中也联合使用这 3 个通知，则@Before 与@Around 的执行顺序取决于它们在 XML 文件中的配置顺序。

5．在注解中使用 bean 表达式

在 Spring 中提供了 bean 表达式来限制切面应用于的目标对象。

创建测试用的项目 aopTest8。

创建切面类 AspectObject.java 的代码如下。

```java
package myaspect;

import org.aspectj.lang.ProceedingJoinPoint;
import org.aspectj.lang.annotation.Around;
import org.aspectj.lang.annotation.Aspect;
```

```
import org.springframework.stereotype.Component;

@Aspect
@Component
public class AspectObject {

    // 表达式也可以这样写:
    // "execution(* service..*(..)) and !bean(service1)"
    // !叹号的意思是取反
    @Around(value = "execution(* service..*(..)) and bean(service1)")
    public void aroundMethod(ProceedingJoinPoint point) {
        try {
            System.out.println("begin 开始");
            point.proceed();
            System.out.println("  end 结束");
        } catch (Throwable e) {
            e.printStackTrace();
        }
    }
}
```

切点表达式 "service.." 中的 .. 代表任意子孙级的包。

表达式 "and bean(service1)" 的作用是切面必须要应用于 bean 的 id 值为 service1 的对象。

创建业务类 UserinfoService1.java 的代码如下。

```
package service;

import org.springframework.stereotype.Service;

@Service(value = "service1")
public class UserinfoService1 {
    public void aroundMethod() {
        System.out.println("UserinfoService1 正在执行业务中！");
    }
}
```

创建业务类 UserinfoService2.java 的代码如下。

```
package service;

import org.springframework.stereotype.Service;

@Service(value = "service2")
public class UserinfoService2 {
    public void aroundMethod() {
        System.out.println("UserinfoService2 正在执行业务中！");
    }
}
```

创建配置类 MyContext.java 的代码如下。

```
package tools;

import org.springframework.context.annotation.ComponentScan;
import org.springframework.context.annotation.Configuration;
import org.springframework.context.annotation.EnableAspectJAutoProxy;
```

```
@Configuration
@ComponentScan(basePackages = { "myaspect", "service" })
@EnableAspectJAutoProxy
public class MyContext {
}
```

创建运行类 Test1.java 的代码如下。

```
package test;

import org.springframework.context.annotation.AnnotationConfigApplicationContext;

import service.UserinfoService1;
import service.UserinfoService2;
import tools.MyContext;

public class Test1 {
    public static void main(String[] args) {
        AnnotationConfigApplicationContext context = new AnnotationConfigApplicationContext(MyContext.class);
        UserinfoService1 service1 = (UserinfoService1) context.getBean(UserinfoService1.class);
        service1.aroundMethod();

        System.out.println();
        System.out.println();

        UserinfoService2 service2 = (UserinfoService2) context.getBean(UserinfoService2.class);
        service2.aroundMethod();
    }
}
```

程序运行结果如图 10-53 所示。

```
begin开始
UserinfoService1正在执行业务中!
  end结束


UserinfoService2正在执行业务中!
```

图 10-53 运行结果

只对 bean 的 id 是 service1 应用了切面。

6. 在 xml 中使用 bean 表达式

创建测试用的项目 aopTest9。
创建切面类 AspectObject.java 的代码如下。

```
package myaspect;

import org.aspectj.lang.ProceedingJoinPoint;

public class AspectObject {
    public void aroundMethod(ProceedingJoinPoint point) {
        try {
            System.out.println("begin 开始");
```

```java
            point.proceed();
            System.out.println(" end结束");
        } catch (Throwable e) {
            e.printStackTrace();
        }
    }
}
```

创建业务类 UserinfoService1.java 的代码如下。

```java
package service;

import org.springframework.stereotype.Service;

public class UserinfoService1 {
    public void aroundMethod() {
        System.out.println("UserinfoService1 正在执行业务中！");
    }
}
```

创建业务类 UserinfoService2.java 的代码如下。

```java
package service;

import org.springframework.stereotype.Service;

public class UserinfoService2 {
    public void aroundMethod() {
        System.out.println("UserinfoService2 正在执行业务中！");
    }
}
```

创建配置文件 applicationContext.xml 的代码如下。

```xml
<?xml version="1.0" encoding="UTF-8"?>
<beans xmlns="http://www.springframework.org/schema/beans"
    xmlns:context="http://www.springframework.org/schema/context"
    xmlns:aop="http://www.springframework.org/schema/aop"
    xmlns:xsi="http://www.w3.org/2001/XMLSchema-instance"
    xsi:schemaLocation="http://www.springframework.org/schema/beans
http://www.springframework.org/schema/beans/spring-beans.xsd
http://www.springframework.org/schema/aop
http://www.springframework.org/schema/aop/spring-aop.xsd
http://www.springframework.org/schema/context
http://www.springframework.org/schema/context/spring-context.xsd
">

    <aop:aspectj-autoproxy></aop:aspectj-autoproxy>

    <bean id="userinfoService1" class="service.UserinfoService1"></bean>
    <bean id="userinfoService2" class="service.UserinfoService2"></bean>

    <bean id="myaspect" class="myaspect.AspectObject"></bean>

    <!-- 也可以这样写： -->
    <!-- execution(* service..*(..)) and !bean(userinfoService1) -->
    <aop:config>
        <aop:aspect ref="myaspect">
            <aop:around method="aroundMethod"
```

```xml
                pointcut="execution(* service..*(..)) and bean(userinfoService1)" />
        </aop:aspect>
    </aop:config>

</beans>
```

创建运行类 Test1.java 的代码如下。

```java
package test;

import org.springframework.context.ApplicationContext;
import org.springframework.context.support.ClassPathXmlApplicationContext;

import service.UserinfoService1;
import service.UserinfoService2;

public class Test1 {
    public static void main(String[] args) {
        ApplicationContext context = new ClassPathXmlApplicationContext("applicationContext.xml");
        UserinfoService1 service1 = (UserinfoService1)
            context.getBean("userinfoService1");
        service1.aroundMethod();
        System.out.println();
        UserinfoService2 service2 = (UserinfoService2)
            context.getBean("userinfoService2");
        service2.aroundMethod();
    }
}
```

程序运行结果如图 10-54 所示。

```
begin开始
UserinfoService1正在执行业务中!
    end结束

UserinfoService2正在执行业务中!
```

图 10-54　运行结果

7．使用注解@Pointcut 定义全局切点

前面章节在多处使用一样的 execution 表达式，可以将 execution 表达式进行全局化，以减少冗余的配置。

创建测试用的项目 aopTest10。

创建切面类 AspectObject.java 的代码如下。

```java
package myaspect;

import org.aspectj.lang.annotation.After;
import org.aspectj.lang.annotation.AfterReturning;
import org.aspectj.lang.annotation.AfterThrowing;
import org.aspectj.lang.annotation.Aspect;
import org.aspectj.lang.annotation.Before;
import org.aspectj.lang.annotation.Pointcut;
import org.springframework.stereotype.Component;

@Aspect
```

```java
@Component
public class AspectObject {

    @Pointcut(value = "execution(* service.UserinfoService.*(..))")
    public void publicPointcut() {
    }

    @Before(value = "publicPointcut()")
    public void beginRun() {
        System.out.println("准备执行了");
    }

    @After(value = "publicPointcut()")
    public void endRun() {
        System.out.println("结束执行了");
    }

    @AfterReturning(value = "publicPointcut()")
    public void afterReturn() {
        System.out.println("返回了值");
    }

    @AfterThrowing(value = "publicPointcut()")
    public void afterThrowing() {
        System.out.println("出错啦！");
    }

}
```

将全局配置@Pointcut(value = "execution(* service.UserinfoService.*(..))")依托于 public void publicPointcut()方法，引用这个全局切点 pointcut 时，引用方法名称即可。

创建业务类 UserinfoService.java 的代码如下。

```java
package service;

import org.springframework.stereotype.Service;

@Service
public class UserinfoService {
    public void save() {
        System.out.println("userinfoService save()");
    }

    public String getUsername() {
        System.out.println("userinfoService getUsername()");
        return "我是返回值";
    }

    public String hasErrorMethod() {
        System.out.println("hasErrorMethod");
        Integer.parseInt("a");
        return "我是返回值";
    }
}
```

创建配置类 MyContext.java 的代码如下。

```java
package tools;
```

```java
import org.springframework.context.annotation.ComponentScan;
import org.springframework.context.annotation.Configuration;
import org.springframework.context.annotation.EnableAspectJAutoProxy;

@Configuration
@ComponentScan(basePackages = { "myaspect", "service" })
@EnableAspectJAutoProxy
public class MyContext {
}
```

创建运行类 Test1.java 的代码如下。

```java
package test;

import org.springframework.context.annotation.AnnotationConfigApplicationContext;

import service.UserinfoService;
import tools.MyContext;

public class Test1 {
    public static void main(String[] args) {
        AnnotationConfigApplicationContext context = new AnnotationConfigApplicationContext(MyContext.class);
        UserinfoService service = (UserinfoService) context.getBean(UserinfoService.class);
        service.save();
        System.out.println();
        System.out.println(service.getUsername());
        System.out.println();
        service.hasErrorMethod();
    }
}
```

程序运行结果如图 10-55 所示。

```
准备执行了
userinfoService save()
结束执行了
返回了值

准备执行了
userinfoService getUsername()
结束执行了
返回了值
我是返回值

准备执行了
hasErrorMethod
结束执行了
出错啦!
Exception in thread "main" java.lang.NumberFormatException: For input string: "a"
        at java.lang.NumberFormatException.forInputString(NumberFormatException.java:65)
        at java.lang.Integer.parseInt(Integer.java:580)
        at java.lang.Integer.parseInt(Integer.java:615)
```

图 10-55 运行结果

8. 使用 xml<aop:pointcut>定义全局切点

创建测试用的项目 aopTest11。

创建切面类 AspectObject.java 的代码如下。

```java
package myaspect;

public class AspectObject {

    public void beginRun() {
        System.out.println("准备执行了");
    }

    public void endRun() {
        System.out.println("结束执行了");
    }

    public void afterReturn() {
        System.out.println("返回了值");
    }

    public void afterThrowing() {
        System.out.println("出错啦！");
    }
}
```

创建业务类 UserinfoService.java 的代码如下。

```java
package service;

public class UserinfoService {
    public void save() {
        System.out.println("userinfoService save()");
    }

    public String getUsername() {
        System.out.println("userinfoService getUsername()");
        return "我是返回值";
    }

    public String hasErrorMethod() {
        System.out.println("hasErrorMethod");
        Integer.parseInt("a");
        return "我是返回值";
    }
}
```

创建配置文件 applicationContext.xml 的代码如下。

```xml
<?xml version="1.0" encoding="UTF-8"?>
<beans xmlns="http://www.springframework.org/schema/beans"
    xmlns:context="http://www.springframework.org/schema/context"
    xmlns:aop="http://www.springframework.org/schema/aop"
    xmlns:xsi="http://www.w3.org/2001/XMLSchema-instance"
    xsi:schemaLocation="http://www.springframework.org/schema/beans
http://www.springframework.org/schema/beans/spring-beans.xsd
http://www.springframework.org/schema/aop
http://www.springframework.org/schema/aop/spring-aop.xsd
http://www.springframework.org/schema/context
http://www.springframework.org/schema/context/spring-context.xsd
">

    <aop:aspectj-autoproxy></aop:aspectj-autoproxy>
```

```xml
<bean id="userinfoService" class="service.UserinfoService"></bean>

<bean id="myaspect" class="myaspect.AspectObject"></bean>


<aop:config>
    <aop:pointcut id="mypointcut"
        expression="execution(* service.UserinfoService.*(..))" />
    <aop:aspect ref="myaspect">
        <aop:before method="beginRun" pointcut-ref="mypointcut" />
        <aop:after method="endRun" pointcut-ref="mypointcut" />
        <aop:after-returning method="afterReturn"
            pointcut-ref="mypointcut" />
        <aop:after-throwing method="afterThrowing"
            pointcut-ref="mypointcut" />
    </aop:aspect>
</aop:config>

</beans>
```

创建运行类 Test1.java 的代码如下。

```java
package test;

import org.springframework.context.ApplicationContext;
import org.springframework.context.support.ClassPathXmlApplicationContext;

import service.UserinfoService;

public class Test1 {
    public static void main(String[] args) {
        ApplicationContext context = new ClassPathXmlApplicationContext("applicationContext.xml");
        UserinfoService service = (UserinfoService) context.getBean("userinfoService");
        service.save();
        System.out.println();
        System.out.println(service.getUsername());
        System.out.println();
        service.hasErrorMethod();
    }
}
```

程序运行结果如图 10-56 所示。

```
准备执行了
userinfoService save()
结束执行了
返回了值

准备执行了
userinfoService getUsername()
结束执行了
返回了值
我是返回值

准备执行了
hasErrorMethod
结束执行了
出错啦!
Exception in thread "main" java.lang.NumberFormatException: For input string: "a"
    at java.lang.NumberFormatException.forInputString(NumberFormatException.java:65)
    at java.lang.Integer.parseInt(Integer.java:580)
    at java.lang.Integer.parseInt(Integer.java:615)
```

图 10-56 运行结果

9. 使用注解向切面传入参数

如果 Service 业务方法有参数，切面类也能接收这个参数，并可以进行预处理。

创建测试用的项目 aopTest12。

创建切面类 AspectObject.java 的代码如下。

```java
package myaspect;

import java.util.Date;

import org.aspectj.lang.annotation.Aspect;
import org.aspectj.lang.annotation.Before;
import org.aspectj.lang.annotation.Pointcut;
import org.springframework.stereotype.Component;

@Aspect
@Component
public class AspectObject {

    // 只能用&&，如果使用 and 则报错
    @Pointcut(value = "execution(* service.UserinfoService.save1(int)) && args(userId)")
    public void publicPointcut1(int userId) {
    }

    @Before(value = "publicPointcut1(userId)")
    public void beginSave1Before(int userId) {
        System.out.println("在切面中取得的 save1()参数值：userId=" + userId);
    }

    @Pointcut(value = "execution(* service.UserinfoService.save2(int,String,String,int,java.util.Date)) && args(userId,username,password,age,insertdate)")
    public void publicPointcut2(int userId, String username, String password, int age, Date insertdate) {
    }

    @Before(value = "publicPointcut2(userId,username,password,age,insertdate)")
    public void beginSave2Before(int userId, String username, String password, int age, Date insertdate) {
        System.out.println("在切面中取得的 save2()参数值：userId=" + userId + " username=" + username + " password=" + password
                + " age=" + age + " insertdate=" + insertdate.getTime());
    }

}
```

关于@Pointcut 和@Before 配置的解释如下。

```java
@Component
@Aspect
public class AspectObject {

    // 下面@Pointcut 注解的解释：
    // (1)@Pointcut 注解的功能是声明 1 个切点表达式
    // (2)由于切面要取得调用方法时传入的参数，
    //    所以要使用 args 表达式：args(xxxxxx)来进行获取
    // (3)与@Pointcut 关联方法的参数名称必须相同，args()表达式中的 xxxxxx 一样
    @Pointcut(value = "execution(* service.UserinfoService.method1(int)) && args(xxxxxx)")
```

10.9 面向切面编程 AOP 的使用

```java
    public void methodAspect(int xxxxxx) {
    }

    // 下面@Before注解的解释:
    // (1)属性value = "methodAspect(ageabc)"是引用方法:
    //    public void methodAspect(int xxxxxx),
    //    引用时参数名称可以不一样, 一个是xxxxxx, 另1个是ageabc
    // (2)与@Before关联的方法:
    //    public void method1Before(int ageabc)
    //    中的参数名称必须和@Before(value = "methodAspect(ageabc)")
    //    配置中方法的参数名称一样
    // (3)@Pointcut和@Before交接关联点在于方法的名称methodAspect,
    //    不包含参数的命名统一性
    @Before(value = "methodAspect(ageabc)")
    public void method1Before(int ageabc) {
        System.out.println("public void method1Before(int ageabc) ageabc=" + ageabc);
    }
}
```

创建业务类 UserinfoService.java 的代码如下。

```java
package service;

import java.util.Date;

import org.springframework.stereotype.Service;

@Service
public class UserinfoService {

    public String save1(int userId) {
        System.out.println("save1Method userId=" + userId);
        return "返回值A";
    }

    public String save2(int userId, String username, String password, int age, Date insertdate) {
        System.out.println("save2Method userId=" + userId + " username=" + username + " password=" + password + " age="
                + age + " insertdate=" + insertdate.getTime());
        return "返回值B";
    }
}
```

创建配置类 MyContext.java 的代码如下。

```java
package tools;

import org.springframework.context.annotation.ComponentScan;
import org.springframework.context.annotation.Configuration;
import org.springframework.context.annotation.EnableAspectJAutoProxy;

@Configuration
@ComponentScan(basePackages = { "myaspect", "service" })
@EnableAspectJAutoProxy
public class MyContext {
}
```

创建运行类 Test1.java 的代码如下。

```java
package test;

import java.util.Date;

import org.springframework.context.annotation.AnnotationConfigApplicationContext;

import service.UserinfoService;
import tools.MyContext;

public class Test1 {
    public static void main(String[] args) {
        AnnotationConfigApplicationContext context = new AnnotationConfigApplicationContext(MyContext.class);
        UserinfoService service = (UserinfoService) context.getBean(UserinfoService.class);
        System.out.println(service.save1(100));
        System.out.println();
        System.out.println(service.save2(200, "中国", "大中国", 300, new Date()));
    }
}
```

程序运行结果如图 10-57 所示。

```
在切面中取得的save1()参数值：userId=100
save1Method userId=100
返回值A

在切面中取得的save2()参数值：userId=200 username=中国 password=大中国 age=300 insertdate=1480860507278
save2Method userId=200 username=中国 password=大中国 age=300 insertdate=1480860507278
返回值B
```

图 10-57　运行结果

10. 使用 xml 向切面传入参数

创建测试用的项目 aopTest13。

创建切面类 AspectObject.java 的代码如下。

```java
package myaspect;

import java.util.Date;

public class AspectObject {

    public void beginSave1Before(int userId) {
        System.out.println("在切面中取得的 save1()参数值：userId=" + userId);
    }

    public void beginSave2Before(int userId, String username, String password, int age, Date insertdate) {
        System.out.println("在切面中取得的 save2()参数值：userId=" + userId + " username=" + username + " password=" + password
                + " age=" + age + " insertdate=" + insertdate.getTime());
    }

}
```

10.9 面向切面编程 AOP 的使用

创建业务类 UserinfoService.java 的代码如下。

```java
package service;

import java.util.Date;

public class UserinfoService {

    public String save1(int userId) {
        System.out.println("save1Method userId=" + userId);
        return "返回值A";
    }

    public String save2(int userId, String username, String password, int age, Date insertdate) {
        System.out.println("save2Method userId=" + userId + " username=" + username + " password=" + password + " age="
                + age + " insertdate=" + insertdate.getTime());
        return "返回值B";
    }
}
```

创建配置文件 applicationContext.xml 的代码如下。

```xml
<?xml version="1.0" encoding="UTF-8"?>
<beans xmlns="http://www.springframework.org/schema/beans"
    xmlns:context="http://www.springframework.org/schema/context"
    xmlns:aop="http://www.springframework.org/schema/aop"
    xmlns:xsi="http://www.w3.org/2001/XMLSchema-instance"
    xsi:schemaLocation="http://www.springframework.org/schema/beans
    http://www.springframework.org/schema/beans/spring-beans.xsd
    http://www.springframework.org/schema/aop
    http://www.springframework.org/schema/aop/spring-aop.xsd
    http://www.springframework.org/schema/context
    http://www.springframework.org/schema/context/spring-context.xsd
    ">

    <aop:aspectj-autoproxy></aop:aspectj-autoproxy>

    <bean id="userinfoService" class="service.UserinfoService"></bean>

    <bean id="myaspect" class="myaspect.AspectObject"></bean>

    <aop:config>
        <aop:pointcut id="mypointcut1"
            expression="execution(* service.UserinfoService.save1(int)) and args(userId)" />
        <aop:pointcut id="mypointcut2"
            expression="execution(* service.UserinfoService.save2(int,String,String,int,java.util.Date)) and args(userId,username,password,age,insertdate)" />
        <aop:aspect ref="myaspect">
            <aop:before method="beginSave1Before" pointcut-ref="mypointcut1" />
            <aop:before method="beginSave2Before" pointcut-ref="mypointcut2" />
        </aop:aspect>
    </aop:config>

</beans>
```

配置代码：
```
<aop:pointcut id="pointCut1"
    expression="execution (* service..method1(int)) and args(xxxxxx)" />
```

中的 args() 里面的参数名称必须和切面类的通知方法的参数名称要一致，通知方法代码如下。

```java
public void method1Before(int xxxxxx) {
    System.out.println("切面: public void method1Before(int xxxxxx) xxxxxx=" + xxxxxx);
}
```

方法 public void method1Before(int xxxxxx)的参数名称必须是 xxxxxx。

创建运行类 Test1.java 的代码如下。

```java
package test;

import java.util.Date;

import org.springframework.context.ApplicationContext;
import org.springframework.context.support.ClassPathXmlApplicationContext;

import service.UserinfoService;

public class Test1 {
    public static void main(String[] args) {
        ApplicationContext context = new ClassPathXmlApplicationContext("applicationContext.xml");
        UserinfoService service = (UserinfoService) context.getBean(UserinfoService.class);
        System.out.println(service.save1(100));
        System.out.println();
        System.out.println(service.save2(200, "中国", "大中国", 300, new Date()));
    }
}
```

程序运行结果如图 10-58 所示。

```
在切面中取得的save1()参数值: userId=100
save1Method userId=100
返回值A

在切面中取得的save2()参数值: userId=200 username=中国 password=大中国 age=300 insertdate=1480860640790
save2Method userId=200 username=中国 password=大中国 age=300 insertdate=1480860640790
返回值B
```

图 10-58 运行结果

11. 使用注解@AfterReturning 和@AfterThrowing 向切面传入参数

创建测试用的项目 aopTest14。

创建切面类 AspectObject.java 的代码如下。

```java
package myaspect;

import java.util.Date;

import org.aspectj.lang.annotation.AfterReturning;
import org.aspectj.lang.annotation.AfterThrowing;
```

```java
import org.aspectj.lang.annotation.Aspect;
import org.aspectj.lang.annotation.Before;
import org.aspectj.lang.annotation.Pointcut;
import org.springframework.stereotype.Component;

@Aspect
@Component
public class AspectObject {

    // 只能用&&，如果使用 and 则报错
    @Pointcut(value = "execution(* service.UserinfoService.save1(int)) && args(userId)")
    public void publicPointcut1(int userId) {
    }

    @Pointcut(value = "execution(* service.UserinfoService.save2(int,String,String,int,java.util.Date)) && args(userId,username,password,age,insertdate)")
    public void publicPointcut2(int userId, String username, String password, int age, Date insertdate) {
    }

    @Pointcut(value = "execution(* service.UserinfoService.*(..))")
    public void publicPointcut3() {
    }

    @Before(value = "publicPointcut1(userId)")
    public void beginSave1Before(int userId) {
        System.out.println("在切面中取得的 save1()参数值：userId=" + userId);
    }

    @Before(value = "publicPointcut2(userId,username,password,age,insertdate)")
    public void beginSave2Before(int userId, String username, String password, int age, Date insertdate) {
        System.out.println("在切面中取得的 save2()参数值：userId=" + userId + " username=" + username + " password=" + password
                + " age=" + age + " insertdate=" + insertdate.getTime());
    }

    @AfterThrowing(value = "publicPointcut3()", throwing = "t")
    public void aopThrowMethod(Throwable t) {
        System.out.println("aopThrowMethod t=" + t.getMessage());
    }

    @AfterReturning(value = "publicPointcut3()", returning = "result")
    public void aopReturnMethod(String result) {
        System.out.println("aopReturnMethod result=" + result);
    }

}
```

创建业务类 UserinfoService.java 的代码如下。

```java
package service;

import java.util.Date;

import org.springframework.stereotype.Service;

@Service
```

```java
public class UserinfoService {
    public String save1(int userId) {
        System.out.println("save1Method userId=" + userId);
        return "我是返回值A";
    }

    public String save2(int userId, String username, String password, int age, Date insertdate) {
        System.out.println("save2Method userId=" + userId + " username=" + username + " password=" + password + " age="
                + age + " insertdate=" + insertdate.getTime());
        Integer.parseInt("a");
        return "我是返回值B";
    }
}
```

创建配置类 MyContext.java 的代码如下。

```java
package tools;

import org.springframework.context.annotation.ComponentScan;
import org.springframework.context.annotation.Configuration;
import org.springframework.context.annotation.EnableAspectJAutoProxy;

@Configuration
@ComponentScan(basePackages = { "myaspect", "service" })
@EnableAspectJAutoProxy
public class MyContext {
}
```

创建运行类 Test1.java 的代码如下。

```java
package test;

import java.util.Date;

import org.springframework.context.annotation.AnnotationConfigApplicationContext;

import service.UserinfoService;
import tools.MyContext;

public class Test1 {
    public static void main(String[] args) {
        AnnotationConfigApplicationContext context = new AnnotationConfigApplicationContext(MyContext.class);
        UserinfoService service = (UserinfoService) context.getBean(UserinfoService.class);
        service.save1(100);
        System.out.println();
        service.save2(200, "中国", "大中国", 300, new Date());
    }
}
```

程序运行结果如图 10-59 所示。

```
在切面中取得的save1()参数值: userId=100
save1Method userId=100
aopReturnMethod result=我是返回值A

在切面中取得的save2()参数值: userId=200 username=中国 password=大中国 age=300 insertdate=1480860930932
save2Method userId=200 username=中国 password=大中国 age=300 insertdate=1480860930932
aopThrowMethod t=For input string: "a"
Exception in thread "main" java.lang.NumberFormatException: For input string: "a"
        at java.lang.NumberFormatException.forInputString(NumberFormatException.java:65)
        at java.lang.Integer.parseInt(Integer.java:580)
        at java.lang.Integer.parseInt(Integer.java:615)
```

图 10-59 运行结果

12. 使用 xml<aop:after-returning>和<aop:after-throwing>向切面传入参数

创建测试用的项目 aopTest15。

创建切面类 AspectObject.java 的代码如下。

```java
package myaspect;

import java.util.Date;

import org.aspectj.lang.annotation.AfterReturning;

public class AspectObject {

    public void beginSave1Before(int userId) {
        System.out.println("在切面中取得的 save1()参数值: userId=" + userId);
    }

    public void beginSave2Before(int userId, String username, String password, int age,
Date insertdate) {
        System.out.println("在切面中取得的 save2()参数值: userId=" + userId + " username=" +
username + " password=" + password
                + " age=" + age + " insertdate=" + insertdate.getTime());
    }

    public void aopReturnMethod(String result) {
        System.out.println("aopReturnMethod result=" + result);
    }

    public void aopThrowMethod(Throwable t) {
        System.out.println("aopThrowMethod t=" + t.getMessage());
    }

}
```

创建业务类 UserinfoService.java 的代码如下。

```java
package service;

import java.util.Date;

public class UserinfoService {

    public String save1(int userId) {
        System.out.println("save1Method userId=" + userId);
        return "我是返回值 A";
```

```java
    }

    public String save2(int userId, String username, String password, int age, Date insertdate) {
        System.out.println("save2Method userId=" + userId + " username=" + username + " password=" + password + " age="
                + age + " insertdate=" + insertdate.getTime());
        Integer.parseInt("a");
        return "我是返回值B";
    }
}
```

创建配置文件 applicationContext.xml 的代码如下。

```xml
<?xml version="1.0" encoding="UTF-8"?>
<beans xmlns="http://www.springframework.org/schema/beans"
    xmlns:context="http://www.springframework.org/schema/context"
    xmlns:aop="http://www.springframework.org/schema/aop"
    xmlns:xsi="http://www.w3.org/2001/XMLSchema-instance"
        xsi:schemaLocation="http://www.springframework.org/schema/beans
    http://www.springframework.org/schema/beans/spring-beans.xsd
    http://www.springframework.org/schema/aop
    http://www.springframework.org/schema/aop/spring-aop.xsd
    http://www.springframework.org/schema/context
    http://www.springframework.org/schema/context/spring-context.xsd
    ">

    <aop:aspectj-autoproxy></aop:aspectj-autoproxy>

    <bean id="userinfoService" class="service.UserinfoService"></bean>

    <bean id="myaspect" class="myaspect.AspectObject"></bean>

    <aop:config>
        <aop:pointcut id="mypointcut1"
            expression="execution(* service.UserinfoService.save1(int)) and args(userId)" />
        <aop:pointcut id="mypointcut2"
            expression="execution(* service.UserinfoService.save2(int,String,String,int,java.util.Date)) and args(userId,username,password,age,insertdate)" />
        <aop:pointcut id="mypointcut3"
            expression="execution(* service.UserinfoService.*(..)) " />
        <aop:aspect ref="myaspect">
            <aop:before method="beginSave1Before" pointcut-ref="mypointcut1" />
            <aop:before method="beginSave2Before" pointcut-ref="mypointcut2" />
            <aop:after-returning method="aopReturnMethod"
                pointcut-ref="mypointcut3" returning="result" />
            <aop:after-throwing method="aopThrowMethod"
                pointcut-ref="mypointcut3" throwing="t" />
        </aop:aspect>
    </aop:config>

</beans>
```

创建运行类 Test1.java 的代码如下。

```java
package test;

import java.util.Date;
```

```
import org.springframework.context.ApplicationContext;
import org.springframework.context.support.ClassPathXmlApplicationContext;

import service.UserinfoService;

public class Test1 {
    public static void main(String[] args) {
        ApplicationContext context = new ClassPathXmlApplicationContext("applicationContext.xml");
        UserinfoService service = (UserinfoService) context.getBean(UserinfoService.class);
        service.save1(100);
        System.out.println();
        service.save2(200, "中国", "大中国", 300, new Date());
    }
}
```

程序运行结果如图 10-60 所示。

```
在切面中取得的save1()参数值：userId=100
save1Method userId=100
aopReturnMethod result=我是返回值A

在切面中取得的save2()参数值：userId=200 username=中国 password=大中国 age=300 insertdate=1480861032916
save2Method userId=200 username=中国 password=大中国 age=300 insertdate=1480861032916
aopThrowMethod t=For input string: "a"
Exception in thread "main" java.lang.NumberFormatException: For input string: "a"
    at java.lang.NumberFormatException.forInputString(NumberFormatException.java:65)
    at java.lang.Integer.parseInt(Integer.java:580)
    at java.lang.Integer.parseInt(Integer.java:615)
```

图 10-60　运行结果

13．使用注解向环绕通知传入参数

本实验将实现向环绕通知传入参数。

创建测试用的项目 aopTest16。

创建切面类 AspectObject.java 的代码如下。

```
package aspect;

import java.util.Date;

import org.aspectj.lang.ProceedingJoinPoint;
import org.aspectj.lang.annotation.AfterReturning;
import org.aspectj.lang.annotation.AfterThrowing;
import org.aspectj.lang.annotation.Around;
import org.aspectj.lang.annotation.Aspect;
import org.aspectj.lang.annotation.Pointcut;
import org.springframework.stereotype.Component;

@Component
@Aspect
public class AspectObject {

    @Pointcut(value = "execution(* service.UserinfoService.method1(int)) && args(xxxxxx)")
    public void methodAspect1(int xxxxxx) {
    }

    @Pointcut(value = "execution(*
```

```java
            service.UserinfoService.method2(String,String,int,java.util.Date)) && args(u,p,a,i)")
    public void methodAspect2(String u, String p, int a, Date i) {
    }

    @Pointcut(value = "execution(* service.UserinfoService.*(..))")
    public void methodAspect3() {
    }

    @Around(value = "methodAspect1(xxxxxx)")
    public void method1Around(ProceedingJoinPoint point, int xxxxxx) throws Throwable {
        System.out.println("切面开始: public void method1Before(ProceedingJoinPoint point, int xxxxxx) xxxxxx=" + xxxxxx);
        point.proceed();
        System.out.println("切面结束: public void method1Before(ProceedingJoinPoint point, int xxxxxx) xxxxxx=" + xxxxxx);
    }

    @Around(value = "methodAspect2(u, p, a, i)")
    public Object method2Around(ProceedingJoinPoint point, String u, String p, int a, Date i) throws Throwable {
        Object returnValue = null;
        System.out.println(
                "切面开始: public void method2Before(ProceedingJoinPoint point, String u, String p, int a, Date i) u=" + u
                        + " p=" + p + " a=" + a + " i=" + i);
        returnValue = point.proceed();
        System.out.println(
                "切面开始: public void method2Before(ProceedingJoinPoint point, String u, String p, int a, Date i) u=" + u
                        + " p=" + p + " a=" + a + " i=" + i);
        return returnValue;
    }

    @AfterReturning(value = "methodAspect3()", returning = "returnParam")
    public void method3AfterReturning(Object returnParam) {
        System.out.println("public void method3AfterReturning(Object returnParam) returnParam=" + returnParam);
    }

    @AfterThrowing(value = "methodAspect3()", throwing = "t")
    public void method4AfterThrowing(Throwable t) {
        System.out.println("public void method4AfterThrowing(Throwable t) t=" + t);
    }

}
```

注意：在切面中不要对异常进行捕获，要将异常抛出给 Spring 框架进行后续处理，这样会使 @AfterThrowing 通知得到执行。

创建业务类 UserinfoService.java 的代码如下。

```java
package service;

import java.util.Date;
import org.springframework.stereotype.Service;

@Service
public class UserinfoService {
```

```java
    public void method1(int ageage) {
        System.out.println("method1 age=" + ageage);
    }

    public String method2(String username, String password, int age, Date insertdate) {
        System.out.println(
                "method2 username=" + username + " password=" + password + " age=" + age
                        + " insertdate=" + insertdate);
        Integer.parseInt("a");
        return "我是返回值method2";
    }
}
```

创建配置类 MyContext.java 的代码如下。

```java
package javaconfig;

import org.springframework.context.annotation.ComponentScan;
import org.springframework.context.annotation.Configuration;
import org.springframework.context.annotation.EnableAspectJAutoProxy;

@Configuration
@EnableAspectJAutoProxy
@ComponentScan(basePackages = { "aspect", "service" })
public class MyContext {
}
```

创建运行类 Test.java 的代码如下。

```java
package test;

import java.util.Date;

import org.springframework.context.ApplicationContext;
import org.springframework.context.annotation.AnnotationConfigApplicationContext;

import javaconfig.MyContext;
import service.UserinfoService;

public class Test {
    public static void main(String[] args) {
        ApplicationContext context = new AnnotationConfigApplicationContext(MyContext.class);
        UserinfoService service = (UserinfoService) context.getBean(UserinfoService.class);
        service.method1(100);
        System.out.println();
        System.out.println();
        System.out.println("main get method2 returnValue=" + service.method2("中国", "大中国", 123, new Date()));
    }
}
```

程序运行结果如下。

```
切面开始: public void method1Before(ProceedingJoinPoint point, int xxxxxx) xxxxxx=100
method1 age=100
切面结束: public void method1Before(ProceedingJoinPoint point, int xxxxxx) xxxxxx=100
public void method3AfterReturning(Object returnParam) returnParam=null
```

```
切面开始：public void method2Before(ProceedingJoinPoint point, String u, String p, int a,
Date i) u=中国 p=大中国 a=123 i=Tue Jan 10 15:34:16 CST 2017
    method2 username=中国 password=大中国 age=123 insertdate=Tue Jan 10 15:34:16 CST 2017
    public void method4AfterThrowing(Throwable t) t=java.lang.NumberFormatException: For
input string: "a"
    Exception in thread "main" java.lang.NumberFormatException: For input string: "a"
        at java.lang.NumberFormatException.forInputString(NumberFormatException.java:65)
        at java.lang.Integer.parseInt(Integer.java:580)
        at java.lang.Integer.parseInt(Integer.java:615)
        at service.UserinfoService.method2(UserinfoService.java:16)
        at service.UserinfoService$$FastClassBySpringCGLIB$$564fa423.invoke(<generated>)
        at org.springframework.cglib.proxy.MethodProxy.invoke(MethodProxy.java:204)
        at org.springframework.aop.framework.CglibAopProxy$CglibMethodInvocation.invokeJoinpoint(CglibAopProxy.java:720)
        at org.springframework.aop.framework.ReflectiveMethodInvocation.proceed(ReflectiveMethodInvocation.java:157)
        at org.springframework.aop.aspectj.MethodInvocationProceedingJoinPoint.proceed(MethodInvocationProceedingJoinPoint.java:85)
        at aspect.AspectObject.method2Around(AspectObject.java:42)
        at sun.reflect.NativeMethodAccessorImpl.invoke0(Native Method)
        at sun.reflect.NativeMethodAccessorImpl.invoke(NativeMethodAccessorImpl.java:62)
        at sun.reflect.DelegatingMethodAccessorImpl.invoke(DelegatingMethodAccessorImpl.java:43)
        at java.lang.reflect.Method.invoke(Method.java:497)
        at org.springframework.aop.aspectj.AbstractAspectJAdvice.invokeAdviceMethodWithGivenArgs(AbstractAspectJAdvice.java:629)
        at org.springframework.aop.aspectj.AbstractAspectJAdvice.invokeAdviceMethod(AbstractAspectJAdvice.java:618)
        at org.springframework.aop.aspectj.AspectJAroundAdvice.invoke(AspectJAroundAdvice.java:70)
        at org.springframework.aop.framework.ReflectiveMethodInvocation.proceed(ReflectiveMethodInvocation.java:168)
        at org.springframework.aop.framework.adapter.AfterReturningAdviceInterceptor.invoke(AfterReturningAdviceInterceptor.java:52)
        at org.springframework.aop.framework.ReflectiveMethodInvocation.proceed(ReflectiveMethodInvocation.java:179)
        at org.springframework.aop.aspectj.AspectJAfterThrowingAdvice.invoke(AspectJAfterThrowingAdvice.java:62)
        at org.springframework.aop.framework.ReflectiveMethodInvocation.proceed(ReflectiveMethodInvocation.java:179)
        at org.springframework.aop.interceptor.ExposeInvocationInterceptor.invoke(ExposeInvocationInterceptor.java:92)
        at org.springframework.aop.framework.ReflectiveMethodInvocation.proceed(ReflectiveMethodInvocation.java:179)
        at
```

```
org.springframework.aop.framework.CglibAopProxy$DynamicAdvisedInterceptor.intercept(CglibA
opProxy.java:655)
    at service.UserinfoService$$EnhancerBySpringCGLIB$$2bbbce8d.method2(<generated>)
    at test.Test.main(Test.java:18)
```

14. 使用 xml 向环绕通知传入参数

创建测试用的项目 aopTest17。

创建切面类 AspectObject.java 的代码如下。

```java
package aspect;

import java.util.Date;
import org.aspectj.lang.ProceedingJoinPoint;

public class AspectObject {

    public void method1Before(ProceedingJoinPoint point, int xxxxxx) throws Throwable {
        System.out.println("切面开始: public void method1Before(ProceedingJoinPoint point, int xxxxxx) xxxxxx=" + xxxxxx);
        point.proceed();
        System.out.println("切面结束: public void method1Before(ProceedingJoinPoint point, int xxxxxx) xxxxxx=" + xxxxxx);
    }

    public Object method2Before(ProceedingJoinPoint point, String u, String p, int a, Date i) throws Throwable {
        Object returnValue = null;
        System.out.println(
                "切面开始: public void method2Before(ProceedingJoinPoint point, String u, String p, int a, Date i) u=" + u
                        + " p=" + p + " a=" + a + " i=" + i);
        returnValue = point.proceed();
        System.out.println(
                "切面开始: public void method2Before(ProceedingJoinPoint point, String u, String p, int a, Date i) u=" + u
                        + " p=" + p + " a=" + a + " i=" + i);
        return returnValue;
    }

    public void method3AfterReturning(Object returnParam) {
        System.out.println("public void method3AfterReturning(Object returnParam) returnParam=" + returnParam);
    }

    public void method4AfterThrowing(Throwable t) {
        System.out.println("public void method4AfterThrowing(Throwable t) t=" + t);
    }

}
```

创建业务类 UserinfoService.java 的代码如下。

```java
package service;

import java.util.Date;
```

```java
public class UserinfoService {
    public void method1(int ageage) {
        System.out.println("method1 age=" + ageage);
    }

    public String method2(String username, String password, int age, Date insertdate) {
        System.out.println(
                "method2 username=" + username + " password=" + password + " age=" + age
                        + " insertdate=" + insertdate);
        Integer.parseInt("a");
        return "我是返回值method2";
    }
}
```

创建配置文件 applicationContext.xml 的代码如下。

```xml
<?xml version="1.0" encoding="UTF-8"?>
<beans xmlns="http://www.springframework.org/schema/beans"
    xmlns:xsi="http://www.w3.org/2001/XMLSchema-instance"
    xmlns:aop="http://www.springframework.org/schema/aop"
    xsi:schemaLocation="http://www.springframework.org/schema/beans
http://www.springframework.org/schema/beans/spring-beans.xsd
       http://www.springframework.org/schema/aop
http://www.springframework.org/schema/aop/spring-aop-4.3.xsd">
    <bean id="userinfoService" class="service.UserinfoService"></bean>
    <bean id="myAspect" class="aspect.AspectObject"></bean>

    <aop:aspectj-autoproxy></aop:aspectj-autoproxy>

    <aop:config>
        <aop:pointcut id="pointCut1"
            expression="execution(* service.UserinfoService.method1(int)) and args(xxxxxx)" />
        <aop:pointcut id="pointCut2"
            expression="execution(* service.UserinfoService.method2(String,String,int,java.util.Date)) and args(u,p,a,i)" />
        <aop:pointcut id="pointCut3"
            expression="execution(* service.UserinfoService.*(..))" />

        <aop:aspect ref="myAspect">
            <aop:around method="method1Before" pointcut-ref="pointCut1" />
            <aop:around method="method2Before" pointcut-ref="pointCut2" />
            <aop:after-returning method="method3AfterReturning"
                pointcut-ref="pointCut3" returning="returnParam" />
            <aop:after-throwing method="method4AfterThrowing"
                pointcut-ref="pointCut3" throwing="t" />
        </aop:aspect>
    </aop:config>

</beans>
```

创建运行类 Test1.java 的代码如下。

```java
package test;

import java.util.Date;
```

```java
import org.springframework.context.ApplicationContext;
import org.springframework.context.support.ClassPathXmlApplicationContext;

import service.UserinfoService;

public class Test {
    public static void main(String[] args) {
        ApplicationContext context = new ClassPathXmlApplicationContext("ac1.xml");
        UserinfoService service = (UserinfoService) context.getBean(UserinfoService.class);
        service.method1(100);
        System.out.println();
        System.out.println();
        System.out.println("main get method2 returnValue=" + service.method2("中国", "大中国", 123, new Date()));
    }
}
```

程序运行结果如下。

```
切面开始: public void method1Before(ProceedingJoinPoint point, int xxxxxx) xxxxxx=100
method1 age=100
切面结束: public void method1Before(ProceedingJoinPoint point, int xxxxxx) xxxxxx=100
public void method3AfterReturning(Object returnParam) returnParam=null


切面开始: public void method2Before(ProceedingJoinPoint point, String u, String p, int a, Date i) u=中国 p=大中国 a=123 i=Tue Jan 10 15:36:58 CST 2017
method2 username=中国 password=大中国 age=123 insertdate=Tue Jan 10 15:36:58 CST 2017
public void method4AfterThrowing(Throwable t) t=java.lang.NumberFormatException: For input string: "a"
Exception in thread "main" java.lang.NumberFormatException: For input string: "a"
    at java.lang.NumberFormatException.forInputString(NumberFormatException.java:65)
    at java.lang.Integer.parseInt(Integer.java:580)
    at java.lang.Integer.parseInt(Integer.java:615)
    at service.UserinfoService.method2(UserinfoService.java:14)
    at service.UserinfoService$$FastClassBySpringCGLIB$$564fa423.invoke(<generated>)
    at org.springframework.cglib.proxy.MethodProxy.invoke(MethodProxy.java:204)
    at org.springframework.aop.framework.CglibAopProxy$CglibMethodInvocation.invokeJoinpoint(CglibAopProxy.java:720)
    at org.springframework.aop.framework.ReflectiveMethodInvocation.proceed(ReflectiveMethodInvocation.java:157)
    at org.springframework.aop.aspectj.MethodInvocationProceedingJoinPoint.proceed(MethodInvocationProceedingJoinPoint.java:85)
    at aspect.AspectObject.method2Before(AspectObject.java:19)
    at sun.reflect.NativeMethodAccessorImpl.invoke0(Native Method)
    at sun.reflect.NativeMethodAccessorImpl.invoke(NativeMethodAccessorImpl.java:62)
    at sun.reflect.DelegatingMethodAccessorImpl.invoke(DelegatingMethodAccessorImpl.java:43)
    at java.lang.reflect.Method.invoke(Method.java:497)
    at org.springframework.aop.aspectj.AbstractAspectJAdvice.invokeAdviceMethodWithGivenArgs(AbstractAspectJAdvice.java:629)
    at
```

```
org.springframework.aop.aspectj.AbstractAspectJAdvice.invokeAdviceMethod(AbstractAspectJAdvice.java:618)
    at
org.springframework.aop.aspectj.AspectJAroundAdvice.invoke(AspectJAroundAdvice.java:70)
    at
org.springframework.aop.framework.ReflectiveMethodInvocation.proceed(ReflectiveMethodInvocation.java:168)
    at
org.springframework.aop.framework.adapter.AfterReturningAdviceInterceptor.invoke(AfterReturningAdviceInterceptor.java:52)
    at
org.springframework.aop.framework.ReflectiveMethodInvocation.proceed(ReflectiveMethodInvocation.java:179)
    at
org.springframework.aop.aspectj.AspectJAfterThrowingAdvice.invoke(AspectJAfterThrowingAdvice.java:62)
    at
org.springframework.aop.framework.ReflectiveMethodInvocation.proceed(ReflectiveMethodInvocation.java:179)
    at
org.springframework.aop.interceptor.ExposeInvocationInterceptor.invoke(ExposeInvocationInterceptor.java:92)
    at
org.springframework.aop.framework.ReflectiveMethodInvocation.proceed(ReflectiveMethodInvocation.java:179)
    at
org.springframework.aop.framework.CglibAopProxy$DynamicAdvisedInterceptor.intercept(CglibAopProxy.java:655)
    at service.UserinfoService$$EnhancerBySpringCGLIB$$55192420.method2(<generated>)
    at test.Test.main(Test.java:17)
```

15. 使用 xml<aop:aspectj-autoproxy></aop:aspectj-autoproxy>应用 AOP 切面

创建测试用的项目 aopTest18。

创建切面类 AspectObject.java 的代码如下。

```java
package myaspect;

import org.aspectj.lang.annotation.After;
import org.aspectj.lang.annotation.AfterReturning;
import org.aspectj.lang.annotation.AfterThrowing;
import org.aspectj.lang.annotation.Aspect;
import org.aspectj.lang.annotation.Before;
import org.aspectj.lang.annotation.Pointcut;
import org.springframework.stereotype.Component;

@Aspect
@Component
public class AspectObject {

    @Pointcut(value = "execution(* service.UserinfoService.*(..))")
    public void publicPointcut() {
    }

    @Before(value = "publicPointcut()")
    public void beginRun() {
        System.out.println("准备执行了");
```

```java
    }

    @After(value = "publicPointcut()")
    public void endRun() {
        System.out.println("结束执行了");
    }

    @AfterReturning(value = "publicPointcut()")
    public void afterReturn() {
        System.out.println("返回了值");
    }

    @AfterThrowing(value = "publicPointcut()")
    public void afterThrowing() {
        System.out.println("出错啦！");
    }
}
```

创建业务类 UserinfoService.java 的代码如下。

```java
package service;

import org.springframework.stereotype.Service;

@Service
public class UserinfoService {
    public void save() {
        System.out.println("userinfoService save()");
    }

    public String getUsername() {
        System.out.println("userinfoService getUsername()");
        return "我是返回值";
    }

    public String hasErrorMethod() {
        System.out.println("hasErrorMethod");
        Integer.parseInt("a");
        return "我是返回值";
    }
}
```

创建配置文件 applicationContext.xml 的代码如下。

```xml
<?xml version="1.0" encoding="UTF-8"?>
<beans xmlns="http://www.springframework.org/schema/beans"
    xmlns:context="http://www.springframework.org/schema/context"
    xmlns:aop="http://www.springframework.org/schema/aop"
    xmlns:xsi="http://www.w3.org/2001/XMLSchema-instance"
    xsi:schemaLocation="http://www.springframework.org/schema/beans
    http://www.springframework.org/schema/beans/spring-beans.xsd
    http://www.springframework.org/schema/context
    http://www.springframework.org/schema/context/spring-context.xsd
    http://www.springframework.org/schema/aop
    http://www.springframework.org/schema/aop/spring-aop.xsd
    ">
    <context:component-scan base-package="service"></context:component-scan>
    <context:component-scan base-package="myaspect"></context:component-scan>
```

```xml
<!-- 不加 aop:aspectj-autoproxy 则 AOP 切面不能进行参与执行的流程 -->
<aop:aspectj-autoproxy></aop:aspectj-autoproxy>
</beans>
```

创建运行类 Test1.java 的代码如下。

```java
package test;

import org.springframework.context.ApplicationContext;
import org.springframework.context.support.ClassPathXmlApplicationContext;

import service.UserinfoService;

public class Test1 {
    public static void main(String[] args) {
        ApplicationContext context = new ClassPathXmlApplicationContext("applicationContext.xml");
        UserinfoService service = (UserinfoService) context.getBean(UserinfoService.class);
        service.save();
        System.out.println();
        System.out.println(service.getUsername());
        System.out.println();
        service.hasErrorMethod();
    }
}
```

程序运行结果如图 10-61 所示。

```
准备执行了
userinfoService save()
结束执行了
返回了值

准备执行了
userinfoService getUsername()
结束执行了
返回了值
我是返回值

准备执行了
hasErrorMethod
结束执行了
出错啦!
Exception in thread "main" java.lang.NumberFormatException: For input string: "a"
    at java.lang.NumberFormatException.forInputString(NumberFormatException.java:65)
    at java.lang.Integer.parseInt(Integer.java:580)
    at java.lang.Integer.parseInt(Integer.java:615)
```

图 10-61　运行结果

10.9.4　Strust 2、Spring 4 整合及应用 AOP 切面

学习完 Spring 框架的核心原理 DI 与 AOP 之后，要将 Struts 2 与 Spring 进行整合，为后面章节的 SSH 整合做好铺垫。

将 Struts 2 与 Spring 整合的目的是将 Struts 2 的控制层交由 Spring 框架进行处理，以对控

10.9 面向切面编程 AOP 的使用

制层进行有效地管理，方便对控制层应用切面。

创建 Web 项目 Struts2_Spring4。

1. 添加 Struts 2 和 Spring 4 的基础 JAR 文件。
2. 添加 Struts 2 与 Spring 整合的插件 jar 包：struts2-spring-plugin-2.5.5.jar。
3. 在 web.xml 文件中添加如下代码。

```xml
<context-param>
    <param-name>contextConfigLocation</param-name>
    <param-value>\WEB-INF\classes\applicationContext.xml</param-value>
</context-param>

<!-- ContextLoaderListener 监听类的主要作用就是在 tomcat 启动后 -->
<!-- 读取指定的参数 contextConfigLocation 对应的 ac.xml 中的配置 -->
<listener>
    <listener-class>org.springframework.web.context.ContextLoaderListener</listener-class>
</listener>
```

4. 更改 applicationContext.xml 配置文件的代码如下。

```xml
<?xml version="1.0" encoding="UTF-8"?>
<beans xmlns="http://www.springframework.org/schema/beans"
    xmlns:xsi="http://www.w3.org/2001/XMLSchema-instance"
    xmlns:aop="http://www.springframework.org/schema/aop"
    xsi:schemaLocation="http://www.springframework.org/schema/beans
http://www.springframework.org/schema/beans/spring-beans.xsd
        http://www.springframework.org/schema/aop
http://www.springframework.org/schema/aop/spring-aop-4.3.xsd">
    <bean id="test" class="controller.Test" scope="prototype"></bean>
</beans>
```

对 Struts 2 的控制层声明成多例模式。

5. 继续更改 web.xml 代码，添加 Struts 2 的过滤器配置，代码如下。

```xml
<filter>
    <filter-name>struts2</filter-name>
    <filter-class>org.apache.struts2.dispatcher.filter.StrutsPrepareAndExecuteFilter</filter-class>
</filter>

<filter-mapping>
    <filter-name>struts2</filter-name>
    <url-pattern>*.action</url-pattern>
</filter-mapping>
```

6. 设计 Struts 2 的控制层代码如下。

```java
package controller;

public class Test {
    public String execute() {
        System.out.println("进入了Struts2的控制层代码, code:" + this.hashCode());
        return null;
    }
}
```

7. 下面开始设计最为关键的文件 struts.xml 设计，代码如下。

```xml
<?xml version="1.0" encoding="UTF-8" ?>
```

```xml
<!DOCTYPE struts PUBLIC "-//Apache Software Foundation//DTD Struts Configuration 2.1//EN"
"http://struts.apache.org/dtds/struts-2.1.dtd">
<struts>
    <package name="struts_spring" extends="struts-default">
        <action name="test" class="springTestBean">
        </action>
    </package>
</struts>
```

与以往的 struts.xml 配置文件不同，在这里需要将 class 的属性值与 applicationContext.xml 配置文件中的 id 值设置为一样。这样做的目的是当前台请求进入 Struts 2 框架后，Struts 2 框架自己不创建控制层类的对象，而是直接到 Spring 的 DI 容器中取得对应的控制层对象，然后再执行内部的 public String execute()方法。

8. 部署项目执行程序，在控制层输出多个控制层对象的信息，结果如下。

```
进入了Struts2的控制层代码，code:1438338512
进入了Struts2的控制层代码，code:244301570
进入了Struts2的控制层代码，code:1982221729
进入了Struts2的控制层代码，code:781861225
进入了Struts2的控制层代码，code:1805175518
进入了Struts2的控制层代码，code:1579095894
进入了Struts2的控制层代码，code:651478447
```

9. 为了证明 Struts 2 的控制层是由 Spring 容器提供的，将 applicationContext.xml 文件代码更改如下。

```xml
<bean id="springTestBean" class="controller.Test"></bean>
```

改成单例模式，让 Struts 2 的控制层由多例变成单例。重启 Tomcat 再执行控制层，控制台输出信息如下。

```
进入了Struts2的控制层代码，code:1094714342
进入了Struts2的控制层代码，code:1094714342
进入了Struts2的控制层代码，code:1094714342
进入了Struts2的控制层代码，code:1094714342
进入了Struts2的控制层代码，code:1094714342
进入了Struts2的控制层代码，code:1094714342
进入了Struts2的控制层代码，code:1094714342
```

Strust 2 的控制层的确变成了单例，说明控制层的对象的确是从 Spring 的 DI 容器中进行创建与取得的。

10. 由于 Struts 2 的控制层具有实例变量,多个线程在访问单例对象的同一个实例变量时会出现"非线程安全"问题，所以还得把 applicationContext.xml 文件更成多例模式，代码更改如下。

```xml
<bean id="springTestBean" class="controller.Test" scope="prototype"></bean>
```

11. 重启 Tomcat 再次执行控制层，多例的效果出现了，不再出现"非线程安全"问题。
12. 至此 Struts 2 和 Spring 4 框架整合完毕。
13. 还可以对 Strust 2 控制层中的 execute()方法应用 AOP 切面。

创建 AspectObject.java 切面类的代码如下。

```java
package myaspect;
```

```java
import org.aspectj.lang.annotation.After;
import org.aspectj.lang.annotation.Aspect;
import org.aspectj.lang.annotation.Before;
import org.aspectj.lang.annotation.Pointcut;
import org.springframework.context.annotation.EnableAspectJAutoProxy;
import org.springframework.stereotype.Component;

@Component
@Aspect
@EnableAspectJAutoProxy
public class AspectObject {

    public AspectObject() {
        System.out.println("public AspectObject()----------------");
    }

    @Pointcut(value = "execution(* controller..*.execute(..))")
    public void struts2ExecuteMethodPointCut() {
    }

    @Before(value = "struts2ExecuteMethodPointCut()")
    public void beforeAspectMethod() {
        System.out.println("开始日志，开始时间：" + System.currentTimeMillis());
    }

    @After(value = "struts2ExecuteMethodPointCut()")
    public void afterAspectMethod() {
        System.out.println("结束日志，结束时间：" + System.currentTimeMillis());
    }

}
```

14. 在 applicationContext.xml 文件中添加扫描的配置代码。

```xml
<context:component-scan base-package="myaspect"></context:component-scan>
```

15. 重启 Tomcat 再运行控制层，输出的结果说明切面成功应用到 execute()方法上。

开始日志，开始时间：1484036253347
进入了 Struts2 的控制层代码，code:121011952
结束日志，结束时间：1484036253362
开始日志，开始时间：1484036393978
进入了 Struts2 的控制层代码，code:1253893552
结束日志，结束时间：1484036393978
开始日志，开始时间：1484036396353
进入了 Struts2 的控制层代码，code:499230446
结束日志，结束时间：1484036396353

第 11 章　Struts 2+Hibernate 5+Spring 4 整合

本章将对 Struts 2+Hibernate 5+Spring 4 进行整合,整合的目的是有效处理 Struts 2 的 Action、Hibernate 5 中的 DAO 及自定义的 Service 业务层 JavaBean 的管理,以及对事务进行自动化处理与控制,增强程序员设计代码的规范度。

11.1　目的

在 Struts 2+Hibernate 5+Spring 4 整合的架构中,Spring 充当了一个 JavaBean 容器的作用。Spring 使用 DI 和 AOP 技术接管了 Hibernate 5 的 DAO 和事务以及 Struts 2 的 Action 对象的创建,还有 Service 业务层的创建与管理,从而能充分地管理事务和代理请求。经过 DI 容器的处理后,针对面向接口的编程使软件项目的分层更明确。

MVC 分别使用
(1) M 层：Spring 4
(2) V 层：JSP
(3) C 层：Struts 2
(4) 持久层：Hibernate 5

Spring 4 对 Hibernate 5 的 DAO 进行了非常好的封装,使程序开发者完全不用理会事务,只需要一心一意的开发业务,在后面的介绍中会一一讲解 Spring 整合方面的知识点。

本章从零起步,一起实现一个 Struts 2+Hibernate 5+Spring 4 整合的项目框架。

11.2　创建数据库环境

在实现 SSH 框架整合之前先要创建实验用的数据库环境,这个过程可以使用前面章节介绍过的 Oracle 客户端工具 Toad 来创建数据表,在此不再重复。

11.2.1　新建数据表 userinfo

在 toad 工具中创建数据表,名称为 userinfo,如图 11-1 所示。

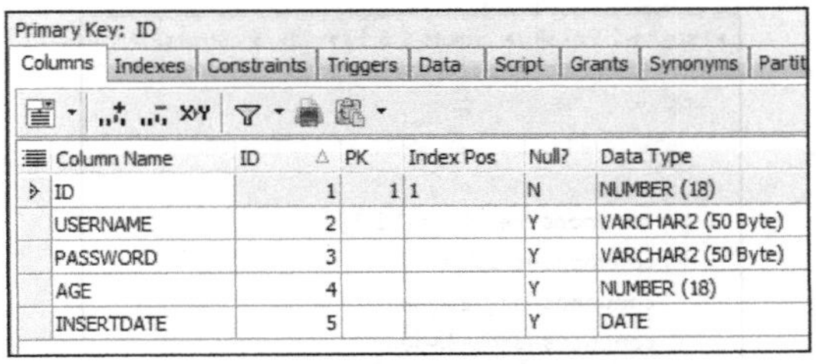

图 11-1　创建数据表 userinfo 结构

11.2.2　创建序列对象

在数据库中还要创建序列对象 idauto，如图 11-2 所示。

图 11-2　创建序列

11.3　新建整合用的 Web 项目

新建 Web 项目 struts2New_Hibernate5New_Spring4New。

在此 Web 项目中存放 Struts 2、Hibernate 5 及 Spring 4 的 Jar 包和配置文件，在 Web 项目中将这 3 个框架进行整合。

Web 项目使用的 Java EE 版本是 5，结果如图 11-3 所示。

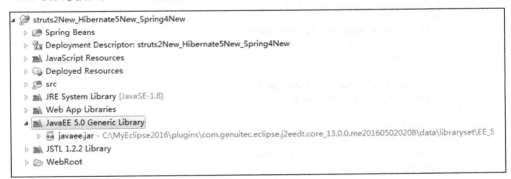

图 11-3　创建 Web 项目

11.4　添加 Struts 2 框架支持

在 Web 项目中添加 Struts 2 框架的主要目的就是使 Web 项目能接收以*.action 为后缀的请求，再把请求交由 Spring 的 DI 容器来进行实例化控制层 Action，继而通过执行 Action 内部的 execute()方法来达到调用业务层的目的。

11.4.1　添加 Struts 2 框架

添加 Struts 2 框架的支持环境，并且把这些 jar 文件复制到项目中的 lib 目录下，如图 11-4 所示。

图 11-4 Struts2 依赖 jar 包

最为关键的文件就是 struts2-spring-plugin-2.5.5.jar，它是 Struts 2 和 Spring 4 框架整合的插件文件。

11.4.2 在 web.xml 文件中注册 Struts 2 的过滤器

在 web.xml 中加入 Struts 2 的过滤器配置，代码如下。

```xml
<filter>
    <filter-name>struts2</filter-name>
    <filter-class>org.apache.struts2.dispatcher.filter.StrutsPrepareAndExecuteFilter</filter-class>
</filter>
<filter-mapping>
    <filter-name>struts2</filter-name>
    <url-pattern>*.jsp</url-pattern>
</filter-mapping>
<filter-mapping>
    <filter-name>struts2</filter-name>
    <url-pattern>*.js</url-pattern>
</filter-mapping>
<filter-mapping>
    <filter-name>struts2</filter-name>
    <url-pattern>*.action</url-pattern>
</filter-mapping>
```

在 web.xml 文件中注册了 Struts 2 的过滤器，拦截以*.action、*.jsp 和*.js 为后缀的请求。

11.4.3 在项目的 src 目录下创建 struts.xml 配置文件

初始代码如下。

```xml
<?xml version="1.0" encoding="UTF-8" ?>
<!DOCTYPE struts PUBLIC "-//Apache Software Foundation//DTD Struts Configuration 2.1//EN"
"http://struts.apache.org/dtds/struts-2.1.dtd">
<struts>
    <package name="ssh" extends="struts-default">
        <action name="test" class="test">
        </action>
```

```
        </package>
        <constant name="struts.devMode" value="true" />
</struts>
```

配置代码<constant name="struts.devMode" value="true" />的作用是使 Struts 2 以开发模式进行运行，这样在报错的时候可以显示更多便于调试的信息。

11.5　添加 Hibernate 5 框架支持

添加 Hibernate 5 框架支持的 jar 包文件，如图 11-5 所示。

图 11-5　添加 Hbiernate 框架支持的 jar 文件

在 src 中创建 hibernate.cfg.xml 配置文件，代码如下。

```
<?xml version='1.0' encoding='UTF-8'?>
<!DOCTYPE hibernate-configuration PUBLIC
        "-//Hibernate/Hibernate Configuration DTD 3.0//EN"
        "http://www.hibernate.org/dtd/hibernate-configuration-3.0.dtd">
<!-- Generated by MyEclipse Hibernate Tools. -->
<hibernate-configuration>
    <session-factory>
        <property name="dialect">
            org.hibernate.dialect.Oracle9Dialect
        </property>
        <mapping resource="orm/Userinfo.hbm.xml" />
    </session-factory>
</hibernate-configuration>
```

在此 hibernate.cfg.xml 文件中并不存在连接数据库的基本信息了,这些信息被转移到 Spring 配置文件中。

11.6 添加 Spring 4 框架支持

添加 Spring 框架支持环境，如图 11-6 所示。

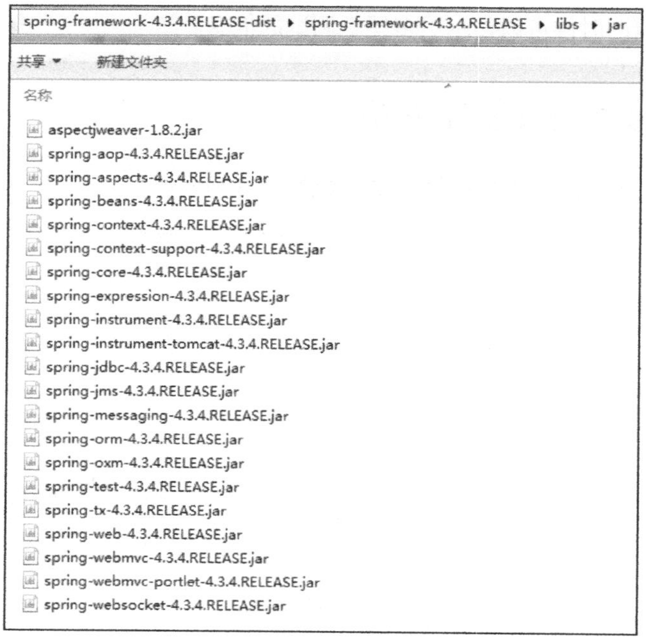

图 11-6 添加 Spring 框架支持的 jar 包

11.7 创建 spring-dao.xml 文件

在 src 路径下创建 XML 文件，代码如下。

```xml
<?xml version="1.0" encoding="UTF-8"?>
<beans xmlns="http://www.springframework.org/schema/beans"
    xmlns:xsi="http://www.w3.org/2001/XMLSchema-instance"
xmlns:p="http://www.springframework.org/schema/p"
    xmlns:aop="http://www.springframework.org/schema/aop"
xmlns:tx="http://www.springframework.org/schema/tx"
    xsi:schemaLocation="http://www.springframework.org/schema/beans
http://www.springframework.org/schema/beans/spring-beans-4.1.xsd
        http://www.springframework.org/schema/aop
http://www.springframework.org/schema/aop/spring-aop-4.3.xsd
        http://www.springframework.org/schema/tx
http://www.springframework.org/schema/tx/spring-tx-4.3.xsd">
    <bean id="userinfoDao" class="dao.UserinfoDao">
        <property name="sessionFactory" ref="sessionFactory"></property>
    </bean>
</beans>
```

11.8 创建 spring-service.xml 文件

在 src 路径下创建 XML 文件，代码如下。

```xml
<?xml version="1.0" encoding="UTF-8"?>
<beans xmlns="http://www.springframework.org/schema/beans"
    xmlns:xsi="http://www.w3.org/2001/XMLSchema-instance"
    xmlns:p="http://www.springframework.org/schema/p"
    xmlns:aop="http://www.springframework.org/schema/aop"
    xmlns:tx="http://www.springframework.org/schema/tx"
    xsi:schemaLocation="http://www.springframework.org/schema/beans
    http://www.springframework.org/schema/beans/spring-beans-4.1.xsd
        http://www.springframework.org/schema/aop
    http://www.springframework.org/schema/aop/spring-aop-4.3.xsd
        http://www.springframework.org/schema/tx
    http://www.springframework.org/schema/tx/spring-tx-4.3.xsd">
        <bean id="userinfoService" class="service.UserinfoServcie">
            <property name="userinfoDao" ref="userinfoDao"></property>
        </bean>
</beans>
```

11.9 创建 spring-controller.xml 文件

在 src 路径下创建 XML 文件，代码如下。

```xml
<?xml version="1.0" encoding="UTF-8"?>
<beans xmlns="http://www.springframework.org/schema/beans"
    xmlns:xsi="http://www.w3.org/2001/XMLSchema-instance"
    xmlns:p="http://www.springframework.org/schema/p"
    xmlns:aop="http://www.springframework.org/schema/aop"
    xmlns:tx="http://www.springframework.org/schema/tx"
    xsi:schemaLocation="http://www.springframework.org/schema/beans
    http://www.springframework.org/schema/beans/spring-beans-4.1.xsd
        http://www.springframework.org/schema/aop
    http://www.springframework.org/schema/aop/spring-aop-4.3.xsd
        http://www.springframework.org/schema/tx
    http://www.springframework.org/schema/tx/spring-tx-4.3.xsd">
        <bean id="test" class="controller.Test" scope="prototype">
            <property name="userinfoService" ref="userinfoService"></property>
        </bean>
</beans>
```

11.10 创建 applicationContext.xml 文件

在 src 目录下手动创建了 1 个 applicationContext.xml 文件，配置代码设计如下。

```xml
<?xml version="1.0" encoding="UTF-8"?>
<beans xmlns="http://www.springframework.org/schema/beans"
    xmlns:xsi="http://www.w3.org/2001/XMLSchema-instance"
    xmlns:p="http://www.springframework.org/schema/p"
    xmlns:aop="http://www.springframework.org/schema/aop"
    xmlns:tx="http://www.springframework.org/schema/tx"
    xsi:schemaLocation="http://www.springframework.org/schema/beans
    http://www.springframework.org/schema/beans/spring-beans-4.1.xsd
        http://www.springframework.org/schema/aop
    http://www.springframework.org/schema/aop/spring-aop-4.3.xsd
        http://www.springframework.org/schema/tx
    http://www.springframework.org/schema/tx/spring-tx-4.3.xsd">
```

```xml
<import resource="spring-controller.xml" />
<import resource="spring-dao.xml" />
<import resource="spring-service.xml" />

<bean id="dataSource"
    class="org.springframework.jdbc.datasource.DriverManagerDataSource">
    <property name="driverClassName" value="oracle.jdbc.OracleDriver"></property>
    <property name="url" value="jdbc:oracle:thin:@localhost:1521:orcl"></property>
    <property name="username" value="y2"></property>
    <property name="password" value="123"></property>
</bean>

<bean id="sessionFactory"
    class="org.springframework.orm.hibernate5.LocalSessionFactoryBean">
    <property name="configLocation" value="classpath:hibernate.cfg.xml"></property>
    <property name="dataSource" ref="dataSource"></property>
</bean>

<bean id="transactionManager"
    class="org.springframework.orm.hibernate5.HibernateTransactionManager">
    <property name="sessionFactory" ref="sessionFactory"></property>
</bean>
<tx:annotation-driven transaction-manager="transactionManager" />
<aop:aspectj-autoproxy proxy-target-class="true"></aop:aspectj-autoproxy>

</beans>
```

在 Hibernate 5 和 Spring 4 整合时，必须要在 applicationContext.xml 文件中定义 dataSource 对象，这点和老版本的整合方式有些差异。如果不定义 dataSource 对象，则启动 Tomcat 时会报异常。

11.11 在 web.xml 文件中注册 Spring 监听器

更改后的 web.xml 代码如下。

```xml
<?xml version="1.0" encoding="UTF-8"?>
<web-app xmlns:xsi="http://www.w3.org/2001/XMLSchema-instance"
    xmlns="http://java.sun.com/xml/ns/javaee"
    xsi:schemaLocation="http://java.sun.com/xml/ns/javaee http://java.sun.com/xml/ns/javaee/web-app_2_5.xsd"
    id="WebApp_ID" version="2.5">
    <welcome-file-list>
        <welcome-file>index.html</welcome-file>
        <welcome-file>index.htm</welcome-file>
        <welcome-file>index.jsp</welcome-file>
        <welcome-file>default.html</welcome-file>
        <welcome-file>default.htm</welcome-file>
        <welcome-file>default.jsp</welcome-file>
    </welcome-file-list>
    <filter>
        <filter-name>struts2</filter-name>
        <filter-class>org.apache.struts2.dispatcher.filter.StrutsPrepareAndExecuteFilter</filter-class>
    </filter>
    <filter-mapping>
        <filter-name>struts2</filter-name>
```

```xml
            <url-pattern>*.jsp</url-pattern>
        </filter-mapping>
        <filter-mapping>
            <filter-name>struts2</filter-name>
            <url-pattern>*.js</url-pattern>
        </filter-mapping>
        <filter-mapping>
            <filter-name>struts2</filter-name>
            <url-pattern>*.action</url-pattern>
        </filter-mapping>

        <listener>
            <listener-class>org.springframework.web.context.ContextLoaderListener</listener-class>
        </listener>
        <context-param>
            <param-name>contextConfigLocation</param-name>
            <param-value>WEB-INF\classes\applicationContext.xml</param-value>
        </context-param>
</web-app>
```

org.springframework.web.context.ContextLoaderListener 类的作用就是启动 Web 容器 Tomcat 时会自动装配 applicationContext.xml 中的 JavaBean 信息。

11.12 加 Spring 4 框架后的 Web 项目结构

项目结构如图 11-7 所示。

图 11-7 现阶段项目结构

11.13 创建 Hibernate 中的实体类与映射文件

在 orm 包中创建实体类 Userinfo.java，代码如下。

```java
package orm;

import java.util.Date;

public class Userinfo implements java.io.Serializable {

    private Long id;
    private String username;
    private String password;
    private Long age;
    private Date insertdate;

    public Userinfo() {
    }

    public Userinfo(String username, String password, Long age, Date insertdate) {
        this.username = username;
        this.password = password;
        this.age = age;
        this.insertdate = insertdate;
    }

    //省略 get 和 set 方法

}
```

继续创建 Hibernate 的 Userinfo.hbm.xml 映射文件，代码如下。

```xml
<?xml version="1.0" encoding="utf-8"?>
<!DOCTYPE hibernate-mapping PUBLIC "-//Hibernate/Hibernate Mapping DTD 3.0//EN"
"http://www.hibernate.org/dtd/hibernate-mapping-3.0.dtd">

<hibernate-mapping>
    <class name="orm.Userinfo" table="USERINFO" schema="Y2">
        <id name="id" type="java.lang.Long">
            <column name="ID" precision="18" scale="0" />
            <generator class="sequence">
                <param name="sequence_name">idauto</param>
            </generator>
        </id>
        <property name="username" type="java.lang.String">
            <column name="USERNAME" length="50" />
        </property>
        <property name="password" type="java.lang.String">
            <column name="PASSWORD" length="50" />
        </property>
        <property name="age" type="java.lang.Long">
            <column name="AGE" precision="18" scale="0" />
        </property>
        <property name="insertdate" type="java.util.Date">
            <column name="INSERTDATE" length="7" />
        </property>
    </class>
```

```
</hibernate-mapping>
```

主键生成策略为 sequence,主键值是来自于数据库中的序列对象 idauto,这个 idauto 对象名称还在要 hbm 映射文件中进行注册,代码如下。

```xml
<id name="id" type="java.lang.Long">
    <column name="ID" precision="18" scale="0" />
    <generator class="sequence">
        <param name="sequence_name">idauto</param>
    </generator>
</id>
```

11.14 创建 Hibernate 5 的 DAO 类

由于还没有创建 DAO 类,所以手动创建 UserinfoDao.java,代码如下。

```java
package dao;

import org.hibernate.SessionFactory;

import orm.Userinfo;

public class UserinfoDao {

    private SessionFactory sessionFactory;

    public SessionFactory getSessionFactory() {
        return sessionFactory;
    }

    public void setSessionFactory(SessionFactory sessionFactory) {
        this.sessionFactory = sessionFactory;
    }

    public void save(Userinfo userinfo) {
        sessionFactory.getCurrentSession().save(userinfo);
    }
}
```

DAO 类的作用就是基本的 CURD 操作,它不继承自任何类,属性 sessionFactory 的值来自于注入。

11.15 创建 UserinfoService.java 服务对象

创建 UserinfoService.java 的代码如下。

```java
package service;

import dao.UserinfoDao;
import orm.Userinfo;

public class UserinfoServcie {

    private UserinfoDao userinfoDao;
```

```java
    public UserinfoDao getUserinfoDao() {
        return userinfoDao;
    }

    public void setUserinfoDao(UserinfoDao userinfoDao) {
        this.userinfoDao = userinfoDao;
    }

    public void saveService() {
        Userinfo userinfo1 = new Userinfo();
        userinfo1.setUsername("中国1");

        Userinfo userinfo2 = new Userinfo();
        userinfo2.setUsername("中国2");

        userinfoDao.save(userinfo1);
        userinfoDao.save(userinfo2);
    }

}
```

在 Service 服务类中创建了一个具有保存功能的业务方法。

11.16 新建一个操作 userinfo 表数据的 Controller 控制层

创建类名称为 Test.java 的 Action 对象，代码如下。

```java
package controller;

import org.springframework.transaction.annotation.Transactional;

import service.UserinfoServcie;

@Transactional
public class Test {

    private UserinfoServcie userinfoService;

    public UserinfoServcie getUserinfoService() {
        return userinfoService;
    }

    public void setUserinfoService(UserinfoServcie userinfoService) {
        this.userinfoService = userinfoService;
    }

    public String execute() {
        userinfoService.saveService();
        return null;
    }

}
```

在 struts.xml 配置文件中注册这个 Action，代码如下。

```xml
<?xml version="1.0" encoding="UTF-8" ?>
```

```xml
<!DOCTYPE struts PUBLIC "-//Apache Software Foundation//DTD Struts Configuration 2.1//EN"
"http://struts.apache.org/dtds/struts-2.1.dtd">
<struts>
    <package name="ssh" extends="struts-default">
        <action name="test" class="test">
        </action>
    </package>
    <constant name="struts.devMode" value="true" />
</struts>
```

在 struts.xml 配置文件中注册 Action 的 class 属性值一定要和 Spring 配置文件中的<bean>标签的 id 属性值一样，通过这个规范 Spring 才可以在 DI 容器中创建指定 Controller 控制层的类对象，结果如图 11-8 所示。

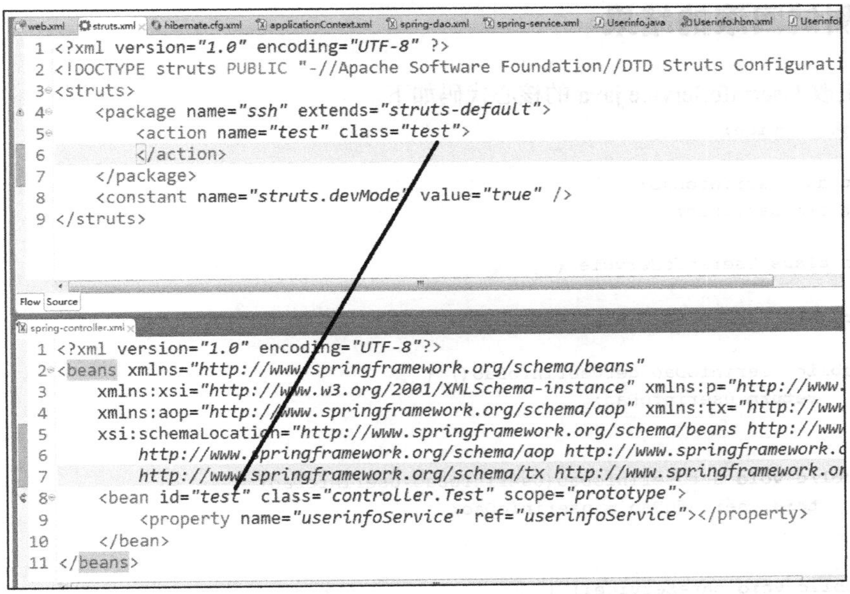

图 11-8　class 属性值与<bean>的 id 属性值一样

另外由于 Struts 2 的 Action 是多例的机制，所以要加上属性 scope，并且值为 prototype，目的是在 Spring 容器中创建多个 Action 实例，这样就不会出现线程安全问题了。

还有重要的一步，就是在 Action 类的上部加入声明式事务，代码如下。

```
@Transactional
public class Test {
```

这样就可以在 Action 中对 Hibernate 操作数据库的事务进行自动化管理。

11.17　测试成功的结果

将项目布署到 Tomcat 中，如果启动后没有出现异常则在 IE 地址栏中输入以下网址。
http://localhost:8081/struts2New_Hibernate5New_Spring4New/test.action
在数据表中成功添加了两条记录，结果如图 11-9 所示。

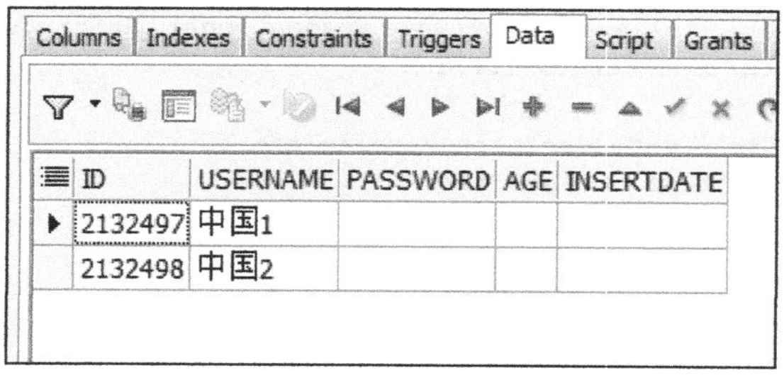

图 11-9 成功添加了两条记录

Struts 2+Hibernate 5+Spring 4 的整合成功！

11.18 测试回滚的结果

继续更改 UserinfoService.java 的核心代码如下。

```java
package service;

import dao.UserinfoDao;
import orm.Userinfo;

public class UserinfoServcie {

    private UserinfoDao userinfoDao;

    public UserinfoDao getUserinfoDao() {
        return userinfoDao;
    }

    public void setUserinfoDao(UserinfoDao userinfoDao) {
        this.userinfoDao = userinfoDao;
    }

    public void saveService() {
        Userinfo userinfo1 = new Userinfo();
        userinfo1.setUsername("中国1");

        Userinfo userinfo2 = new Userinfo();
        userinfo2.setUsername(
                "中国2中国2中国2中国2中国2中国2中国2中国2中国2中国2中国2中国2中国2中国2中国2中国2中国2中国2中国2中国2中国2中国2中国2中国2中国2中国2中国2中国2中国2中国2中国2中国2中国2中国2中国2中国2中国2中国2中国2中国2中国2中国2中国2中国2中国2中国2中国2中国2中国2中国2中国2中国2中国2中国2中国2中国2中国2中国2中国2中国2中国2中国2");

        userinfoDao.save(userinfo1);
        userinfoDao.save(userinfo2);
    }

}
```

重启 Tomcat 后，再次运行在控制台出现报错信息，如图 11-10 所示。

11.18 测试回滚的结果

```
Caused by: java.sql.SQLException: ORA-12899: 列 "Y2"."USERINFO"."USERNAME" 的值太大 (实际值: 290, 最大值: 50)
    at oracle.jdbc.driver.T4CTTIoer.processError(T4CTTIoer.java:447)
    at oracle.jdbc.driver.T4CTTIoer.processError(T4CTTIoer.java:396)
    at oracle.jdbc.driver.T4C8Oall.processError(T4C8Oall.java:951)
    at oracle.jdbc.driver.T4CTTIfun.receive(T4CTTIfun.java:513)
    at oracle.jdbc.driver.T4CTTIfun.doRPC(T4CTTIfun.java:227)
    at oracle.jdbc.driver.T4C8Oall.doOALL(T4C8Oall.java:531)
    at oracle.jdbc.driver.T4CPreparedStatement.doOall8(T4CPreparedStatement.java:208)
    at oracle.jdbc.driver.T4CPreparedStatement.executeForRows(T4CPreparedStatement.java:1046)
    at oracle.jdbc.driver.OracleStatement.doExecuteWithTimeout(OracleStatement.java:1336)
    at oracle.jdbc.driver.OraclePreparedStatement.executeInternal(OraclePreparedStatement.java:3613)
    at oracle.jdbc.driver.OraclePreparedStatement.executeUpdate(OraclePreparedStatement.java:3694)
    at oracle.jdbc.driver.OraclePreparedStatementWrapper.executeUpdate(OraclePreparedStatementWrapper.java:1354)
    at org.hibernate.engine.jdbc.internal.ResultSetReturnImpl.executeUpdate(ResultSetReturnImpl.java:204)
    ... 113 more
```

图 11-10　报错了

这里人为地创建了一个异常。继续查看数据表，还是两条记录，看来 Spring 对事务的回滚是成功的，数据表中的数据如图 11-11 所示。

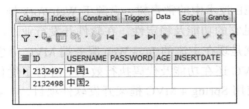

图 11-11　数据表中还是 4 条记录

到此 Struts 2+Hibernate 5+Spring 4 整合的步骤就结束了。

第 12 章　Spring 4 MVC 核心技能

Struts 2 的确是 Java EE 技术中的常青树，它源于 WebWork 框架，发展多年以来一直被 Java 程序员所使用，可见它的功能及稳定性。在软件开发行业，旧的技术会被新的技术所替代，Spring MVC 框架就是现在主流的 Java Web 服务端 MVC 分层框架，它避免了 Struts 2 的种种缺点，Spring 4 MVC 还在功能及代码执行效率上进行了优化，进行了增强。现阶段越来越多的软件公司在使用 Spring 4 MVC 框架开发软件项目，本章内容也较多，读者应该掌握如下内容：

- 使用 Spring 4 MVC 实现登录功能；
- Spring 4 MVC 转发与重定向；
- Spring 4 MVC 如何进行多模块的开发；
- Spring 4 MVC 对 URL 的实体封装；
- 在 Spring 4 MVC 中使用 jQuery+JSON 开发无刷新的应用；
- 在 Spring 4 MVC 中开发上传与下载文件的功能。

12.1　Spring 4 MVC 介绍

随着 Spring 4 框架的发布，其中包含的 MVC 模块越来越受到软件公司的关注。此技术在未来会有更多的程序员去使用，也就是说它具有非常大的优点。从本章开始就和大家一起学习 Spring 4 MVC 框架的使用。

Spring 的官方网站是 http://www.springsource.org/spring-framework，打开网页后界面如图 12-1 所示。

单击 Download 即可下载 Spring 的框架文件，里面包含开发文档、源代码以及示例。

12.1 Spring 4 MVC 介绍

图 12-1 Spring 的官方网站

12.1.1 Spring 4 MVC 核心控制器

与 Struts2 的核心过滤器：

```
org.apache.struts2.dispatcher.filter.StrutsPrepareAndExecuteFilter
```

不同，Spring 4 MVC 使用的是基于 Servlet 的 DispatcherServlet 核心控制器来作为 request 对象的处理，它的父类是真正的 HttpServlet 对象，继承关系如图 12-2 所示。

```
Object - java.lang
  GenericServlet - javax.servlet
    HttpServlet - javax.servlet.http
      HttpServletBean - org.springframework.web.servlet
        FrameworkServlet - org.springframework.web.servlet
          DispatcherServlet - org.springframework.web.servlet
```

图 12-2 DispatcherServlet 的继承关系

并且使用时还要在 web.xml 文件中声明它，代码如下。

```
<servlet>
    <servlet-name>springMVC</servlet-name>
    <servlet-class>org.springframework.web.servlet.DispatcherServlet</servlet-class>
    <load-on-startup>1</load-on-startup>
</servlet>

<servlet-mapping>
    <servlet-name>springMVC</servlet-name>
    <url-pattern>*.spring</url-pattern>
</servlet-mapping>
```

上面的代码只要在 URL 上以 .spring 为后缀访问服务器，就把这个请求交给 DispatcherServlet

对象来进行处理。上面的代码将 DispatcherServlet 设置为别名 springMVC，而 Spring 4 MVC 框架还要在项目中的 WEB-INF 文件夹中找 springMVC-servlet.xml 配置文件。也就是以配置代码 <servlet-name>springMVC</servlet-name> 中的 springMVC 别名做为 XML 文件名的前缀，后面连接 "-servlet.xml" 做为文件名，所有 Bean 和相关的配置定义都要声明在这个 springMVC-servlet.xml 文件中。

通过使用 DispatcherServlet 对象可以将请求分发到不同的功能组件上，比如 handler mappings、view resolution、locale 和 uploading files 等。

类 DispatcherServlet 也被称为前端控制器，所有的请求都要经过前端控制器。DispatcherServlet 的任务是将请求交给控制层，由于系统中存在多个控制层，所以 DispatcherServlet 会根据 URL 的映射关系，来将 URL 的路径与控制层配置的 URL mapping 映射进行对比，从而找到目的控制层对象执行控制层中的方法，进而执行 Service 业务层中的功能。

12.1.2 基于注解的 Spring 4 MVC 开发

Spring 4 MVC 也像 Struts 2 一样支持 POJO 式的控制层，也就是说不需要实现 Spring 4 中的任何接口或类，只需要在类中使用@Controller 和@RequestMapping 注解即可。这样做可以极大地提高组件的松耦合，有利于软件模块间的设计，最重要的是减少了在 XML 配置文件中的代码量。

12.2 Spring 4 MVC 第一个登录测试

软件开发、实践才是真正的学习方法！
本示例将和大家一起学习一下如何用 Spring 4 MVC 开发一个最经典的登录案例。

12.2.1 添加 Spring 4 MVC 的依赖 jar 文件

新建名称为 springlogin 的项目，添加 Spring 4 MVC 框架依赖的 jar 包文件，如图 12-3 所示。

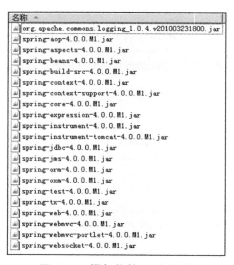

图 12-3 添加依赖 jar 包文件

注意：由于 Spring 官方在 jar 包中并没有提供 org.apache.commons.logging.jar 文件，所以还得自己去搜索然后添加到 lib 文件夹中。

12.2.2 在 web.xml 中配置核心控制器

在 web.xml 文件中注册 Spring 4 MVC 核心控制器 DispatcherServlet，代码如下。

```xml
<?xml version="1.0" encoding="UTF-8"?>
<web-app version="2.5" xmlns="http://java.sun.com/xml/ns/javaee"
    xmlns:xsi="http://www.w3.org/2001/XMLSchema-instance"
    xsi:schemaLocation="http://java.sun.com/xml/ns/javaee
    http://java.sun.com/xml/ns/javaee/web-app_2_5.xsd">
    <servlet>
        <servlet-name>springMVC</servlet-name>
        <servlet-class>org.springframework.web.servlet.DispatcherServlet</servlet-class>
        <load-on-startup>1</load-on-startup>
    </servlet>
    <servlet-mapping>
        <servlet-name>springMVC</servlet-name>
        <url-pattern>*.spring</url-pattern>
    </servlet-mapping>
    <welcome-file-list>
        <welcome-file>index.jsp</welcome-file>
    </welcome-file-list>
</web-app>
```

12.2.3 新建 springMVC-servlet.xml 配置文件

还要在项目中的 WEB-INF 文件夹中创建配置文件 springMVC-servlet.xml，代码如下。

```xml
<?xml version="1.0" encoding="UTF-8"?>
<beans xmlns="http://www.springframework.org/schema/beans"
    xmlns:xsi="http://www.w3.org/2001/XMLSchema-instance"
    xmlns:p="http://www.springframework.org/schema/p"
    xmlns:context="http://www.springframework.org/schema/context"
    xsi:schemaLocation="
        http://www.springframework.org/schema/beans
        http://www.springframework.org/schema/beans/spring-beans-3.0.xsd
        http://www.springframework.org/schema/context
        http://www.springframework.org/schema/context/spring-context-3.0.xsd">
    <context:component-scan base-package="controller" />
</beans>
```

代码 `<context:component-scan base-package="controller" />` 的作用是在包中扫描带 @Controller 的控制层 Java 文件。如果找到，则说明该 Java 文件是控制层，参与处理 request 和 response 对象。

12.2.4 新建相关的 JSP 文件

新建 3 个 JSP 文件，其中 login.jsp 文件代码如下。

```jsp
<%@ page language="java" import="java.util.*" pageEncoding="utf-8"%>
<!DOCTYPE HTML PUBLIC "-//W3C//DTD HTML 4.01 Transitional//EN">
<html>
```

```html
        <head>
        </head>
        <body>
            <form action="login.spring" method="post">
                username:
                <input type="text" name="username">
                <br />
                password:
                <input type="text" name="password">
                <br />
                <input type="submit" value="submit">
                <br />
            </form>
        </body>
</html>
```

登录成功界面 ok.jsp 的代码如下。

```
<%@ page language="java" import="java.util.*" pageEncoding="utf-8"%>
<!DOCTYPE HTML PUBLIC "-//W3C//DTD HTML 4.01 Transitional//EN">
<html>
    <head>
    </head>
    <body>
        ok page!welcome:${username}
    </body>
</html>
```

登录失败界面 no.jsp 的代码如下。

```
<%@ page language="java" import="java.util.*" pageEncoding="utf-8"%>
<!DOCTYPE HTML PUBLIC "-//W3C//DTD HTML 4.01 Transitional//EN">
<html>
    <head>
    </head>
    <body>
        no page!
    </body>
</html>
```

12.2.5 新建控制层 Java 类文件

下面开始设计最为关键的组件：控制层。

新建 Login.java 文件，代码如下。

```java
package controller;

import org.springframework.stereotype.Controller;
import org.springframework.ui.Model;
import org.springframework.web.bind.annotation.RequestMapping;
import org.springframework.web.bind.annotation.RequestParam;

//@Controller 注解代表本 Java 类是 controller 控制层
@Controller
public class Login {
    // 通过@RequestMapping 注解可以
    // 用指定的 URL 路径访问本控制层
    @RequestMapping("/login")
```

```
// @RequestParam 注解的功能从 url 根据参数名取得参数值
// Model model 的作用相当于往 request 对象中放入数据
public String login(@RequestParam("username") String username,
        @RequestParam("password") String password, Model model) {
    if (username.equals("ghy") && password.equals("123")) {
        model.addAttribute("username", username);
        return "ok.jsp";
    } else {
        return "no.jsp";
    }
}
```

12.2.6 部署项目并运行

项目中的代码设计完毕后,部署项目到 Tomcat,输入网址 http://localhost:8081/springlogin/login.jsp,显示登录界面,如图 12-4 所示。

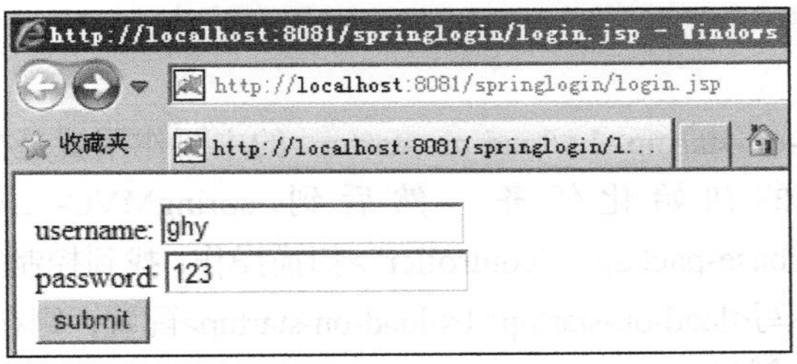

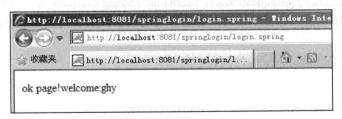

图 12-4 显示登录界面

在 username 处填写 ghy,在 password 处填写 123 以便实现成功登录的效果。单击 "submit" 按钮,显示成功登录界面如图 12-5 所示。

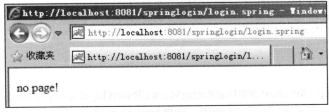

图 12-5 登录成功界面

重新登录,在 username 和 password 处什么也不填写,直接单击 "submit" 按钮,出现结果如图 12-6 所示。

图 12-6 登录失败界面

12.2.7　第一个示例的总结

通过本示例，要掌握 Spring 4 MVC 的基本使用流程，着重掌握以下知识点。
- web.xml 中配置 DispatcherServlet 核心控制器。
- WEB-INF 文件夹中要添加 XML 配置文件。
- @Controller、@RequestMapping、@RequestParam 以及 Model model 对象的作用和使用。
- 将 login.jsp 中的代码改成 get 提交。

```
<form action="login.spring" method="get">
```

控制层 Login.java 代码也能正确无误地运行，说明控制层在默认的情况下可以处理 get 和 post 请求。
- 在 web.xml 中有如下配置代码。

```xml
<servlet>
    <servlet-name>springMVC</servlet-name>
    <servlet-class>org.springframework.web.servlet.DispatcherServlet</servlet-class>
    <load-on-startup>1</load-on-startup>
</servlet>
```

其中配置代码<load-on-startup>1</load-on-startup>的主要作用就是在 Tomcat 容器启动时执行 Spring MVC 框架的初始化任务，然后到 springMVC-servlet.xml 文件中进行<context:component-scan base-package="controller">扫描操作，找到控制层 JavaBean 并进行实例化。如果在 web.xml 中不写<load-on-startup>1</load-on-startup>配置，则在第一次执行 Controller，该 Controller 的实例才被创建。

12.2.8　Spring MVC 取参还能更加方便

前面的示例通过代码@RequestParam("username") String username 取得 URL 中 username 的参数值。其实在 Spring 4 MVC 中还可以更加方便地取参，演示代码在 urlParamMethodParam 项目中。

创建控制层文件 Login.java 的代码如下。

```java
package controller;

import org.springframework.stereotype.Controller;
import org.springframework.web.bind.annotation.RequestMapping;

@Controller
public class Login {
    @RequestMapping("/login")
    public String login(String username, String password) {
        System.out.println("username=" + username + " passwword=" + password);
        return "index.jsp";
    }
}
```

在 IE 中打开 http://localhost:8081/urlParamMethodParam/login.spring?username=a&password=b，在控制台输出了 username 和 password 的值，结果如图 12-7 所示。

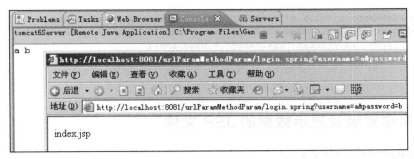

图 12-7　更加方便地取得 URL 参数值

通过此示例可以知道，URL 中同名的参数将要自动传给控制层方法中同名的参数，并且不再需要@RequestParam("username")注解了。

12.3　执行控制层与限制提交的方式

前面的章节使用 Spring 4 MVC 开发一个简单的登录功能，有时候需要在浏览器的地址栏中直接输入 URL 地址而去访问控制层来达到实现一些功能上的目的，比如列表。本节就来实现这个功能。

另外，前面的示例中控制层对 get 和 post 都可以进行处理，但有些情况下如果想让控制层只处理指定的<form method="xxx">提交请求该怎么办呢？这两个问题都在项目 callController 中与大家一起学习。

12.3.1　新建控制层 ListUsername.java 文件

新建 ListUsername.java 文件的代码如下。

```java
package controller;

import java.util.ArrayList;
import java.util.List;

import org.springframework.stereotype.Controller;
import org.springframework.ui.Model;
import org.springframework.web.bind.annotation.RequestMapping;
import org.springframework.web.bind.annotation.RequestMethod;
import org.springframework.web.bind.annotation.RequestParam;

@Controller
public class ListUsername {
    @RequestMapping("/listUsername")
    public String listUsername(Model model) {
        List listUsername = new ArrayList();
        for (int i = 0; i < 10; i++) {
            listUsername.add("username" + (i + 1));
        }
        model.addAttribute("listUsername", listUsername);
        return "listUsername.jsp";
    }
}
```

```
    @RequestMapping(value = "/login", method = RequestMethod.POST)
    public String login(@RequestParam("username") String username) {
        System.out.println("username=" + username);
        return "test.jsp";
    }
}
```

12.3.2 新建登录及显示数据的 JSP 文件

文件 login.jsp 的代码如下。

```
<%@ page language="java" import="java.util.*" pageEncoding="utf-8"%>
<!DOCTYPE HTML PUBLIC "-//W3C//DTD HTML 4.01 Transitional//EN">
<html>
    <head>
    </head>
    <body>
        <form action="login.spring" method="post">
            username:
            <input type="text" name="username">
            <input type="submit" value="submit">
        </form>
    </body>
</html>
```

文件 listUsername.jsp 的代码如下。

```
<%@ page language="java" import="java.util.*" pageEncoding="utf-8"%>
<%@ taglib uri="http://java.sun.com/jsp/jstl/core" prefix="c"%>
<!DOCTYPE HTML PUBLIC "-//W3C//DTD HTML 4.01 Transitional//EN">
<html>
    <head>
    </head>
    <body>
        <c:forEach var="eachUsername" items="${listUsername}">
        ${eachUsername}<br />
        </c:forEach>
    </body>
</html>
```

文件 test.jsp 的代码如下。

```
<%@ page language="java" import="java.util.*" pageEncoding="utf-8"%>
<%@ taglib uri="http://java.sun.com/jsp/jstl/core" prefix="c"%>
<!DOCTYPE HTML PUBLIC "-//W3C//DTD HTML 4.01 Transitional//EN">
<html>
    <head>
    </head>
    <body>
        test.jsp
    </body>
</html>
```

12.3.3 部署项目并测试

部署项目到 Tomcat，输入网址 http://localhost:8081/callController/login.jsp，打开网页并输

入 username 为 ghy，结果如图 12-8 所示。

图 12-8 打开的 login.jsp 页面

单击"submit"按钮进行登录提交，结果如图 12-9 所示。

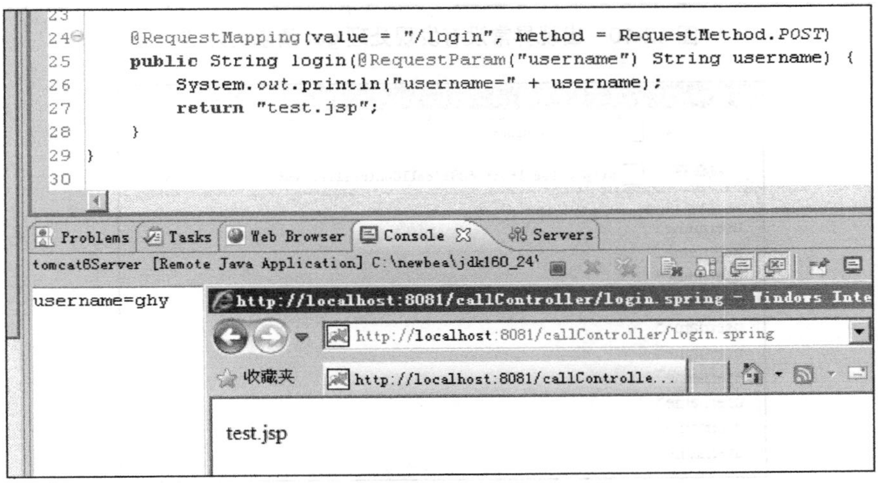

图 12-9 登录后的界面效果

从图 12-9 中可以看到，在控制层正确输出了以 post 提交的 username 值，并且成功转发到 test.jsp 页面中。

那如果将<form>的属性 method 改成 get 会是什么样子呢？

文件 login.jsp 代码的更改如下。

```
<form action="login.spring" method="get">
    username:
    <input type="text" name="username">
    <input type="submit" value="submit">
</form>
```

重新部署项目并刷新浏览器，当在 username 的表单中输入 abc 后单击"submit"按钮，结果如图 12-10 所示。

通过上面的示例可以发现，使用注解 @RequestMapping(value = "/login", method = RequestMethod.POST)可以限制接收提交的方式，有利于代码的规范。

本示例还提供另外一个功能，就是可以直接访问控制层，然后在 JSP 文件中显示 List 中的数据。输入网址 http://localhost:8081/callController/listUsername.spring，在 IE 中输入数据内容，如图 12-11 所示。

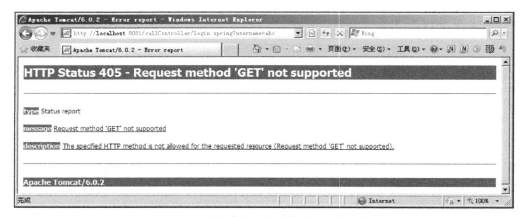

图 12-10　出现异常没有发现处理 get 请求的方法

图 12-11　直接访问控制层输出列表中的数据

12.4　解决多人开发路径可能重复的问题

在开发 Java EE 项目时，分组开发、分工协作是软件公司常用的工作方式，这时就会出现一些问题。比如 A 开发前台登录，路径为 login.spring，B 开发后台登录，路径也是 login.spring，这种情况就会出现一些错误。本实验在项目 springdoublelogin 中进行测试。

另外需要注意的是，在使用 Spring 4 MVC 框架中，不要出现在不同的包中有相同类名的情况。这样在 Web 容器启动时会报错，出错信息如下所示。

```
org.springframework.context.annotation.ConflictingBeanDefinitionException: Annotation-
specified bean name 'userinfoController' for bean class [b.controller.UserinfoController]
conflicts with existing, non-compatible bean definition of same name and class
[a.controller.UserinfoController]
```

12.4.1　错误的情况

A 开发的前台登录代码。

```
package a.controller;

import org.springframework.stereotype.Controller;
import org.springframework.web.bind.annotation.RequestMapping;

@Controller
public class AModule {
    @RequestMapping("/login")
    public String login() {
        System.out.println("aModule");
        return "index.jsp";
    }
}
```

B 开发的后台登录代码。

```
package b.controller;

import org.springframework.stereotype.Controller;
import org.springframework.web.bind.annotation.RequestMapping;

@Controller
public class BModule {
    @RequestMapping("/login")
    public String login() {
        System.out.println("bModule");
        return "index.jsp";
    }
}
```

在这个示例中需要扫描两个包，代码如下。

```
<context:component-scan base-package="a.controller"></context:component-scan>
<context:component-scan base-package="b.controller"></context:component-scan>
```

把这个项目部署到 Tomcat 中，启动时已经出现异常，信息如下。

```
java.lang.IllegalStateException: Cannot map handler 'BModule' to URL path [/login]: There is already handler of type [class a.controller.AModule] mapped.
```

从出错提示可以看到，路径/login 已经被注册，不能重复，如果遇到这种情况该怎么办呢？

12.4.2 解决办法

解决办法就是限定各模块的访问路径。
将文件 AModule.java 代码更改如下。

```
@Controller
@RequestMapping("/a")
public class AModule {
    @RequestMapping("/login")
    public String login() {
        System.out.println("aModule");
        return "../index.jsp";
    }
}
```

文件 BModule.java 代码更改如下。

```
@Controller
```

```
@RequestMapping("/b")
public class BModule {
    @RequestMapping("/login")
    public String login() {
        System.out.println("bModule");
        return "../index.jsp";
    }
}
```

如果在类的上方使用@RequestMapping注解，表示首先定义了相对的父路径，然后在方法上定义的路径是相对于类级别上的。

重新启动Tomcat并没有出现异常，并且在console控制台启动日志中已经看到不同的login被划分到不同模块的工作路径中。

```
信息: Mapped URL path [/a/login] onto handler 'AModule'
2012-6-30 15:23:09 org.springframework.web.servlet.handler.AbstractUrlHandlerMapping registerHandler
信息: Mapped URL path [/a/login.*] onto handler 'AModule'
2012-6-30 15:23:09 org.springframework.web.servlet.handler.AbstractUrlHandlerMapping registerHandler
信息: Mapped URL path [/a/login/] onto handler 'AModule'
2012-6-30 15:23:09 org.springframework.web.servlet.handler.AbstractUrlHandlerMapping registerHandler
信息: Mapped URL path [/b/login] onto handler 'BModule'
2012-6-30 15:23:09 org.springframework.web.servlet.handler.AbstractUrlHandlerMapping registerHandler
信息: Mapped URL path [/b/login.*] onto handler 'BModule'
2012-6-30 15:23:09 org.springframework.web.servlet.handler.AbstractUrlHandlerMapping registerHandler
信息: Mapped URL path [/b/login/] onto handler 'BModule'
```

在浏览器输入网址http://localhost:8081/springdoublelogin/a/login.spring，正确执行a模块的登录，结果如图12-12所示。

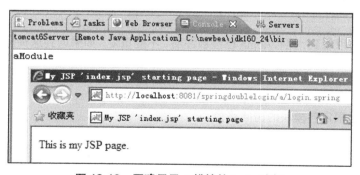

图12-12 正确显示a模块的login路径

继续输入网址http://localhost:8081/springdoublelogin/b/login.spring，成功执行b模块的登录，结果如图12-13所示。

通过在类的上方加入下述代码可以在Spring 4 MVC中进行模块化开发。

```
@RequestMapping("/a")
public class AModule {
```

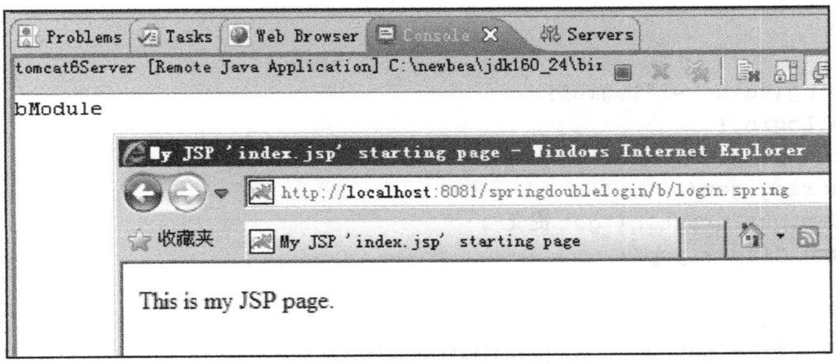

图 12-13 正确显示 b 模块的 login 路径

但是，现在的状态虽然正确实现了多人开发的问题，但控制层中代码 return "../index.jsp";感觉处理的不太美观。因为如果路径过多，则../符号会很多，如何解决呢？在 springMVC-servlet.xml 配置文件中加入如下配置即可。

```
<bean
    class="org.springframework.web.servlet.view.InternalResourceViewResolver"
    p:prefix="/" />
```

它的功能就是限定默认访问资源的路径是/根路径，也就是相对于 WebRoot 路径。

改后的控制层代码如下所示。

```
return "index.jsp";
```

再次运行网址也能在控制台中正确输出想要输出的字符串，并且在 IE 中显示 index.jsp 文件中的内容，如图 12-14 所示。

图 12-14 显示 index.jsp 中的内容

12.5 在控制层中处理指定的提交 get 或 post 方式

可以在控制层中使用指定的方式来处理 get 或 post 提交方式，实验的代码在项目 getpostDefault 中。

12.5.1 控制层代码

文件 Login.java 的代码如下。

```
package controller;

import org.springframework.stereotype.Controller;
import org.springframework.web.bind.annotation.RequestMapping;
import org.springframework.web.bind.annotation.RequestMethod;
```

```
@Controller
@RequestMapping(value = "login")
public class Login {
    @RequestMapping(method = RequestMethod.GET)
    public String get() {
        System.out.println("get 提交");
        return "index.jsp";
    }

    @RequestMapping(method = RequestMethod.POST)
    public String post() {
        System.out.println("post 提交");
        return "index.jsp";
    }
}
```

与前面示例稍有不同，在 Java 文件中，代码@RequestMapping(value = "login")在类 Login.java 的上方进行声明，而类 Login.java 的方法 get()和 post()分别用于处理 get 和 post 的提交方式。

12.5.2 新建 JSP 文件并运行

登录 login.jsp 的代码如下。

```
<body>
    <form action="login.spring" method="get">
        <input type="submit" value="submit">
    </form>
    <br />
    <br />
    <form action="login.spring" method="post">
        <input type="submit" value="submit">
    </form>
</body>
```

文件 index.jsp 的代码如下。

```
<body>
    index.jsp
</body>
```

输入网址 http://localhost:8081/getpostDefault/login.jsp，单击上面的"submit"按钮进行 get 方式的提交，结果如图 12-15 所示。

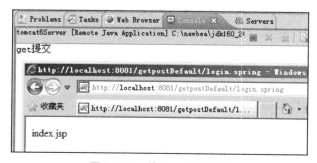

图 12-15　处理 get 的提交

单击下面的"submit"按钮进行 post 方式的提交,运行结果如图 12-16 所示。

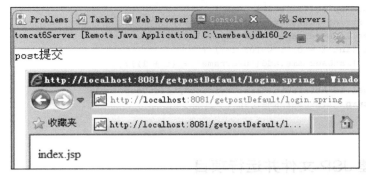

图 12-16 处理 post 的提交

12.6 控制层重定向到控制层——无参数传递

控制层重定向到控制层典型的情况就是登录成功后重定向 List 显示数据。本示例在项目名称为 controllertocontroller 的项目中进行实验。

12.6.1 新建控制层 Java 文件

文件 Login.java 的代码如下。

```
package controller;

import org.springframework.stereotype.Controller;
import org.springframework.web.bind.annotation.RequestMapping;
import org.springframework.web.bind.annotation.RequestParam;

@Controller
public class Login {
    @RequestMapping("/login")
    public String listUsername(@RequestParam("username") String username) {
        System.out.println("username=" + username);
        return "redirect:/listUsername.spring";
    }
}
```

在 Spring 4 MVC 中重定向的关键代码就是在返回字符串中要加入"redirect:/"前缀,代表这个操作是重定向。

文件 ListUsername.java 的代码如下。

```
package controller;

import java.util.ArrayList;
import java.util.List;

import org.springframework.stereotype.Controller;
import org.springframework.ui.Model;
import org.springframework.web.bind.annotation.RequestMapping;
```

```
@Controller
public class ListUsername {
    @RequestMapping("/listUsername")
    public String listUsername(Model model) {
        List listUsername = new ArrayList();
        for (int i = 0; i < 10; i++) {
            listUsername.add("username" + (i + 1));
        }
        model.addAttribute("listUsername", listUsername);
        return "listUsername.jsp";
    }
}
```

12.6.2 创建 JSP 文件并运行项目

文件 listUsername.jsp 的核心代码如下。

```
<body>
    <c:forEach var="eachUsername" items="${listUsername}">
${eachUsername}<br />
    </c:forEach>
</body>
```

文件 login.jsp 的核心代码如下。

```
<body>
    <form action="login.spring" method="get">
        username:
        <input type="text" name="username">
        <input type="submit" value="submit">
    </form>
</body>
```

部署项目，在浏览器中输入网址 http://localhost:8081/controllertocontroller/login.jsp，在 username 表单中输入 123。单击"提交"按钮，在控制台输出 123，并且浏览器的地址栏发生变化，证明成功重定向到另外的控制层，结果如图 12-17 所示。

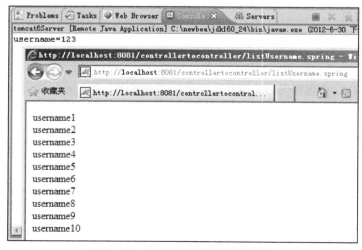

图 12-17　控制层彼此重定向无传参

12.7 控制层重定向到控制层——有参数传递

本示例也要实现控制层重定向到控制层，但是两者之间要传参的示例。代码非常简单，只需要在项目 controllertocontroller 的基础上进行更改即可，当然新建一个项目是更好的复习及学习新知识的方式。

12.7.1 创建两个控制层 Java 文件

新建名称为 controllertocontrollerParam 的 Web 项目。

文件 Login.java 的代码如下。

```java
package controller;

import org.springframework.stereotype.Controller;
import org.springframework.web.bind.annotation.RequestMapping;
import org.springframework.web.bind.annotation.RequestParam;

@Controller
public class Login {
    @RequestMapping("/login")
    public String listUsername(@RequestParam("username") String username) {
        System.out.println("username=" + username);
        return "redirect:/listUsername.spring?username=" + username;
    }
}
```

文件 ListUsername.java 的代码如下。

```java
package controller;

import java.util.ArrayList;
import java.util.List;

import org.springframework.stereotype.Controller;
import org.springframework.ui.Model;
import org.springframework.web.bind.annotation.RequestMapping;
import org.springframework.web.bind.annotation.RequestParam;

@Controller
public class ListUsername {
    @RequestMapping("/listUsername")
    public String listUsername(@RequestParam("username") String username,
            Model model) {
        System.out.println("在 listUsername 中接收 username=" + username);
        List listUsername = new ArrayList();
        for (int i = 0; i < 10; i++) {
            listUsername.add("username" + (i + 1));
        }
        model.addAttribute("listUsername", listUsername);
        return "listUsername.jsp";
    }
}
```

12.7.2 部署项目并运行

部署到 Tomcat 并输入网址 http://localhost:8081/controllertocontrollerParam/login.jsp，在 username 处添写 123，单击"提交"按钮，结果如图 12-18 所示。

图 12-18　控制层重定向到控制层有传参的效果

12.8　匹配 URL 路径执行指定控制层

Spring 4 MVC 框架还提供一种非常类似于 Android 中的 Intent 的技术，也就是将指定 URL 模式地址与指定访问控制层的路径进行关联与匹配，如果某一个访问控制层的 URL 与该 URL 模式进行匹配，则匹配的控制层即被调用。

12.8.1　新建控制层文件

新建名称为 urlMatchTest 测试项目。
文件 URLMatchTest.java 的代码如下。

```java
package controller;

import org.springframework.stereotype.Controller;
import org.springframework.web.bind.annotation.PathVariable;
import org.springframework.web.bind.annotation.RequestMapping;

@Controller
public class URLMatchTest {

    // 对配置变量进行另起别名为 userIdParam
    @RequestMapping("/gaohongyan2/{userId}")
    public String test2(@PathVariable("userId") String userIdParam) {
        System.out.println("run test2 userIdParam=" + userIdParam);
        return "/index.jsp";
    }
}
```

```java
// 直接使用 Path 路径配置变量{userId}
@RequestMapping("/gaohongyan1/{userId}")
public String test1(@PathVariable String userId) {
    System.out.println("run test1 userId=" + userId);
    return "/index.jsp";
}

@RequestMapping("/gaohongyan/{userId}/age/{ageValue}")
public String test3(@PathVariable String userId, @PathVariable int ageValue) {
    System.out.println("run test3 userId=" + userId + " ageValue="
            + ageValue);
    return "/index.jsp";
}
```

另外一种 URL 模式匹配的测试文件 URLMatchTest2.java 的代码如下。

```java
package controller;

import org.springframework.stereotype.Controller;
import org.springframework.web.bind.annotation.PathVariable;
import org.springframework.web.bind.annotation.RequestMapping;

@Controller
@RequestMapping("/gaohongyan100/{userId}")
public class URLMatchTest2 {
    @RequestMapping("/age/{ageValue}")
    public String test100(@PathVariable String userId,
            @PathVariable int ageValue) {
        System.out.println("run test100 userId=" + userId + " ageValue="
                + ageValue);
        return "/index.jsp";
    }
}
```

12.8.2 部署项目并运行

部署项目，重启 Tomcat，输入网址 http://localhost:8081/urlMatchTest/gaohongyan1/123.spring，结果如图 12-19 所示。

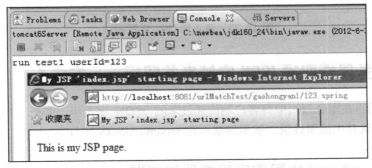

图 12-19 第一种 URL 匹配测试

输入网址 http://localhost:8081/urlMatchTest/gaohongyan2/123.spring，结果如图 12-20 所示。

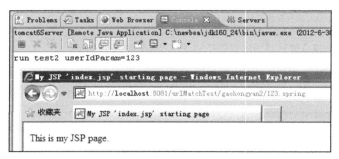

图 12-20　第二种 URL 匹配测试

输入网址 http://localhost:8081/urlMatchTest/gaohongyan/1/age/100.spring，结果如图 12-21 所示。

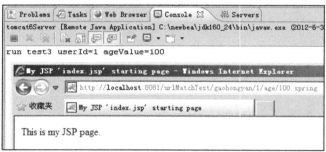

图 12-21　第三种 URL 匹配测试

输入网址 http://localhost:8081/urlMatchTest/gaohongyan100/100/age/200.spring，结果如图 12-22 所示。

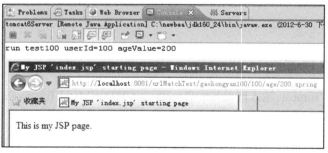

图 12-22　第 4 种 URL 匹配测试

12.9　在服务器端取得 JSON 字符串并解析——方式 1

下面的示例将在服务器端取得客户端传递过来的 JSON 字符串，并把 JSON 字符串转成 JSON 对象并取得其中的属性值，该示例演示在 serverGetJSON 项目中。

12.9.1　在 web.xml 中配置字符编码过滤器

在 web.xml 中添加 Spring 的编码过滤器，目的是防止中文乱码，完整的 web.xml 代码如下。

```xml
<?xml version="1.0" encoding="UTF-8"?>
<web-app version="2.5" xmlns="http://java.sun.com/xml/ns/javaee"
    xmlns:xsi="http://www.w3.org/2001/XMLSchema-instance"
    xsi:schemaLocation="http://java.sun.com/xml/ns/javaee
    http://java.sun.com/xml/ns/javaee/web-app_2_5.xsd">

    <filter>
        <filter-name>encodingFilter</filter-name>
        <filter-class>org.springframework.web.filter.CharacterEncodingFilter</filter-class>
        <init-param>
            <param-name>encoding</param-name>
            <param-value>utf-8</param-value>
        </init-param>
    </filter>
    <filter-mapping>
        <filter-name>encodingFilter</filter-name>
        <url-pattern>/*</url-pattern>
    </filter-mapping>

    <servlet>
        <servlet-name>springMVC</servlet-name>
        <servlet-class>org.springframework.web.servlet.DispatcherServlet</servlet-class>
        <load-on-startup>1</load-on-startup>
    </servlet>
    <servlet-mapping>
        <servlet-name>springMVC</servlet-name>
        <url-pattern>*.spring</url-pattern>
    </servlet-mapping>
    <welcome-file-list>
        <welcome-file>index.jsp</welcome-file>
    </welcome-file-list>
</web-app>
```

12.9.2 新建 JSP 文件

新建名称为 sendAjax.jsp 的 JSP 文件，代码如下。

```jsp
<%@ page language="java" import="java.util.*" pageEncoding="utf-8"%>
<!DOCTYPE HTML PUBLIC "-//W3C//DTD HTML 4.01 Transitional//EN">
<html>
    <head>
        <script src="jquery-1.3.2.js">
        </script>
        <script src="json2.js">
        </script>
        <script>
            function userinfo(username, password){
                this.username = username;
                this.password = password;
            }
            function sendAjax(){
                var userinfoRef = new userinfo('高洪岩', '123');
                var jsonStringRef = JSON.stringify(userinfoRef);
                $.post("getJSONString.spring?t=" + new Date().getTime(), {
                    jsonString: jsonStringRef
                });
            }
        </script>
```

```
</head>
<body>
    <input type="button" onclick="sendAjax()" 1>
</body>
</html>
```

12.9.3　新建控制层 Java 文件

新建控制层 GetJSONString.java 的代码如下。

```java
package controller;

import net.sf.json.JSONObject;

import org.springframework.stereotype.Controller;
import org.springframework.web.bind.annotation.RequestMapping;
import org.springframework.web.bind.annotation.RequestParam;

@Controller
public class GetJSONString {

    @RequestMapping(value = "getJSONString")
    public String getJSONString(@RequestParam("jsonString") String jsonString) {

        JSONObject object = JSONObject.fromObject(jsonString);
        System.out.println(object.get("username"));
        System.out.println(object.get("password"));

        return "test.jsp";
    }

}
```

12.9.4　添加依赖的 jar 包文件

并且还要在项目中添加必要的依赖 jar 文件，jar 文件列表如图 12-23 所示。

图 12-23　添加必要的 jar 文件列表

注意：JSONLIB 库需要依赖其它第三方的 jar 文件，一定要进行添加，不然会出现错误。

12.9.5 运行项目

程序运行后，单击 IE 界面中的"Button"按钮，正确在 Eclipse 控制台中输出 JSON 对象中的属性值，结果如图 12-24 所示。

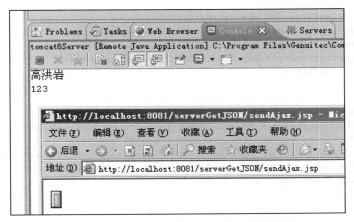

图 12-24 正确打印出 JSON 对象中的属性值

12.10 在服务器端取得 JSON 字符串并解析——方式 2

在前面的章节中，通过使用代码 **public** String getJSONString(@RequestParam("jsonString") String jsonString)来取得从客户端传递过来的 JSON 字符串并进行解析。解析 JSON 字符串使用的是 json-lib-2.3-jdk15.jar，其实在 Spring 4 MVC 中已经提供了一种从 JSON 字符串自动转成实体的技术。

12.10.1 新建封装 JSON 对象属性的实体类

新建名称为 getJSONStringToObject 的 Web 项目，新建实体类 Userinfo.java，代码如下。

```java
package entity;

public class Userinfo {
    private String username;
    private String password;
    public String getUsername() {
        return username;
    }
    public void setUsername(String username) {
        this.username = username;
    }
    public String getPassword() {
        return password;
    }
    public void setPassword(String password) {
        this.password = password;
```

```
        }
    }
```

12.10.2 新建控制层

新建控制层 GetJSONStringToObject.java，代码如下。

```java
package controller;

import org.springframework.stereotype.Controller;
import org.springframework.web.bind.annotation.RequestBody;
import org.springframework.web.bind.annotation.RequestMapping;
import org.springframework.web.bind.annotation.RequestMethod;
import entity.Userinfo;

@Controller
public class GetJSONStringToObject {
    @RequestMapping(value = "createJSONObjectURL", method = RequestMethod.POST)
    public String createJSON(@RequestBody Userinfo userinfo) {
        System.out.println("username value=" + userinfo.getUsername());
        System.out.println("password value=" + userinfo.getPassword());
        return "test.jsp";
    }
}
```

使用@RequestBody 注解后前台只需要向控制层提交一段符合 JSON 格式的 request body 体，Spring 4 MVC 就会自动将其拼装成 JavaBean。

注意：在这里需要留意的是，如果 Userinfo.java 类里具有有参的构造函数，则一定要创建一个无参的构造函数。因为 createJSON()方法是需要调用一个无参的构造函数来实例化一个 Userinfo 实体，进而调用 username 和 password 属性所对应的方法。如果没有无参的构造函数，则通过 HttpWatch 工具查看是报错的结果。

12.10.3 在配置文件中添加<mvc:annotation-driven />注解

配置文件 springMVC-servlet.xml 的代码如下。

```xml
<?xml version="1.0" encoding="UTF-8"?>
<beans xmlns="http://www.springframework.org/schema/beans"
    xmlns:mvc="http://www.springframework.org/schema/mvc"
    xmlns:xsi="http://www.w3.org/2001/XMLSchema-instance"
    xmlns:p="http://www.springframework.org/schema/p"
    xmlns:context="http://www.springframework.org/schema/context"
    xsi:schemaLocation="
        http://www.springframework.org/schema/beans
        http://www.springframework.org/schema/beans/spring-beans-3.0.xsd
        http://www.springframework.org/schema/context
        http://www.springframework.org/schema/context/spring-context-3.0.xsd
        http://www.springframework.org/schema/mvc
        http://www.springframework.org/schema/mvc/spring-mvc-3.1.xsd">
    <context:component-scan base-package="controller" />
    <mvc:annotation-driven />
</beans>
```

添加注解的目的可以使 JSON 字符串自动转成实体类。

在 Spring 4 MVC 中 DefaultAnnotationHandlerMapping 对象是负责处理类级别上的 @RequestMapping 注解，而 AnnotationMethodHandlerAdapter 是负责处理方法级别上的 @RequestMapping 注解，但如果使用 <mvc:annotation-driven /> 注解会自动注册 DefaultAnnotationHandlerMapping 与 AnnotationMethodHandlerAdapter 两个 Bean。

12.10.4 新建 JSP 文件

新建发送 Ajax 请求的文件 sendAjax.jsp，代码如下。

```jsp
<%@ page language="java" import="java.util.*" pageEncoding="utf-8" %>
<!DOCTYPE HTML PUBLIC "-//W3C//DTD HTML 4.01 Transitional//EN">
<html>
    <head>
        <script src="jquery-1.3.2.js">
        </script>
        <script src="json2.js">
        </script>
        <script>
            function userinfo(username, password){
                this.username = username;
                this.password = password;
            }

            function sendAjax(){
                var userinfoRef = new userinfo('高洪岩new123', '123new');
                var jsonStringRef = JSON.stringify(userinfoRef);
                $.ajax({
                    type: "POST",
                    data: jsonStringRef,
                    url: "createJSONObjectURL.spring?t=" + new Date().getTime(),
                    contentType: "application/json"
                });
            }
        </script>
    </head>
    <body>
        <input type="button" onclick="sendAjax()"/>
    </body>
</html>
```

12.10.5 添加 jacksonJSON 解析处理类库并运行

其中最为重要的是要添加 jacksonJSON 支持的 jar 文件 jackson-all-1.9.8.jar。

程序运行后，在控制台输出从 JSON 字符串转换成 Userinfo 实体后的属性值，如图 12-25 所示。

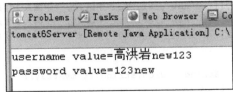

图 12-25 打印 JSON 字符串转成 Userinfo 的属性值

12.10.6 解析不同格式的 JSON 字符串示例

前面的知识点仅仅将 JSON 字符串转成 Userinfo 实体类，其实还可以以多种传输 JSON 字符串类型来在服务器端进行解析，示例前台 HTML 代码如下。

```jsp
<%@ page language="java" import="java.util.*" pageEncoding="utf-8" %>
<!DOCTYPE HTML PUBLIC "-//W3C//DTD HTML 4.01 Transitional//EN">
<html>
    <head>
        <script src="js/jquery/jquery-1.4.2.js">
        </script>
        <script src="js/json/json2.js">
        </script>
        <script>
            function userinfo(username, password){
                this.username = username;
                this.password = password;
            }

            function sendAjax1(){
                var userinfoRef = new userinfo("中国", "中国人");
                var jsonString = JSON.stringify(userinfoRef);
                $.ajax({
                    type: "POST",
                    data: jsonString,
                    url: "getJSON1.spring?t=" + new Date().getTime(),
                    contentType: "application/json"
                });
            }

            function sendAjax2(){
                var myArray = new Array();
                myArray[0] = "中国1";
                myArray[1] = "中国2";
                myArray[2] = "中国3";
                myArray[3] = "中国4";

                var jsonString = JSON.stringify(myArray);
                $.ajax({
                    type: "POST",
                    data: jsonString,
                    url: "getJSON2.spring?t=" + new Date().getTime(),
                    contentType: "application/json"
                });
            }

            function sendAjax3(){
                var myArray = new Array();
                myArray[0] = new userinfo("中国1", "中国人1");
                myArray[1] = new userinfo("中国2", "中国人2");
                myArray[2] = new userinfo("中国3", "中国人3");
                myArray[3] = new userinfo("中国4", "中国人4");


                var jsonString = JSON.stringify(myArray);
                $.ajax({
                    type: "POST",
```

12.10 在服务器端取得 JSON 字符串并解析——方式 2

```
                    data: jsonString,
                    url: "getJSON3.spring?t=" + new Date().getTime(),
                    contentType: "application/json"
                });
            }

            function sendAjax4(){
                var jsonObject = {
                    "username": "accp",
                    "work": [{
                        "address": "address1"
                    }, {
                        "address": "address2"
                    }],
                    "school": {
                        "name": "tc",
                        "address": "pjy"
                    }
                }

                var jsonString = JSON.stringify(jsonObject);
                $.ajax({
                    type: "POST",
                    data: jsonString,
                    url: "getJSON4.spring?t=" + new Date().getTime(),
                    contentType: "application/json"
                });
            }
        </script>
    </head>
    <body>
        <input type="button" onclick="sendAjax1()" value="sendAjax1"/>
        <br/>
        <input type="button" onclick="sendAjax2()" value="sendAjax2"/>
        <br/>
        <input type="button" onclick="sendAjax3()" value="sendAjax3"/>
        <br/>
        <input type="button" onclick="sendAjax4()" value="sendAjax4"/>
        <br/>
    </body>
</html>
```

后台 Java 代码如下。

```
package controller;

import java.util.ArrayList;
import java.util.List;
import java.util.Map;

import org.springframework.stereotype.Controller;
import org.springframework.web.bind.annotation.RequestBody;
import org.springframework.web.bind.annotation.RequestMapping;

import entity.Userinfo;

@Controller
public class Test {
```

```java
@RequestMapping(value = "getJSON1")
public void getJSON1(@RequestBody Userinfo userinfo) {
    System.out.println(userinfo.getUsername());
    System.out.println(userinfo.getPassword());
}

@RequestMapping(value = "getJSON2")
public void getJSON2(@RequestBody ArrayList<String> list) {
    for (int i = 0; i < list.size(); i++) {
        System.out.println(list.get(i));
    }
}

@RequestMapping(value = "getJSON3")
public void getJSON3(@RequestBody List<Map> list) {
    for (int i = 0; i < list.size(); i++) {
        Map map = list.get(i);
        System.out.println(map.get("username") + " " + map.get("password"));
    }
}

@RequestMapping(value = "getJSON4")
public void getJSON4(@RequestBody Map map) {
    System.out.println(map.get("username"));
    List<Map> workList = (List) map.get("work");
    for (int i = 0; i < workList.size(); i++) {
        Map eachAddressMap = workList.get(i);
        System.out.println("address=" + eachAddressMap.get("address"));
    }

    Map schoolMap = (Map) map.get("school");
    System.out.println(schoolMap.get("name"));
    System.out.println(schoolMap.get("address"));
}
}
```

12.11 将 URL 中的参数转成实体的示例

在 Spring 4 MVC 中还可以将 URL 中的参数转成实体,此实验在项目 paramToEntity 中进行实现。

12.11.1 新建控制层文件

新建名称为 ParamToEntity.java 的控制层文件，代码如下。

```java
@Controller
public class ParamToEntity {
    @RequestMapping(value = "paramToEntity", method = RequestMethod.POST)
    public String paramToEntity(Userinfo userinfo) {
        System.out.println("username value=" + userinfo.getUsername());
        System.out.println("password value=" + userinfo.getPassword());
        return "test.jsp";
    }
}
```

12.11.2 新建登录用途的 JSP 文件

新建 index.jsp 文件，核心代码如下。

```
<form action="paramToEntity.spring" method="post">
    username:
    <input type="text" name="username">
    <br />
    password:
    <input type="text" name="password">
    <br />
    <input type="submit" name="submit">
    <br />
</form>
```

12.11.3 在 web.xml 中注册编码过滤器

在 web.xml 中注册编码过滤器防止提交中文时出现乱码，配置如下。

```
<filter>
    <filter-name>encodingFilter</filter-name>
    <filter-class>org.springframework.web.filter.CharacterEncodingFilter</filter-class>
    <init-param>
        <param-name>encoding</param-name>
        <param-value>utf-8</param-value>
    </init-param>
</filter>
<filter-mapping>
    <filter-name>encodingFilter</filter-name>
    <url-pattern>/*</url-pattern>
</filter-mapping>
```

12.11.4 运行结果

在 IE 地址栏中输入网址 http://localhost:8081/paramToEntity/，打开登录界面，如图 12-26 所示。

单击"提交"按钮后，在控制台输出转成 Userinfo 实体的属性值，如图 12-27 所示。

图 12-26 打开登录界面

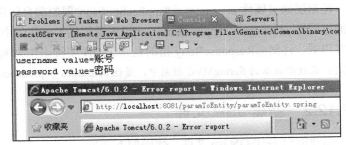

图 12-27 打印 Userinfo 属性值

12.12 在控制层返回 JSON 对象示例

有时候需要在控制层以响应的方式返回 JSON 对象，比如返回学生信息列表等信息。本示例就在 createJSON 项目中进行测试。

12.12.1 新建控制层文件

新建控制层文件 CreateJSON.java，代码如下。

```java
package controller;

import java.util.ArrayList;

import org.springframework.stereotype.Controller;
import org.springframework.web.bind.annotation.RequestBody;
import org.springframework.web.bind.annotation.RequestMapping;
import org.springframework.web.bind.annotation.RequestMethod;
import org.springframework.web.bind.annotation.ResponseBody;
import entity.Userinfo;

@Controller
public class CreateJSON {
    @RequestMapping(value = "createJSONURL", method = RequestMethod.POST, consumes = "application/json", produces = "application/json")
    @ResponseBody
    public Userinfo createJSON(@RequestBody Userinfo userinfoParam) {
        System.out.println("param username=" + userinfoParam.getUsername()
                + " password=" + userinfoParam.getPassword());

        Userinfo userinfo = new Userinfo();
        userinfo.setUsername("高洪岩中文");
        userinfo.setPassword("我的密码中文");
        userinfo.setStudyList(new ArrayList<String>());
        userinfo.getStudyList().add("Java");
        userinfo.getStudyList().add("Delphi");
        userinfo.getStudyList().add("HTML");
        userinfo.getStudyList().add("JQuery+JavaScript");
        return userinfo;
    }

}
```

控制层通过属性 consumes = "application/json"限制前台传递过来的数据格式必须是 JSON，又通过属性 produces = "application/json"设置返回的 Userinfo 中的数据要转成 JSON 对象并回传给客户端。注解@ResponseBody 指的是将 JSON 字符串作为响应处理。

注意：在最新版本的 SpringMVC 框架中，属性 consumes 和 produces 都要写。

12.12.2 新建 JSP 文件

新建前台 getJSON.jsp 文件，代码如下。

```jsp
<%@ page language="java" import="java.util.*" pageEncoding="utf-8" %>
<!DOCTYPE HTML PUBLIC "-//W3C//DTD HTML 4.01 Transitional//EN">
<html>
    <head>
        <script src="jquery-1.3.2.js">
        </script>
        <script src="json2.js">
```

```
        </script>
        <script>
            function userinfo(username, password){
                this.username = username;
                this.password = password;
            }

            function sendAjax(){
                var userinfoRef = new userinfo('ghy', '678');
                var jsonStringRef = JSON.stringify(userinfoRef);

                $.ajax({
                    type: "POST",
                    data: jsonStringRef,
                    url: "createJSONURL.spring?t=" + new Date().getTime(),
                    contentType: "application/json",
                    dataType: "json",
                    success: function(data, type){
                        alert(data.username + " " + data.password);
                        var aihaoArray = data.studyList;
                        for (var i = 0; i < aihaoArray.length; i++) {
                            alert(aihaoArray[i]);
                        }
                    }
                });
            }
        </script>
    </head>
    <body>
        <input type="button" onclick="sendAjax()"/>
    </body>
</html>
```

在这里需要注意的是,尽量不要使用如下的 jQuery 里 Ajax 调用方法,因为同步调用会影响效率。

```
//建议不要使用,因为此方式使用同步调用,影响效率
function notUseIt(){
    var returnJSONObject = $.ajax({
        type: "POST",
        url: "getJSON.spring?t=" + new Date().getTime(),
        data: jsonString,
        contentType: "application/json",
        dataType: "json",
        async: false
    }).responseText;
    var jsonObject = JSON.parse(returnJSONObject);
    alert(jsonObject.username);
}
```

12.12.3 部署项目并运行

布署项目,在 IE 中输出 URL 网址 http://localhost:8081/createJSON/getJSON.jsp,单击 IE 浏览器中的"button"按钮,控制台输出接收到的参数值,如图 12-28 所示。

前台输出返回 JSON 对象的属性,如图 12-29 所示。

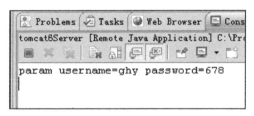

图 12-28　输出接收到的参数值

图 12-29　输出返回 JSON 对象的属性值

12.13　在控制层传回 JSON 字符串示例

在前面小节中实现的是在 response 对象中返回 JSON 对象，本示例要在 response 对象中返回 JSON 字符串。其实返回 JSON 对象和 JSON 字符串都可以在前端进行处理，只是每个程序员的习惯不一样，但结果都是相同的。

12.13.1　新建控制层文件

新建 Web 项目，名称为 returnJSONString。
创建控制层文件 ReturnJSONString.java，代码如下。

```java
package controller;

import java.util.ArrayList;

import net.sf.json.JSONObject;

import org.springframework.stereotype.Controller;
import org.springframework.web.bind.annotation.RequestMapping;
import org.springframework.web.bind.annotation.ResponseBody;

import entity.Userinfo;
@Controller
public class ReturnJSONString {
    @RequestMapping(value = "returnJSONString", produces = "text/html;charset=utf-8")
    @ResponseBody
    public String createJSON() {
        Userinfo userinfo = new Userinfo();
        userinfo.setUsername("我是中文高洪岩");
        userinfo.setPassword("123");
        userinfo.setStudyList(new ArrayList<String>());
        userinfo.getStudyList().add("Java");
        userinfo.getStudyList().add("Delphi");
        userinfo.getStudyList().add("HTML");
        userinfo.getStudyList().add("JQuery+JavaScript");
        return JSONObject.fromObject(userinfo).toString();
    }
}
```

12.13.2　新建 JSP 文件及在配置文件中注册 utf-8 编码处理

创建名称为 getJSONString.jsp 的 JSP 文件，代码如下。

12.14 在控制层取得 HttpServletRequest 和 HttpServletResponse 对象

```
<%@ page language="java" import="java.util.*" pageEncoding="utf-8" %>
<!DOCTYPE HTML PUBLIC "-//W3C//DTD HTML 4.01 Transitional//EN">
<html>
    <head>
        <script src="jquery-1.3.2.js">
        </script>
        <script src="json2.js">
        </script>
        <script>
            function sendAjax(){
                $.ajax({
                    type: "GET",
                    url: "returnJSONString.spring?t=" + new Date().getTime(),
                    success: function(data, textStatus){
                        var jsonObject = JSON.parse(data);
                        alert("returnJSONString=" + data + "  value=" + jsonObject.username + " " + jsonObject.password + " " + jsonObject.studyList[0] + " " + jsonObject.studyList[3]);
                    }
                });
            }
        </script>
    </head>
    <body>
        <input type="button" onclick="sendAjax()"/>
    </body>
</html>
```

需要注意的是在 springMVC-servlet.xml 中要配置 mvc:annotation，不然中文问题无法解决。

12.13.3 运行项目

部署项目并运行，输入网址 http://localhost:8081/returnJSONString/getJSONString.jsp，单击 "button" 按钮后，弹出 JSON 对象中的属性值，如图 12-30 所示。

图 12-30 程序运行结果

12.14 在控制层取得 HttpServletRequest 和 HttpServletResponse 对象

有时候需要在控制层取得 HttpServletRequest 和 HttpServletResponse 对象，进而使用这两个对象的 API 方法。

新建名称为 getHttpServletRequest 的 Web 项目。

12.14.1 新建控制层

新建控制层文件 GetRequest.java 的代码如下。

```java
package controller;

import javax.servlet.http.HttpServletRequest;
import javax.servlet.http.HttpServletResponse;

import org.springframework.stereotype.Controller;
import org.springframework.web.bind.annotation.RequestMapping;

@Controller
public class GetRequest {

    @RequestMapping("/getRequest1")
    public String getRequest1(HttpServletRequest request,
            HttpServletResponse response) {
        System.out.println("request1=" + request);
        System.out.println("response1=" + response);
        request.setAttribute("username", "高洪岩1");
        return "index.jsp";
    }
}
```

12.14.2 JSP 文件中的 EL 代码及运行结果

文件 index.jsp 的代码如下。

```
<body>
    ${username }
</body>
```

程序运行后，在控制台及 IE 界面输出相关的数据信息，如图 12-31 所示。

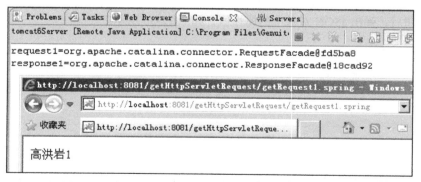

图 12-31 输出结果

12.14.3 直接使用 HttpServletResopnse 对象输出响应字符

通过此实验可以得到另外一个开发方式，就是使用 HttpServletResopnse 对象来输出字符串。把字符串数据做为 Ajax 请求的响应字符，从而在客户端进行进一步处理。此实验在项目 useResponsePrintString 中进行测试。

12.14　在控制层取得 HttpServletRequest 和 HttpServletResponse 对象

创建前台 login.jsp 文件的代码如下。

```jsp
<%@ page language="java" import="java.util.*" pageEncoding="utf-8" %>
<!DOCTYPE HTML PUBLIC "-//W3C//DTD HTML 4.01 Transitional//EN">
<html>
    <head>
        <script src="jquery-1.3.2.js">
        </script>
        <script src="json2.js">
        </script>
        <script>
            function loginUserinfo(username, password){
                this.username = username;
                this.password = password;
            }

            function loginMethod(){
                var usernameValue = "ghy";
                var passwordValue = "789";

                var loginUserinfoRef = new loginUserinfo(usernameValue, passwordValue);
                var jsonLoginString = JSON.stringify(loginUserinfoRef);
                $.post("login.spring", {
                    jsonLoginString: jsonLoginString
                }, function(data){
                    alert(data);
                });
            }
        </script>
    </head>
    <body>
        <input type="button" value="登录" onclick="loginMethod()">
    </body>
</html>
```

控制层文件 Login.java 的代码如下。

```java
package controller;

import java.io.IOException;
import java.io.PrintWriter;

import javax.servlet.http.HttpServletResponse;

import net.sf.json.JSONObject;

import org.springframework.stereotype.Controller;
import org.springframework.web.bind.annotation.RequestMapping;

@Controller
public class Login {
    @RequestMapping("/login")
    public void login(String jsonLoginString, HttpServletResponse response) {
        try {
            JSONObject jsonObject = JSONObject.fromObject(jsonLoginString);
            String username = jsonObject.getString("username");
            String password = jsonObject.getString("password");
            System.out.println("username=" + username + " passwword="
                    + password);
```

```
                response.setContentType("text/html");
                response.setCharacterEncoding("utf-8");
                PrintWriter out = response.getWriter();
                out.println("我是中文高洪岩");
                out.flush();
                out.close();
            } catch (IOException e) {
                // TODO Auto-generated catch block
                e.printStackTrace();
            }
        }
    }
```

程序运行后,单击 IE 界面上的"Button"按钮,运行结果如图 12-32 所示。

图 12-32　取参与响应结果

12.15　通过 URL 参数访问指定的业务方法

在 Spring 4 MVC 中可以实现,在访问同一个 URL 地址的同时以传递参数的方式来调用指定控制层中指定方法的目的。使用@RequestMapping 注解很容易实现这样的需求,此实验在项目 oneURLaccessMethoFromParam 中进行实现。

12.15.1　新建控制层文件 List.java

文件 List.java 的代码如下。

```
package controller;

import org.springframework.stereotype.Controller;
import org.springframework.web.bind.annotation.RequestMapping;

@Controller
public class List {
    @RequestMapping(value = "/list", params = "type=a")
    public String listA() {
        System.out.println("listA");
        return "index.jsp";
    }

    @RequestMapping(value = "/list", params = "type=b")
```

```
    public String listB() {
        System.out.println("listB");
        return "index.jsp";
    }
}
```

控制层中有两个业务方法，如何调用这两个业务方法呢？

12.15.2 运行结果

使用 URL：http://localhost:8081/oneURLaccessMethoFromParam/list.spring?type=a 来进行访问，输出结果如图 12-33 所示。

图 12-33 输出结果 A

使用 URL：http://localhost:8081/oneURLaccessMethoFromParam/list.spring?type=b 来进行访问，输出结果如图 12-34 所示。

图 12-34 输出结果 B

12.16 Spring 4 MVC 单文件上传——写法 1

Spring 4 MVC 也支持文件的上传，演示的项目名称为 springOneFileUpload。

12.16.1 新建控制层

控制层文件 FileUpload.java 的代码如下。

```
package controller;

import java.io.File;
import java.io.FileOutputStream;
```

```java
import java.io.IOException;
import java.io.InputStream;

import javax.servlet.http.HttpServletResponse;

import org.apache.commons.io.IOUtils;
import org.springframework.stereotype.Controller;
import org.springframework.web.bind.annotation.RequestMapping;
import org.springframework.web.multipart.MultipartFile;
import org.springframework.web.multipart.MultipartHttpServletRequest;

@Controller
public class FileUpload {
    @RequestMapping("/oneFileUpload")
    public void oneFileUpload(MultipartHttpServletRequest request,
            HttpServletResponse response) {
        try {
            System.out.println("username=" + request.getParameter("username"));
            MultipartFile file = request.getFile("uploadFile");
            String uploadFileName = file.getOriginalFilename();

            InputStream isRef = file.getInputStream();

            String targetDir = request.getSession().getServletContext()
                    .getRealPath("/upload");
            File targetFile = new File(targetDir, uploadFileName);
            FileOutputStream fosRef = new FileOutputStream(targetFile);
            IOUtils.copy(isRef, fosRef);

        } catch (IOException e) {
            // TODO Auto-generated catch block
            e.printStackTrace();
        }

    }
}
```

12.16.2 在配置文件 springMVC-servlet.xml 中声明上传请求

```xml
<bean id="multipartResolver"
    class="org.springframework.web.multipart.commons.CommonsMultipartResolver">
    <property name="maxUploadSize" value="2048000000" />
</bean>
```

文件上传最大为 2GB，如果不在 XML 中添加上面的配置则在运行时会出现异常。
java.lang.IllegalStateException: Current request is not of type [org.springframework.web.multipart.MultipartHttpServletRequest]: org.apache.catalina.connector.RequestFacade@7ac8095f

另外为了解决上传文件时单行文本域有乱码的问题，所以还需要在 web.xml 中声明编码过滤器，代码如下。

```xml
<filter>
    <filter-name>encodingFilter</filter-name>
    <filter-class>org.springframework.web.filter.CharacterEncodingFilter</filter-class>
    <init-param>
        <param-name>encoding</param-name>
        <param-value>utf-8</param-value>
    </init-param>
```

```
    </filter>
    <filter-mapping>
        <filter-name>encodingFilter</filter-name>
        <url-pattern>/*</url-pattern>
    </filter-mapping>
```

12.16.3 创建前台 JSP 文件

前台 JSP 文件的代码如下。

```
<form action="oneFileUpload.spring" method="post"
    enctype="multipart/form-data">
    <input type="text" name="username">
    <input type="file" name="uploadFile">
    <input type="submit" value="开始上传">
</form>
```

12.16.4 程序运行结果

程序运行后的结果如图 12-35 所示。

图 12-35　程序运行结果

12.17　Spring 4 MVC 单文件上传——写法 2

上传文件还有另外一种写法，演示的项目名称为 springOneFileUpload_Other_A。

新建控制层

新建控制层 FileUpload.java 文件，核心代码如下。

```
@Controller
public class FileUpload {
    @RequestMapping("/oneFileUpload")
    public void oneFileUpload(@RequestParam String username,
            @RequestParam MultipartFile uploadFile, HttpServletRequest request) {
        try {
            System.out.println("username=" + username);
```

```java
            MultipartFile file = uploadFile;
            String uploadFileName = file.getOriginalFilename();

            InputStream isRef = file.getInputStream();

            String targetDir = request.getSession().getServletContext()
                    .getRealPath("/upload");
            File targetFile = new File(targetDir, uploadFileName);
            FileOutputStream fosRef = new FileOutputStream(targetFile);
            IOUtils.copy(isRef, fosRef);
        } catch (IOException e) {
            // TODO Auto-generated catch block
            e.printStackTrace();
        }

    }
}
```

注意：使用此写法时：

```
@RequestParam MultipartFile uploadFile
```

一定要注意前台<input type="file">文件域的 name 名称一定要和参数名称 uploadFile 一样。还有必须在 MultipartFile uploadFile1 对象前要加@RequestParam 注解，不加会出现异常。

```
org.springframework.beans.BeanInstantiationException: Could not instantiate bean class
[org.springframework.web.multipart.MultipartFile]: Specified class is an interface
```

程序运行后也可以正确地上传文件。

12.18　Spring 4 MVC 多文件上传

Spring 4 MVC 中还可以支持多文件上传，示例代码在 springMultiFileUpload 项目中。

12.18.1　新建控制层及 JSP 文件

创建控制层 FileUpload.java 文件，核心代码如下。

```java
@Controller
public class FileUpload {
    @RequestMapping("/moreFileUpload1")
    public String oneFileUpload(MultipartHttpServletRequest request,
            HttpServletResponse response) {
        try {
            System.out.println("username=" + request.getParameter("username"));

            Map<String, MultipartFile> fileMap = request.getFileMap();
            System.out.println("文件个数为: " + fileMap.size());
            Set<String> fileSet = fileMap.keySet();
            Iterator<String> fileNameIterator = fileSet.iterator();
            while (fileNameIterator.hasNext()) {
                String uploadFileName = fileNameIterator.next();
                System.out.println(uploadFileName);
                MultipartFile file = fileMap.get(uploadFileName);
                uploadFileName = file.getOriginalFilename();
                InputStream isRef = file.getInputStream();
```

```
            String targetDir = request.getSession().getServletContext()
                .getRealPath("/upload");
            SimpleDateFormat sdf = new SimpleDateFormat(
                "yyyy_MM_dd_hh_mm_ss");
            String getDateString = sdf.format(new Date());
            File targetFile = new File(targetDir, "" + getDateString + "_"
                + System.nanoTime() + "_" + uploadFileName);
            FileOutputStream fosRef = new FileOutputStream(targetFile);
            IOUtils.copy(isRef, fosRef);
        }
    } catch (IOException e) {
        // TODO Auto-generated catch block
        e.printStackTrace();
    }
    return "index.jsp";
}
```

前台 index.jsp 文件的代码如下。

```
<form action="moreFileUpload1.spring" method="post"
    enctype="multipart/form-data">
    <input type="text" name="username">
    <br />
    <input type="file" name="uploadFile1">
    <br />
    <input type="file" name="uploadFile2">
    <br />
    <input type="file" name="uploadFile3">
    <br />
    <input type="file" name="uploadFile4">
    <br />
    <input type="submit" value="开始上传">
</form>
```

12.18.2 运行结果

程序运行后的结果如图 12-36 所示。

图 12-36 程序运行结果

12.19 Spring 4 MVC 支持中文文件名的文件下载

创建测试用的项目 springMVCDownloadFile，创建下载文件的控制层，核心代码如下。

```java
package controller;

import java.io.File;
import java.io.FileInputStream;
import java.io.FileNotFoundException;
import java.io.IOException;
import java.io.UnsupportedEncodingException;
import java.net.URLEncoder;

import javax.servlet.ServletOutputStream;
import javax.servlet.http.HttpServletRequest;
import javax.servlet.http.HttpServletResponse;

import org.apache.commons.io.IOUtils;
import org.springframework.stereotype.Controller;
import org.springframework.web.bind.annotation.RequestMapping;

@Controller
public class UserinfoController {
    @RequestMapping(value = "downloadFile")
    public void testA(String fileName, HttpServletRequest request, HttpServletResponse response)
            throws UnsupportedEncodingException {
        try {
            String downPath = request.getSession().getServletContext().getRealPath("/");
            fileName = fileName.replace("_", "%");
            fileName = java.net.URLDecoder.decode(fileName, "utf-8");
            System.out.println(fileName);
            String downfileName = "";
            if (request.getHeader("USER-AGENT").toLowerCase().indexOf("msie") > 0) {// IE
                fileName = URLEncoder.encode(fileName, "UTF-8");
                downfileName = fileName.replace("+", "%20");// 处理空格变"+"的问题
            } else {// FF
                downfileName = new String(fileName.getBytes("UTF-8"), "ISO-8859-1");
            }
            System.out.println(downPath + fileName);
            File downloadFile = new File(downPath + fileName);
            response.setContentType("application/octet-stream;");
            response.setHeader("Content-disposition", String.format("attachment;filename=\"%s\"", downfileName)); // 文件名外的双引号处理firefox的空格截断问题
            response.setHeader("Content-Length", String.valueOf(downloadFile.length()));
            FileInputStream fis = new FileInputStream(downloadFile);
            ServletOutputStream out = response.getOutputStream();
            IOUtils.copy(fis, out);
        } catch (FileNotFoundException e) {
            e.printStackTrace();
        } catch (IOException e) {
            e.printStackTrace();
        }
    }
}
```

前台 JSP 文件的代码如下。

```
<body>
    <a
        href="downloadFile.spring?fileName=<%=java.net.URLEncoder.encode("中国.zip", "utf-8").toString().replace("%", "_")%>">中国.zip</a>
    <br />
```

```html
    <a href="downloadFile.spring?fileName=a.zip">a.zip</a>
    <br />
</body>
```

12.20 控制层返回 List 对象及实体的效果

在 Spring 4 MVC 中的控制层还可以返回 Java 的数据类型，比如返回 List 或 Userinfo.java 自定义实体等，而且还可以自动转发到 JSP 页面。本测试在项目 returnListPutRequestScope 中进行。

12.20.1 新建控制层文件

新建控制层文件 ListTest.java，核心代码如下。

```java
@Controller
public class ListTest {

    @RequestMapping(value = "listUserinfo")
    public List<String> listUserinfo() {
        List<String> listString = new ArrayList<String>();
        for (int i = 0; i < 10; i++) {
            listString.add("username" + (i + 1));
        }
        return listString;
        // 返回 List<String>数据类型
        // 所以要在项目中自动找名称为 listUserinfo.jsp 的 JSP 文件
        // List<String> listString 对象还要自动放入 request
        // 作用域中，key 的名称为 stringList，因为
        // List 中存放的是 string
    }

    @RequestMapping(value = "getUserinfo")
    public Userinfo getUserinfo() {
        Userinfo userinfo = new Userinfo("a", "aa");
        return userinfo;
    }

}
```

12.20.2 新建 JSP 文件

创建 JSP 文件 listUserinfo.jsp，核心代码如下。

```jsp
<body>
    ${stringList }
    <br />
    <br />
    <%
        Enumeration enum1 = request.getAttributeNames();
        while (enum1.hasMoreElements()) {
            out.println(enum1.nextElement() + "<br/>");
        }
    %>
</body>
```

创建 JSP 文件 getUserinfo.jsp，核心代码如下。

```
<body>
    ${userinfo }
    <br />
    <br />
    <%
        Enumeration enum1 = request.getAttributeNames();
        while (enum1.hasMoreElements()) {
            out.println(enum1.nextElement() + "<br/>");
        }
    %>
</body>
```

12.20.3　更改 springMVC-servlet.xml 配置文件

配置文件 springMVC-servlet.xml 配置更改如下。

```
<bean
    class="org.springframework.web.servlet.view.InternalResourceViewResolver"
    p:prefix="/" p:suffix=".jsp" />
<context:component-scan base-package="controller" />
```

12.20.4　程序运行结果

部署项目运行程序，输入 http://localhost:8081/returnListPutRequestScope/listUserinfo.spring，程序运行结果如图 12-37 所示。

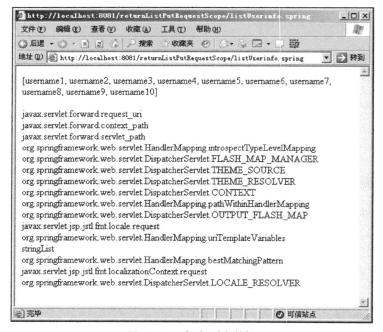

图 12-37　打印列表数据

继续输入 http://localhost:8081/returnListPutRequestScope/getUserinfo.spring，程序运行结果如图 12-38 所示。

12.21 控制层 ModelMap 对象

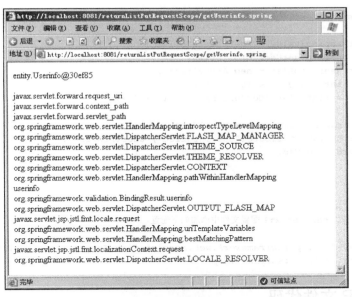

图 12-38 打印实体数据

当然，如果觉得自动放入 request 的 key 并不是自定义的不太好，还可以将代码更改如下。

```
@ModelAttribute(value="abc")
@RequestMapping(value = "listUserinfo")
public List<String> listUserinfo() {
    List<String> listString = new ArrayList<String>();
    for (int i = 0; i < 10; i++) {
        listString.add("username" + (i + 1));
    }
    return listString;
    // 返回 List<String>数据类型
    // 所以要在项目中自动找名称为 listUserinfo.jsp 的 JSP 文件
    // List<String> listString对象还要自动放入 request
    // 作用域中，key 的名称为 stringList，因为
    // List 中存放的是 string
}
```

加入注解@ModelAttribute(value="abc")相当于代码 request.setAttribute("abc",list);将对象 list 以 key 为 abc 放入 request 作用域中，那么前台 JSP 也要改成${abc}来取得对应的值。

12.21 控制层 ModelMap 对象

前面的章节介绍了控制层返回 List 或 Userinfo.java 自定义实体，其实还可以返回 ModelMap 对象。此对象也是自动往 request 中存放数据，与 Model 对象有些相似，只不过一个是返回值，一个是方法的参数。演示的代码在项目 returnModelMap 中。

12.21.1 新建控制层

创建控制层文件 ListTest.java，核心代码如下。

```
@Controller
```

```java
public class ListTest {

    @RequestMapping(value = "getUserinfoList")
    public ModelMap getUserinfoList() {
        List userinfoList = new ArrayList();
        for (int i = 0; i < 10; i++) {
            userinfoList.add("username" + (i + 1));
        }
        ModelMap modelMap = new ModelMap();
        modelMap.addAttribute("userinfoList", userinfoList);
        return modelMap;
    }

}
```

还要在 springMVC-servlet.xml 配置文件中添加如下配置。

```xml
<bean
    class="org.springframework.web.servlet.view.InternalResourceViewResolver"
    p:prefix="/" p:suffix=".jsp" />
```

12.21.2 JSP 文件代码

创建默认转发的 JSP 文件 getUserinfoList.jsp 的代码如下。

```jsp
<body>
    ${userinfoList }
    <br />
    <br />
    <%
        Enumeration enum1 = request.getAttributeNames();
        while (enum1.hasMoreElements()) {
            out.println(enum1.nextElement() + "<br/>");
        }
    %>
</body>
```

12.21.3 运行效果

输入 http://localhost:8081/returnModelMap/getUserinfoList.spring，在 IE 输出相关的数据信息，如图 12-39 所示。

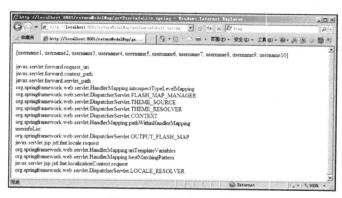

图 12-39　输出相关信息

12.22 Spring 4 MVC 提交的表单进行手动数据验证

虽然 Spring 4 MVC 框架提供了验证框架来作为前台参数的数据有效性验证，但由于在操作性上如果存在业务型的验证还得需要手动写代码的方式来进行处理，那么本示例演示如何用手动的方式来验证前台传递过来的登录信息。

新建名称为 validateNull 的 Web 项目。

12.22.1 创建控制层文件

创建控制层文件 Login.java，核心代码如下。

```java
@Controller
public class Login {

    private Map validateLogin(LoginInfo loginInfo) {
        Map errorMap = new HashMap();
        if (loginInfo.getUsername() == null
                || "".equals(loginInfo.getUsername())) {
            errorMap.put("usernameIsNull", "账号为空!");
        }
        if (loginInfo.getPassword() == null
                || "".equals(loginInfo.getPassword())) {
            errorMap.put("passwordIsNull", "密码为空!");
        }
        return errorMap;
    }

    @RequestMapping("/login")
    public String login(LoginInfo loginInfo, Model model) {
        Map errorMap = validateLogin(loginInfo);
        if (errorMap.size() > 0) {
            model.addAttribute("errorMap", errorMap);
            return "login.jsp";
        }

        return "index.jsp";
    }
}
```

12.22.2 创建 JSP 文件

文件 login.jsp 的核心代码如下。

```jsp
<body>
    <form action="login.spring" method="get">
        username:
        <input type="text" name="username">${errorMap.usernameIsNull}
        <br />
        password:
        <input type="text" name="password">${errorMap.passwordIsNull}
        <br />
        <input type="submit" value="submit">
        <br />
    </form>
```

```
        <br />
        <br />
</body>
```

12.22.3 运行结果

如果账号和密码不输入,则返回 login.jsp 页面,显示出错信息,结果如图 12-40 所示。

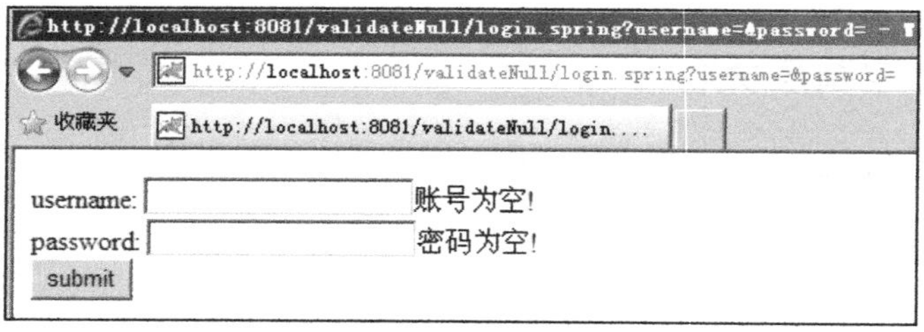

图 12-40 显示出错信息

第 13 章 Spring 4 MVC 必备知识

本章也是 Spring 4 MVC 的扩展内容，代码也较常用，而且出现率较多，比如查看公司同事写的代码时，学习网上 Spring 4 MVC 的资料时。掌握本章内容有助于使用 Spring 4 MVC 开发 Web 软件项目。

在本章读者应该掌握如下内容：
- 对 Spring 4 MVC 的控制层注入业务对象；
- Spring 4 MVC 中如何操作 HttpSession 对象；
- 方法返回 void 的情况；
- 多种使用配置文件的情况。

13.1 web.xml 中的不同配置方法

在前面的章节中都是在 web.xml 中添加如下的 Spring 4 MVC 配置映射。

```
<servlet>
    <servlet-name>springMVC</servlet-name>
    <servlet-class>org.springframework.web.servlet.DispatcherServlet</servlet-class>
    <load-on-startup>1</load-on-startup>
</servlet>
<servlet-mapping>
    <servlet-name>springMVC</servlet-name>
    <url-pattern>*.spring</url-pattern>
</servlet-mapping>
```

然后 Spring 框架自动在 WEB-INF 文件夹中寻找 springMVC-servlet.xml 配置文件，这种是默认的使用方式，示例代码在项目 mvc1 中。

其实在 Spring 4 MVC 中还有其他几种 web.xml 配置映射的方法。

13.1.1 存放于 src 资源路径中

将配置文件存放于 src 资源路径中的配置如图 13-1 所示。

图 13-1　配置文件放入 src 路径中

示例源代码在项目 mvc2 中。

13.1.2　指定存放路径

还可以将 Spring 4 MVC 的配置文件放入指定的路径中，如图 13-2 所示。

图 13-2　Spring 4 MVC 配置文件存放在指定路径中

13.1.3　指定多个配置文件

还可以将大的配置文件拆分成小的配置文件，然后在 web.xml 文件中一一进行注册，使用方式如图 13-3 所示。

13.2 路径中添加通配符的功能

图 13-3 多个配置文件的使用

示例源代码在项目 mvc3 中。

13.2 路径中添加通配符的功能

还可以在访问映射路径中添加通配符的功能，实验代码在项目 url_xing 中。
控制层核心代码如下。

```java
@Controller
public class FindUserinfo {
    @RequestMapping(value = "/findUserinfo_*")
    public String findUserinfo(HttpServletRequest request) {
        System.out.println("URI=" + request.getRequestURI());
        String URI = request.getRequestURI();
        int beginIndex = request.getRequestURI().lastIndexOf("_");
        int endIndex = request.getRequestURI().lastIndexOf(".");
        String param = URI.substring(beginIndex + 1, endIndex);
        System.out.println("id=" + param);
        return "index.jsp";
    }
}
```

可以通过如下的 URL 进行访问。

- http://localhost:8081/url_xing/findUserinfo_1.spring。
- http://localhost:8081/url_xing/findUserinfo_99.spring。

控制台输出的结果如图 13-4 所示。

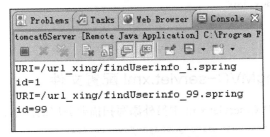

图 13-4 控制台输出通配符的效果

13.3 业务逻辑层在控制层中进行注入

前面的示例都是在控制层中进行业务的处理，下面的代码将实现在控制层中通过注入的方式注入业务逻辑层来实现功能上的处理，示例代码在项目 setService 中。

13.3.1 新建业务逻辑层

创建文件 UserinfoService.java，代码如下。

```java
package service;

import java.util.ArrayList;
import java.util.List;

import org.springframework.stereotype.Service;

@Service
public class UserinfoService {
    public List getAllUserinfo() {
        List userinfoList = new ArrayList();
        for (int i = 0; i < 10; i++) {
            userinfoList.add("userinfo" + (i + 1));
        }
        return userinfoList;
    }
}
```

使用注解@Service 代表本类是一个业务对象。

13.3.2 创建控制层文件

文件 List.java 的核心代码如下。

```java
@Controller
public class List {

    @Autowired
    private UserinfoService userinfoService;

    @RequestMapping(value = "getUserinfoList")
    public String getUserinfoList(Model model) {
        java.util.List userinfoList = userinfoService.getAllUserinfo();
        model.addAttribute("userinfoList", userinfoList);
        return "index.jsp";
    }
}
```

13.3.3 设计 springMVC-servlet.xml 配置文件

在配置文件 springMVC-servlet.xml 中另外添写扫描业务层的注解，完整代码如下。

```xml
<context:component-scan base-package="controller" />
<context:component-scan base-package="service" />
```

13.3.4 程序运行结果

程序运行后，结果如图 13-5 所示。

图 13-5　列出用户信息

13.3.5 多个实现类的情况

前面的示例仅仅实现的是一个实现类，在项目中基于业务经常变化的情况，多个实现类也是经常出现的。演示的代码在项目 setService2 中。

新建业务接口及实现类，代码如图 13-6 所示。

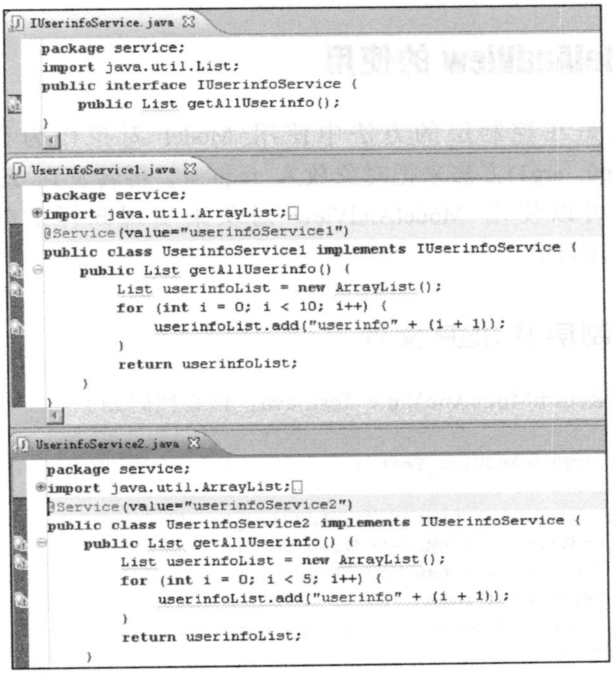

图 13-6　业务接口及两个实现类

在控制层明确注明引用的业务对象别名,核心代码如下。

```
@Controller
public class List {

    @Resource(name = "userinfoService2")
    private IUserinfoService userinfoService;

    @RequestMapping(value = "getUserinfoList")
    public String getUserinfoList(Model model) {
        java.util.List userinfoList = userinfoService.getAllUserinfo();
        model.addAttribute("userinfoList", userinfoList);
        return "index.jsp";
    }

}
```

输入网址 http://localhost:8081/setService2/getUserinfoList.spring,输出具有 5 个元素的 List 对象中的数据,结果如图 13-7 所示。

图 13-7　输出 5 个元素的 List 对象

13.4　对象 ModelAndView 的使用

前面的章节都是在控制层的方法中使用 Model 对象作为参数,再使用它的 model.addAttribute(arg0, arg1)方法来把对象放入 request,再转发到 JSP 文件中。其实在 Spring 4 MVC 中还可以使用 ModelAndView 对象来实现同样的效果。示例代码在 returnModelAndView 项目中。

13.4.1　创建控制层及 JSP 文件

创建控制层文件 ReturnModeAndView_Test.java,核心代码如下。

```
@Controller
public class ReturnModeAndView_Test {

    @RequestMapping(value = "returnModeAndView_Test")
    public ModelAndView returnModeAndView_Test() {
        ModelAndView mvRef = new ModelAndView();
        mvRef.setViewName("index.jsp");
        mvRef.addObject("usernameKey", "usernameValue");
        return mvRef;
    }
```

}
JSP 文件 index.jsp 核心代码如下。
```
<body>
    ${usernameKey }
</body>
```

13.4.2 程序运行结果

部署项目,执行 URL:http://localhost:8081/returnModelAndView/returnModeAndView_Test.spring。程序运行的结果如图 13-8 所示。

图 13-8 程序运行结果

13.5 控制层返回 void 数据的情况

前面示例的控制层大多数返回的是 String 数据类型,代表转发到指定名称的 JSP 文件。控制层还可以返回 void 数据类型,存在两种情况:
- 使用默认的 JSP;
- 通过 HttpServletResponse 打印输出。

此示例在项目 controllerReturnNull 中进行演示。

13.5.1 创建控制层及 index.jsp 文件

创建控制层文件,核心代码如下。
```
@Controller
public class Index {
    @RequestMapping(value = "index")
    public void gotoIndexJSP() {
//此方法的参数不能添加 HttpServletResponse response 类型
//如果一旦添加,则转发 forward 失效
//则可以使用 response 方式输出最终信息
//如果没有 HttpServletResponse response 参数,则可以转发
        System.out.println("自动转到 index.jsp 文件");
    }

    @RequestMapping(value = "outTest")
    public void outTest(HttpServletResponse response) {
        try {
            response.setContentType("text/html");
            response.setCharacterEncoding("utf-8");
```

```
                PrintWriter out = response.getWriter();
                out.println("我是高洪岩，此字符串通过 PrintWriter 输出！");
                out.flush();
                out.close();
            } catch (IOException e) {
                // TODO Auto-generated catch block
                e.printStackTrace();
            }
        }
    }
```

文件 index.jsp 核心代码如下。

```
<body>
    哈哈，来吧，我是 index.jsp 文件，欢迎你！
</body>
```

13.5.2 更改配置文件

更改配置文件 springMVC-servlet.xml 的核心代码如下。

```xml
<bean id="viewResolver"
    class="org.springframework.web.servlet.view.InternalResourceViewResolver">
    <property name="viewClass">
        <value>org.springframework.web.servlet.view.JstlView</value>
    </property>
    <property name="prefix">
        <value>/</value>
    </property>
    <property name="suffix">
        <value>.jsp</value>
    </property>
</bean>
<context:component-scan base-package="controller" />
```

加入一个 JSP 视图文件解析器，它的作用是根据 URL 的路径进行参考，然后以.jsp 为后缀的文件做为视图层，不然出现 404 出错异常。

13.5.3 部署项目运行程序

部署项目并运行程序，输入 http://localhost:8081/controllerReturnNull/outTest.spring，程序运行结果如图 13-9 所示。

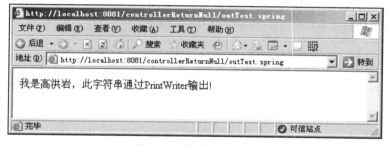

图 13-9 程序运行结果

13.6 使用 Spring 4 MVC 中的注解来操作 HttpSession 中的对象

再次输入 http://localhost:8081/controllerReturnNull/index.spring，程序运行结果如图 13-10 所示。

图 13-10　默认转发到 index.jsp 文件中

13.6 使用 Spring 4 MVC 中的注解来操作 HttpSession 中的对象

在 Servlet 技术中可以使用如下代码来操作 HttpSession 对象。

```
HttpServletRequest request = null;
HttpSession session = request.getSession();
```

在 Spring 4 MVC 中可以使用注解来进行操作 HttpSession 对象，示例代码在 springMVCSession 项目中。

13.6.1　创建控制层文件 PutGetSession.java

文件 PutGetSession.java 的核心代码如下。

```
@Controller
@SessionAttributes(value = "usernameInSession")
// 还可以定义多个放入 session 的对象：
// @SessionAttributes({"a","b"})
public class PutGetSession {
    @RequestMapping(value = "putSession")
    public String putSession(Model model) {
        model.addAttribute("usernameInSession", "usernameInSessionValue");
        return "index.jsp";
    }

    @RequestMapping(value = "getSession")
    public String getSession(
            @ModelAttribute("usernameInSession") String getUsernameFromSession) {
        System.out.println("getUsernameFromSession==" + getUsernameFromSession);
        return "pp.jsp";
    }

}
```

在类的上方使用注解@SessionAttributes(value = "usernameInSession")的含义是如果在 Model 中存在名称为 usernameInSession 的 key 时，就把 key 及其对应的值放入 HttpSession 对象中。

在方法 getSession()使用注解@ModelAttribute("usernameInSession")的含义是将 HttpSession 中 key 为 usernameInSession 的值注入进 getUsernameFromSession 参数中。

13.6.2　创建显示不同作用域中的值的 JSP 文件

文件 index.jsp 的核心代码如下。

```
<body>
    request======${requestScope.usernameInSession }
    <br />
    session======${sessionScope.usernameInSession }
    <br />
</body>
```

13.6.3　部署项目并运行程序

执行 http://localhost:8081/springMVCSession/putSession.spring，在 IE 界面上输出信息，如图 13-11 所示。

图 13-11　输出 request 及 HttpSession 作用域中的值

再继续调用 http://localhost:8081/springMVCSession/getSession.spring，从 HttpSession 对象中取出字符串的值并输出，结果如图 13-12 所示。

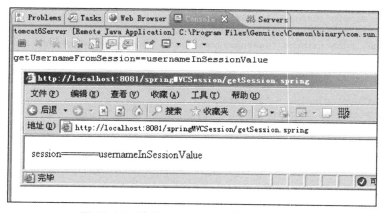

图 13-12　输出 HttpSession 中字符串的值

第 14 章　Spring 4 MVC+MyBatis 3 +Spring 4 整合

学习完 MyBatis 3 和 Spring 4 MVC 框架后，我们就可以将它们进行整合，整合的目的可以使事务自动地提交或回滚。可以生成 MyBatis 3 操作数据库接口的实现类，使程序员在开发基于 Spring 4 MVC+MyBatis 3+Spring 4 的项目时写法更加统一，便于维护。

14.1　准备 Spring 4 的 JAR 包文件

使用 Spring 4 版本进行整合，从官方网站下载的 Spring 4 完整 JAR 文件列表如图 14-1 所示。

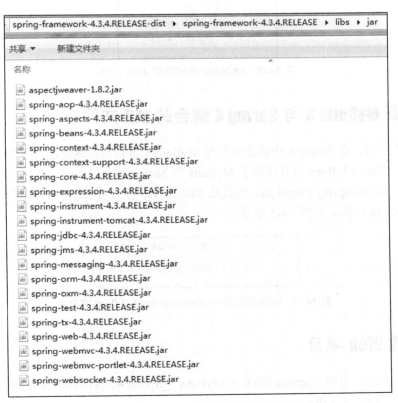

图 14-1　Spring 4 整合需要的 JAR 文件

在这些 JAR 文件中就包含 Spring 4 MVC 框架所需的 JAR 文件，因为 Spring 4 框架包含 Spring 4 MVC 模块。

14.2 准备 MyBatis 的 JAR 包文件

添加 ORM 框架 MyBatis 的 JAR 包文件，如图 14-2 所示。

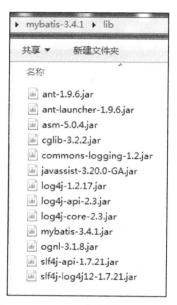

图 14-2　MyBatis 框架所需 JAR 文件

14.3 准备 MyBatis 3 与 Spring 4 整合的 JAR 文件

在撰写本书时，在 Spring 4 中还是不能与 MyBatis 3 进行直接地整合，需要借助于第三方的 JAR 文件，所以 MyBatis 官方发布了 MyBatis 与 Spring 整合的插件。此插件就是一个 JAR 文件，类似于 struts2-spring-plugin.jar。通过此 JAR 文件即可与 Spring 4 进行整合，此文件名是 mybatis_spring_xxxx.jar，如图 14-3 所示。

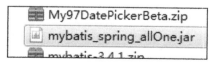

图 14-3　整合需要的 mybatis-spring-1.2.0.jar 文件

14.4 创建 Web 项目

创建 Web 项目，名称为 spring MVC 4_MyBatis 3_Spring 4，将前面所有涉及的 JAR 文件放入 lib 文件夹，如图 14-4 所示。

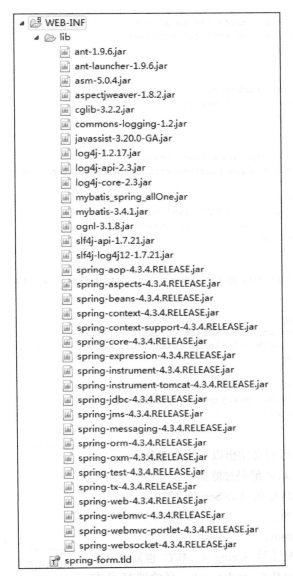

图 14-4　Web 项目中 lib 的 JAR 文件

14.5　配置 web.xml 文件

项目中的 web.xml 配置的代码如下。

```xml
<?xml version="1.0" encoding="UTF-8"?>
<web-app version="2.5" xmlns="http://java.sun.com/xml/ns/javaee"
    xmlns:xsi="http://www.w3.org/2001/XMLSchema-instance"
    xsi:schemaLocation="http://java.sun.com/xml/ns/javaee
    http://java.sun.com/xml/ns/javaee/web-app_2_5.xsd">
    <filter>
        <filter-name>charFilter</filter-name>
        <filter-class>org.springframework.web.filter.CharacterEncodingFilter</filter-class>
        <init-param>
```

```xml
            <param-name>encoding</param-name>
            <param-value>utf-8</param-value>
        </init-param>
    </filter>
    <filter-mapping>
        <filter-name>charFilter</filter-name>
        <url-pattern>/*</url-pattern>
    </filter-mapping>

    <servlet>
        <servlet-name>springMVC</servlet-name>
        <servlet-class>org.springframework.web.servlet.DispatcherServlet</servlet-class>
    </servlet>
    <servlet-mapping>
        <servlet-name>springMVC</servlet-name>
        <url-pattern>*.spring</url-pattern>
    </servlet-mapping>

    <listener>
        <listener-class>org.springframework.web.context.ContextLoaderListener</listener-class>
    </listener>

    <context-param>
        <param-name>contextConfigLocation</param-name>
        <param-value>\WEB-INF\classes\applicationContext.xml</param-value>
    </context-param>

    <welcome-file-list>
        <welcome-file>index.jsp</welcome-file>
    </welcome-file-list>
</web-app>
```

在 web.xml 文件中可以总结出以下几个部分。
- 使用 Spring 的 utf-8 编码过滤器。
- 配置 Spring 4 MVC 的核心 Servlet 控制器。
- 注册 listener 监听。
- 注册 context-param。

在上述第 2 部分中说明 Spring 4 MVC 也需要一个 XML 配置文件 springMVC-servlet.xml，存放路径是在 WEB-INF 文件夹中，如图 14-5 所示。

图 14-5 springMVC-servlet.xml 存放路径

14.6 配置 springMVC-servlet.xml 文件

文件 springMVC-servlet.xml 的代码如下。

```xml
<?xml version="1.0" encoding="UTF-8"?>
<beans xmlns="http://www.springframework.org/schema/beans"
    xmlns:xsi="http://www.w3.org/2001/XMLSchema-instance"
    xmlns:p="http://www.springframework.org/schema/p"
    xmlns:aop="http://www.springframework.org/schema/aop"
    xmlns:context="http://www.springframework.org/schema/context"
    xmlns:tx="http://www.springframework.org/schema/tx"
    xsi:schemaLocation="http://www.springframework.org/schema/beans
```

```
http://www.springframework.org/schema/beans/spring-beans.xsd
        http://www.springframework.org/schema/context
http://www.springframework.org/schema/context/spring-context.xsd
        http://www.springframework.org/schema/aop
http://www.springframework.org/schema/aop/spring-aop-4.3.xsd
        http://www.springframework.org/schema/tx
http://www.springframework.org/schema/tx/spring-tx-4.3.xsd">
    <context:component-scan base-package="controller" />
    <aop:aspectj-autoproxy proxy-target-class="true"></aop:aspectj-autoproxy>
</beans>
```

使用<context:component-scan base-package="controller" />配置代码扫描 controller 包中带有@Controller 注解的控制层类。

使用<aop:aspectj-autoproxy proxy-target-class="true" />配置代码对控制层进行事务代理 AOP 支持。

14.7 配置 MyBatis 配置文件

创建 MyBatis 框架的配置文件 mybatis-3-config.xml，代码如下。

```
<?xml version="1.0" encoding="UTF-8" ?>
<!DOCTYPE configuration PUBLIC "-//mybatis.org//DTD Config 3.0//EN" "mybatis-3-config.dtd">
<configuration>
</configuration>
```

由于连接数据库等信息是在 Spring 4 配置文件中进行定义，所以此配置文件中的配置代码极少。但要保持 MyBatis 的一些默认行为，还是可以在此文件中进行声明定义。

14.8 创建 MyBatis 与映射有关文件

在 Web 项目中创建 mapping 和 entity 包，包中的内容如图 14-6 所示。

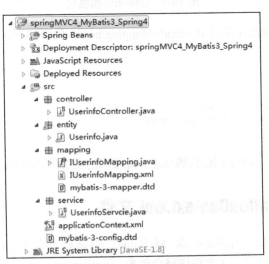

图 14-6 ssm.orm 包内容

关键的 SQL 映射文件 IUserinfoMapping.xml 的代码如下。

```xml
<?xml version="1.0" encoding="UTF-8" ?>
<!DOCTYPE mapper
PUBLIC "-//mybatis.org//DTD Mapper 3.0//EN"
"mybatis-3-mapper.dtd">
<mapper namespace="mapping.IUserinfoMapping">
    <insert id="save" parameterType="entity.Userinfo">
        <selectKey order="BEFORE" resultType="java.lang.Long"
            keyProperty="id">
            select idauto.nextval from dual
        </selectKey>
        insert into userinfo(id,username)
        values(#{id},#{username})
    </insert>
</mapper>
```

实体类 Userinfo.java 类结构如图 14-7 所示。

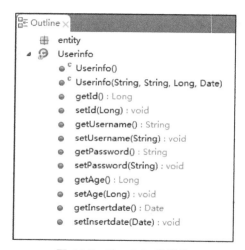

图 14-7 Userinfo 类结构

再创建 ORM 映射的操作接口 IUserinfoMapping.java，代码如下。

```java
package mapping;

import entity.Userinfo;

public interface IUserinfoMapping {
    public void save(Userinfo userinfo);
}
```

接口仅仅是定义，并不能实现任何的功能，此接口的实现类是由 Spring 4 框架生成的代理类。

14.9 配置 applicationContext.xml 文件

现在，Spring 4 MVC、Spring 4 及 MyBatis 3 的基础环境已经就绪，现在就要使用 applicationContext.xml 文件把这 3 个模块整合起来，applicationContext.xml 文件配置的代码如下。

```xml
<?xml version="1.0" encoding="UTF-8"?>
```

14.9 配置 applicationContext.xml 文件

```xml
<beans xmlns="http://www.springframework.org/schema/beans"
    xmlns:xsi="http://www.w3.org/2001/XMLSchema-instance"
xmlns:p="http://www.springframework.org/schema/p"
    xmlns:aop="http://www.springframework.org/schema/aop"
xmlns:tx="http://www.springframework.org/schema/tx"
    xmlns:context="http://www.springframework.org/schema/context"
    xsi:schemaLocation="http://www.springframework.org/schema/beans
http://www.springframework.org/schema/beans/spring-beans-4.1.xsd
        http://www.springframework.org/schema/context
http://www.springframework.org/schema/context/spring-context-4.3.xsd
        http://www.springframework.org/schema/aop
http://www.springframework.org/schema/aop/spring-aop-4.3.xsd
        http://www.springframework.org/schema/tx
http://www.springframework.org/schema/tx/spring-tx-4.3.xsd">

    <context:component-scan base-package="service"></context:component-scan>

    <bean id="dataSource"
        class="org.springframework.jdbc.datasource.DriverManagerDataSource">
        <property name="driverClassName" value="oracle.jdbc.OracleDriver"></property>
        <property name="url" value="jdbc:oracle:thin:@localhost:1521:orcl"></property>
        <property name="username" value="y2"></property>
        <property name="password" value="123"></property>
    </bean>

    <bean id="sqlSessionFactory" class="org.mybatis.spring.SqlSessionFactoryBean">
        <property name="dataSource" ref="dataSource"></property>
    </bean>

    <bean id="mapperScannerConfigurer" class="org.mybatis.spring.mapper.MapperScannerConfigurer">
        <property name="basePackage" value="mapping"></property>
        <property name="sqlSessionFactory" ref="sqlSessionFactory"></property>
    </bean>

    <bean id="transactionManager"
        class="org.springframework.jdbc.datasource.DataSourceTransactionManager">
        <property name="dataSource" ref="dataSource"></property>
    </bean>
    <tx:annotation-driven transaction-manager="transactionManager" />
</beans>
```

在上面的配置中使用代码<context:component-scan base-package="*service*"></context:component-scan>对服务层中的包进行扫描。

使用代码：

```xml
<bean id="dataSource" class="org.springframework.jdbc.datasource.DriverManagerDataSource">
```

创建数据源 dataSource 对象，再将数据源 dataSource 对象注入进 SqlSessionFactoryBean 对象中：

```xml
<bean id="sqlSessionFactory"
class="org.mybatis.spring.SqlSessionFactoryBean">
```

还要将 dataSource 对象关联事务的功能，配置代码如下。

```xml
<bean id="transactionManager"
    class="org.springframework.jdbc.datasource.DataSourceTransactionManager">
    <property name="dataSource" ref="dataSource" />
</bean>
```

```xml
<tx:annotation-driven transaction-manager="transactionManager" />
```

最后还需要将 sqlSessionFactory 对象注入进 UserinfoMapper.java 接口的代理实现类中，这就需要告诉 Spring 这些映射接口在什么位置。使用如下代码进行定义。

```xml
<bean id="mapperScannerConfigurer" class="org.mybatis.spring.mapper.MapperScannerConfigurer">
    <property name="basePackage" value="mapping"></property>
    <property name="sqlSessionFactory" ref="sqlSessionFactory"></property>
</bean>
```

14.10 创建 Service 对象

创建 service 包并创建 UserinfoService.java 类，代码如下。

```java
package service;

import org.springframework.beans.factory.annotation.Autowired;
import org.springframework.stereotype.Service;

import entity.Userinfo;
import mapping.IUserinfoMapping;

@Service
public class UserinfoServcie {

    @Autowired
    private IUserinfoMapping userinfoMapping;

    public IUserinfoMapping getUserinfoMapping() {
        return userinfoMapping;
    }

    public void setUserinfoMapping(IUserinfoMapping userinfoMapping) {
        this.userinfoMapping = userinfoMapping;
    }

    public void saveService() {
        Userinfo userinfo1 = new Userinfo();
        userinfo1.setUsername("中国1");

        Userinfo userinfo2 = new Userinfo();
        userinfo2.setUsername("中国2");

        userinfoMapping.save(userinfo1);
        userinfoMapping.save(userinfo2);
    }
}
```

14.11 创建 Controller 对象

创建 controller 包，并创建 controller 控制层 UserinfoController.java，代码如下。

```java
package controller;
```

```java
import org.springframework.beans.factory.annotation.Autowired;
import org.springframework.stereotype.Controller;
import org.springframework.transaction.annotation.Transactional;
import org.springframework.web.bind.annotation.RequestMapping;

import service.UserinfoServcie;

@Transactional
@Controller
public class UserinfoController {

    @Autowired
    private UserinfoServcie userinfoService;

    public UserinfoServcie getUserinfoService() {
        return userinfoService;
    }

    public void setUserinfoService(UserinfoServcie userinfoService) {
        this.userinfoService = userinfoService;
    }

    @RequestMapping(value = "test")
    public String test() {
        userinfoService.saveService();
        return "index.jsp";
    }
}
```

至此，整合的步骤全部结束。

14.12 测试正常的效果

将项目部署到 Tomcat 中，执行控制层后，在控制台并没有输出异常。

数据表 Userinfo 中的内容如图 14-8 所示。

图 14-8 userinfo 数据表新添加的记录

14.13 测试回滚的效果

在 ssm 整合的过程中考虑到操作数据库如果出现异常事务应该回滚，所以更改 UserinfoService.java 类中的代码如下。

```java
public void saveService() {
    Userinfo userinfo1 = new Userinfo();
```

```
        userinfo1.setUsername("中国1");

        Userinfo userinfo2 = new Userinfo();
        userinfo2.setUsername(
            "中国2中国2中国2中国2中国2中国2中国2中国2中国2中国2中国2中国2中国2中国2中国2中国2中国2中国2中国2中国2中国2中国2中国2中国2中国2中国2中国2中国2中国2中国2中国2中国2中国2中国2中国2中国2中国2中国2中国2中国2中国2中国2中国2中国2中国2中国2中国2中国2中国2");

        userinfoMapping.save(userinfo1);
        userinfoMapping.save(userinfo2);
    }
```

程序运行后在控制台输出异常信息如下。

```
java.sql.SQLException: ORA-12899: 列 "USERNAME" 的值太大
```

再来查看 userinfo 表中的数据是否还是两条记录,如图 14-9 所示。

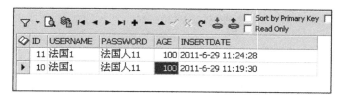

图 14-9 事务成功回滚

本章演示的是 Spring MVC 4_MyBatis 3_Spring 4 框架的整合,关于其他框架的整合,比如,Struts 2_Hibernate 5_Spring 4、Spring MVC 4_Hibernate 5_Spring 4 和 Struts 2_MyBatis 3_Spring 4 的整合示例请查看本章附带的源代码。

欢迎来到异步社区！

异步社区的来历

异步社区（www.epubit.com.cn）是人民邮电出版社旗下IT专业图书旗舰社区，于2015年8月上线运营。

异步社区依托于人民邮电出版社20余年的IT专业优质出版资源和编辑策划团队，打造传统出版与电子出版和自出版结合、纸质书与电子书结合、传统印刷与POD按需印刷结合的出版平台，提供最新技术资讯，为作者和读者打造交流互动的平台。

社区里都有什么？

购买图书

我们出版的图书涵盖主流IT技术，在编程语言、Web技术、数据科学等领域有众多经典畅销图书。社区现已上线图书1000余种，电子书400多种，部分新书实现纸书、电子书同步出版。我们还会定期发布新书书讯。

下载资源

社区内提供随书附赠的资源，如书中的案例或程序源代码。

另外，社区还提供了大量的免费电子书，只要注册成为社区用户就可以免费下载。

与作译者互动

很多图书的作译者已经入驻社区，您可以关注他们，咨询技术问题；可以阅读不断更新的技术文章，听作译者和编辑畅聊好书背后有趣的故事；还可以参与社区的作者访谈栏目，向您关注的作者提出采访题目。

灵活优惠的购书

您可以方便地下单购买纸质图书或电子图书，纸质图书直接从人民邮电出版社书库发货，电子书提供多种阅读格式。

对于重磅新书，社区提供预售和新书首发服务，用户可以第一时间买到心仪的新书。

用户帐户中的积分可以用于购书优惠。100积分=1元，购买图书时，在 使用积分 里填入可使用的积分数值，即可扣减相应金额。

特别优惠

购买本书的读者专享异步社区购书优惠券。

使用方法：注册成为社区用户，在下单购书时输入 S4XC5 使用优惠码，然后点击"使用优惠码"，即可在原折扣基础上享受全单9折优惠。（订单满39元即可使用，本优惠券只可使用一次）

纸电图书组合购买

社区独家提供纸质图书和电子书组合购买方式，价格优惠，一次购买，多种阅读选择。

社区里还可以做什么？

提交勘误

您可以在图书页面下方提交勘误，每条勘误被确认后可以获得100积分。热心勘误的读者还有机会参与书稿的审校和翻译工作。

写作

社区提供基于Markdown的写作环境，喜欢写作的您可以在此一试身手，在社区里分享您的技术心得和读书体会，更可以体验自出版的乐趣，轻松实现出版的梦想。

如果成为社区认证作译者，还可以享受异步社区提供的作者专享特色服务。

会议活动早知道

您可以掌握IT圈的技术会议资讯，更有机会免费获赠大会门票。

加入异步

扫描任意二维码都能找到我们：

异步社区	微信服务号	微信订阅号	官方微博	QQ群：436746675

社区网址：www.epubit.com.cn

投稿 & 咨询：contact@epubit.com.cn